ENVIRONMENTAL RESTORATION of METALS-CONTAMINATED SOILS

ENVIRONMENTAL RESTORATION of METALS-CONTAMINATED SOILS

Edited by

I.K. Iskandar

CRC Press
Taylor & Francis Group
Boca Raton   London   New York

CRC Press is an imprint of the
Taylor & Francis Group, an informa business

Library of Congress Cataloging-in-Publication Data

Environmental restoration of metals-contaminated soils / edited by I.K. Iskandar.
    p. cm.
Includes bibliographical references and index.
ISBN 1-56670-457-X (alk. paper)
1. Metals—Environmental aspects. 2. Soil remediation. I. Iskandar, I. K. (Iskandar Karam), 1938–

TD879.M47 E58 2001
628.5'5—dc21

00-030172
CIP

Visit the CRC Press Web site at www.crcpress.com

© 2001 by CRC Press LLC
Lewis Publishers is an imprint of CRC Press LLC

No claim to original U.S. Government works
International Standard Book Number 0-56670-457-X
Library of Congress Card Number 00-030172

Preface

During recent decades, phenomenal progress has been made in several areas of biology, ecology, health, and environmental geochemistry of heavy metals in soils. Prior to the 1960s, research was focused on enhancing the plant uptake or availability of selected heavy metals or minor elements (also referred to as micronutrients) from the soil. More recently, concerns regarding heavy metals contamination in the environment affecting all ecosystem components, including aquatic and terrestrial systems, have been identified with increasing efforts on limiting their bioavailability in the vadose zone.

Recently, many sites have been identified as hazardous waste sites because of the presence of elevated concentrations of heavy metals. In some cases, contamination of groundwater with metals that have potential health effects has also been discovered. The total mass of metals in surface soils is an important factor influencing their migration in the soil to the groundwater. However, soil environmental conditions and physical, chemical, and biological processes are also important factors affecting the fate of metals in soils.

Unlike organic contaminants that can be destroyed (or mineralized) through treatment technologies, such as bioremediation or incineration, metal contaminants cannot. Once a metal has contaminated soil, it will remain a threat to the environment until it is removed or immobilized. The cleanup techniques most used for the remediation of heavy metals contamination are excavation and subsequent landfilling of the heavy metals-contaminated soil or waste (commonly referred to as "dig and haul"). Dig and haul does not remove the contaminant from the waste but simply transfers the contamination from one area to another. Usually, no effort is made to reduce the mobility of the heavy metals beyond containment in a secured landfill.

Because of the concerns regarding the role of heavy metals in the environment, a series of international conferences was held to explore the emerging issues of the biogeochemistry of trace elements in the environment. In June 1997, the Fourth International Conference on the Biogeochemistry of Trace Elements was held in Berkeley, CA. The contributions in this book were presented in part at this conference. This book, *Environmental Restoration of Metals-Contaminated Soils*, follows earlier titles: *Engineering Aspects of Metal-Waste Management*, 1992, and *Remediation of Soils Contaminated with Metals*, 1997.

The contributors are a multidisciplinary group of scientists and engineers; the book was written to update current information on environmentally accepted methods for site restoration.

The book is organized in 14 chapters. The first eight chapters deal with the physical and chemical methods and processes for soil remediation. The other six chapters focus on selected biological methods and processes for remediation.

Chapters 1 and 2 describe physical-chemical processes for *in situ* remediation by adding amendments for stabilization.

Chapter 3 considers the immobilization of lead. Chapter 4 describes the mechanics of metal retention and release from soils. Chapter 5 describes a chemical remediation method for soil contaminated with cadmium and lead.

Chapter 6 examines the effect of soil pH on the distribution of metals among soil fractions. Chapters 7 and 8 describe physical and electrical separation methods for soil remediation. The relationship between the phytoavailability and the extractability of heavy metals in contaminated soils is discussed in Chapter 9, while Chapter 10 provides an overview on environmental restoration of selenium-contaminated soils. Chapter 11 discusses trace elements in soil-plant systems under tropical environment. The process of metal removal by chelation using amino acids is presented in Chapter 12. Chapter 13 examines the effects of natural zeolite and bentonite on the phytoavailability of heavy metals. Chapter 14 discusses metal uptake by agricultural crops from sewage sludge-treated soils.

I thank the authors for their contributions. I am also grateful for their patience, valuable time, and effort in preparing and critiquing the various chapters and in keeping the focus on the main theme of soil remediation. I gratefully acknowledge the technical reviews provided by Dr. A.L. Page of the University of California, Riverside.

Without the support of the Center for Environmental Engineering, Science and Technology (CEEST) at the University of Massachusetts, Lowell, and the U.S. Army Cold Regions Research and Engineering Laboratory (CRREL), this project could not have been achieved. I thank Edmund A. Wright (since deceased) and Donna Harp of CRREL for their editing and typing support. Finally, I thank my wife, Bonnie Iskandar, for her encouragement and support and for allowing me to work at home many hours to complete this volume.

This volume is dedicated to the memory of the late Edmund A. Wright, who over the past 25 years provided me with technical editing.

Iskandar K. Iskandar

Editor

Iskandar K. Iskandar earned his Ph.D. degree in soil science and water chemistry at the University of Wisconsin–Madison, in 1972. He is a research physical scientist at the Cold Regions Research and Engineering Laboratory (CRREL) and a Distinguished Research Professor at the University of Massachusetts, Lowell. He developed a major research program on land treatment of municipal wastewater and coordinated a number of research areas including transformation and transport of nitrogen, phosphorus, and heavy metals. He also developed the Cold Regions Environmental Quality Program at CRREL which he managed from 1985 to 1997. His recent research efforts have focused on the fate and transformation of toxic chemicals, development of non-destructive methods for site assessments, and evaluation of *in situ* and on-site remediation alternatives. Dr. Iskandar has edited several books and published numerous technical papers. He organized several national and international workshops, conferences, and symposia. He received a number of awards including the Army Science Conference Award, CRREL Research and Development Award, and CRREL Technology Transfer Award. Dr. Iskandar is a fellow of both the Soil Science Society of America and the American Society of Agronomy, and vice president of the International Society of Trace Element Biogeochemistry.

Contributors

Sultana Ahmed   Bangladesh Institute of Nuclear Agriculture, Bangladesh

Akram N. Alshawabkeh   Northeastern University, Department of Civil and Environmental Engineering, Boston, Massachusetts 02115, U.S.A.

Herbert E. Allen   Department of Civil and Environmental Engineering, University of Delaware, Newark, Delaware 19716, U.S.A.

Alain Bermond   Institut National Agronomique Laboratoire de Chimie Analytique, Paris, France

R. Mark Bricka   U.S. Army Corps of Engineers Engineering Research & Development Center, Vicksburg, Mississippi 39180, U.S.A.

Lenom J. Cajuste   Colegio de Postgraduados, Chapingo Montecillo, Mexico

Zueng-Sang Chen   Graduate Institute of Agricultural Chemistry, National Taiwan University, Taipei, Taiwan

K.S. Dhillon   Department of Soils, Punjab Agricultural University, Ludhiana, India

S.K. Dhillon   Department of Soils, Punjab Agricultural University, Ludhiana, India

James A. Holcombe   Department of Chemistry and Biochemistry, University of Texas, Austin, Texas 78712, U.S.A.

Maury Howard   Department of Chemistry and Biochemistry, University of Texas, Austin, Texas 78712, U.S.A.

Achim Kayser   Swiss Federal Institute of Technology, Institute of Terrestrial Ecology, Soil Protection Group, Zürich, Switzerland

Armin Keller   Swiss Federal Institute of Technology, Institute of Terrestrial Ecology, Soil Protection Group, Zürich, Switzerland

Catherine Keller   Swiss Federal Institute of Technology, Institute of Terrestrial Ecology, Soil Protection Group, Zürich, Switzerland

A.S. Knox (formerly A. Chlopecka)   Savannah River Ecology Laboratory, University of Georgia, Aiken, South Carolina 29802, U.S.A.

Reggie J. Laird   Colegio de Postgraduados, Chapingo Montecillo, Mexico

Valérie Laperche   Centre National de Recherche sur les Sites et Sols Pollués, Douai, France

Geng-Jauh Lee   Graduate Institute of Agricultural Chemistry, National Taiwan University, Taipei, Taiwan

Suen-Zone Lee   Department of Environmental Engineering and Health, Chia Nan College of Pharmacy and Science, Tainan, Taiwan

Jen-Chyi Liu   Department of Agricultural Chemistry, Taiwan Agricultural Research Institute, Council of Agriculture, Taichung, Taiwan

M.J. Mench   Centre Bordeaux-Aquitaine, INRA Agronomy Unit, Villenave d'Ornon, France

S.M. Rahman   Bangladesh Institute of Nuclear Agriculture, Mymensingh, Bangladesh

Rainer Schulin   Swiss Federal Institute of Technology, Institute of Terrestrial Ecology, Soil Protection Group, Zürich, Switzerland

J.C. Seaman   Savannah River Ecology Laboratory Advanced Analytical Center for Environmental Science, University of Georgia, Aiken, South Carolina 29802, U.S.A.

László Simon   College of Agriculture, Gödöllő University of Agricultural Sciences, Nyiregyhaza, Hungary

C.D. Tsadilas   Institute of Soil Classification and Mapping, National Agricultural Research Foundation, Larissa, Greece

J. Vangronsveld   Limburgs Universitair Centrum, Environmental Biology Universitaire Campus, Diepenbeek, Belgium

Clint W. Williford, Jr.   Department of Chemical Engineering, University of Mississippi, Oxford, Mississippi

Yujun Yin   Department of Civil and Environmental Engineering, University of Delaware, Newark, Delaware 19716, U.S.A.

Sun-Jae You   Department of Marine Environmental Engineering, Kunsan National University, Republic of Korea

Isabelle Yousfi (deceased)   IPSN\DPRE\SERGD\Laboratoire d'Étude des Stockages de Surface, Fontenay-aux-Roses, France

Contents

---

* Formerly A. Chlopecka

Section I

Physical and Chemical Methods and Processes

1

Physical-Chemical Approach to Assess the Effectiveness of Several Amendments Used for In Situ Remediation of Trace Metals-Contaminated Soils by Adding Solid Phases

Isabelle Yousfi (deceased) and Alain Bermond (corresponding author)

CONTENTS

1.1 Introduction

The increasing use of trace metal-contaminated sewage sludge as agricultural fertilizer or the storage of polluted wastes can cause toxic elements to accumulate in soils. After complex processes, these elements may pass into the soil solution where plant uptake or leaching to groundwater can contaminate the food chain.

Thus, increasing contamination of agricultural soil by toxic compounds such as heavy metals, metalloids, or organic pollutants has important health and economic implications. With this in mind, several cleanup methods have been investigated, which can be divided into two groups: those that remove contaminants, and those that transform pollutants into harmless forms (immobilization) by fixation, oxidation, etc. These cleanup technologies can be applied on- or off-site, utilizing three kinds of remediation treatments: biological, physical, and chemical techniques.

1.1.1 Remediation Techniques

Biological treatments generally consist of *in situ* remediation and biodegradation, relying on microbiological activities to remove contaminants, which in most cases are organic pollutants. This technique is being used to remove HAPs or PBCs (Hamby, 1996) or to transform pollutants to harmless forms. It is an immobilization technique involving, for example, photolysis of organic compounds. Another biological cleanup approach is the cultivation of plants, called hyperaccumulators, that are able to accumulate one or several toxic elements. This method has been successfully used for different toxic trace elements, for instance Cd and Zn (Keller et al., 1997) or nickel (Homer et al., 1991).

Physical treatments remove contaminants through physical means. Among these approaches are thermal desorption or vapor stripping of semi-volatile or volatile compounds. Electrokinetic migration, useful for anionic and cationic compound extraction, can also be mentioned (Chapter 8).

Chemical treatments include all techniques involving reagents or external compounds. Soil washing is one. In this technique different reagents able to solubilize toxic elements are coupled with the removal of leaching solutions and added chelators that make cations less labile, thus allowing them to immobilize toxic elements in a less bioavailable form. For instance, EDTA is one of the proposed complexing reagents (Li and Shuman, 1996). In addition, adding solid phases should also be mentioned. Solid phases can fix (irreversibly) toxic elements, and they can sometimes be used to reduce remediation costs.

Among these different remediation techniques, this chapter focuses on this last method. Adding solid phases is an interesting and low-cost technique.

1.1.2 *In Situ* Remediation by Adding Solid Phase

The question here is what will the effect be of adding the following different kinds of amendments to contaminated soils: clay minerals, hydrous iron oxides (HFO), hydrous manganese oxides (HMO), fly ashes, etc. Two kinds of studies have been carried out. The first type consists of plant studies, in which some authors conducted experiments on the effect of amendments on the uptake of trace elements by plants. They did this by comparing the amounts of cadmium, lead, copper, or zinc in plants grown on contaminated soils with or without amendments (Gworek, 1992a, 1992b; Wong and Wong, 1990; Petruzzelli et al., 1987; Chlopecka and Adriano, 1997). For instance, Gworek (1992a) studied the effect of adding synthetic zeolites on the amount of Cd in lettuce, oats, and ryegrass. She noted a significant decrease in Cd in plant tissue in 85% of lettuce leaves and about 45% of lettuce roots.

Chlopecka and Adriano (1997) worked with four different soils and studied the effect of two kinds of zeolite and apatite on the amount of Cd, Zn, and Pb in maize tissue. Among the results shown, they found a decrease of about 50% of Pb in maize leaves grown on a zeolite-amended soil, and a 20% reduction in Cd content in maize grown on apatite-amended soil.

Fly ashes are used as agricultural amendments, too, particularly because they are low-cost wastes. Petruzzelli et al. (1987) have added from 0 to 5% (w/w) fly ash to soils and studied the amount of Cu, Ni, Cr, and Zn in wheat seedlings. Amendments seem to have a useful effect and result, for instance, in a decrease of about 65% of Zn in roots when 5% fly ash is added. Wong and Wong (1990) have also studied fly ashes and mixed coal fly ashes at rates of 0, 3, 6, and 12% in two soils. If they divided the amount of Zn in vegetables (cabbages) by a factor of two, the authors then noticed an increase of soil pH and a decrease in several cases of the vegetal yield. They concluded that there was a possible phytotoxic effect of fly ashes. However, the liming effect involved adding fly ashes, implying that the lowered bioavailability of oligo-elements can be another assumption, as Waren et al. (1993) and Davies et al. (1993) have assumed.

Although the range of effects varied with the plants studied, in most cases the amendments brought about a reduction of trace metal concentrations in plant tissues. However, in the case of fly ashes, an important variation of soil pH after application was noticed, and a correlation between trace element concentrations in the plant and the pH of the amended soil was found (Wong and Wong, 1990).

In another approach to remediation, some authors studied the correlation between trace elements in plants and soils. In this case, the quantities of trace metals in soils were determined using chemical fractionation procedures, before and after the amendments. A few examples can be given. Pierynski and Schwab (1993) have studied the effect of limestone, N Viro Soil, or phosphates on the content of Zn, Cd, and Pb in soybean tissues. In other tests, chemical extraction procedures have been carried out on soils amended or not. By means of these protocols, they observed a redistribution of trace elements between soil fractions; in addition, several phases seem to improve plant quality. Thus, they noted that limestone can decrease about 65% of Zn and 33% of Cd in plant tissue. Adding N Viro Soil resulted in a decrease of 33% of Pb in leaves.

Sappin-Didier et al. (1994), after adding 1% (w/w) of steel shot to two different soil samples, have studied the amount of Ni, Cr, and Pb in ryegrass and tobacco. Among their results, they showed that Cd can be decreased by 40% in tobacco. Simultaneously, the fractions of extractable Cd H_2O and $Ca(NO_3)_2$ decrease about 75 and 85%, respectively. With the same plants, Mench et al. (1994) have studied the effect of lime, HFO, HMO, and zeolites on different types of soils. Among their results, they found that the amount of Cd decreases by 75% in ryegrass roots grown on HMO-amended soils and by 50% in tobacco grown under the same conditions. In addition, lead content in plant tissue significantly decreases with HMO (–75%) or lime (–65%).

Davies et al. (1993) studied the effect of adding agricultural limestone to soils contaminated with different ranges of trace metals. The EDTA extractable fractions of Cd, Pb, and Zn were studied before and after amendment; in addition, the total content of trace elements in radishes was investigated. For single application, Davies et al. (1993) noticed, under the best conditions, a decrease of Zn and Pb in vegetative tissue of about 20%, while the EDTA-extractable fraction of zinc and lead in soil also decreased.

From a general point of view, chemical extractions showed that the amendments resulted in changes in the fractionation of trace elements available to crops.

Amendments with solid phases such as fly ashes (Wong and Wong, 1990; Petruzzelli et al., 1987; Pierzynski and Schwab, 1993), limestone, K_2PO_4, or hydrous manganese oxides and hydrous iron oxides (Wong and Wong, 1990; Petruzzelli et al., 1987; Pierzynski and Schwab, 1993; Sims and Kline, 1991; Mench et al., 1994) and zeolites (Gworek, 1992a, 1992b; Chlopecka and Adriano, 1997) can influence plant uptake, but unfortunately, these amendments can also influence the physicochemical parameters of the soil, particularly its pH. These changes influence the effectiveness of amendments, but they are rarely taken into account.

Several studies mention the effect on pH value of adding solid phases. For instance, Wong and Wong (1990) show that the pH value can increase from 1.5 to 2 units by adding alkaline fly ashes. With fly ashes, too, Petruzzelli et al. (1987) observed a decrease by a third of Ni in roots, and they also observed simultaneous pH increases from 6.1 to 8.5. Davies et al. (1993) underline the accepted liming effect of agricultural limestone and note an increase of pH from 5.9 before amendment to 7.5 after adding limestone. Waren et al. (1993), studying the bioavailability of B, Ca, Cr, Zn, Cu, and Ni for wheat seedlings grown on fly ash-amended soils, have seen an increase in pH, accompanied by a decrease in crop yield. They explain their results by a decreased availability of plant essential elements.

Although solid phases in most cases change pH value, the liming effect is not general. For instance, Juste and Soldâ (1988) have studied the effect of ammonium sulphate, calcium carbonate, acid peat, HFO, and steel shot on the total content of Cd, Ni, Zn, and Mn in

TABLE 1.1

Effect of Different Amendments on Soil pH Value and Amount of Zn, Cd, and Ni in Ryegrass
(First cut in $\mu g \cdot kg^{-1}$; the effect is calculated as (soil – soil + amendment)/soil)

	pH	[Zn] (effect)	[Cd] (effect)	[Ni] (effect)
Soil	5.5	130	18	61
+ Ammonium sulfate	4.9	204 (+57%)	27 (+50%)	99 (+62%)
+ CaCO₃	7.4	97 (–25%)	12 (–33%)	39 (–36%)
+ Acid peat	5.4	137 (+3%)	19 (+5%)	65 (+6%)
+ HFO	5.2	117 (–17%)	16 (–18%)	55 (–10%)
+ Steel shot	6.4	65 (–50%)	6 (–66%)	26 (–57%)

Source: Juste, C. and P. Soldâ. Influence de l'addition de différentes matières fertilisantes sur la biodisponibilité du cadmium, du manganèse, du nickel et du zinc contenus dans un sol sableux amendé par des boues de station d'épuration. *Agronomie*, 8(10) 897–904, 1988. With permission.

ryegrass. They observed, for example, Zn in the first cut of ryegrass was increased by a factor of 1.6 by adding ammonium sulfate while the pH value decreased from 5.5 to 4.9, but Zn decreased about 25% by adding lime up to pH 7.4. Similar results were obtained for Cd and Ni. These results are summarized in Table 1.1.

These possible changes of the physicochemical parameters make it difficult to interpret the influence of amendment action on the immobilization of trace metals. It is not easy, for instance, to distinguish the true fixation of trace elements onto the added solid phase from a lower availability of the element due to the effect of an increasing pH that enhances the fixation of trace metals onto the soil voids. Actually, the role of soil pH on the solubility of trace metals such as lead or cadmium is well known: the trace metals released in solution increase when the pH decreases (Sims, 1986). Moreover, comparisons of the effects of different amendments are quite difficult, if made simultaneously, for the same sample, pH, and added phase. Finally, it can be assumed that the plant studies do not give any informa-

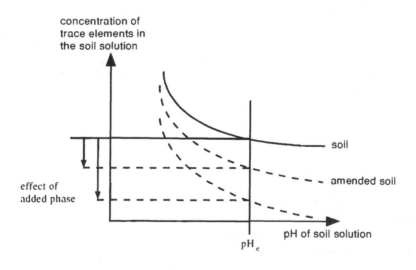

FIGURE 1.1
Schematic representation of the determination of the amendment effect at a given pH (pH$_e$). Curves correspond to the possible relative position of the extracted trace metal amounts vs. pH without and with amendments, respectively. Solid line represents extracted trace metal amounts from soil without amendment vs. pH; it is a "reference curve" to assess amendment effect. Dotted lines represent extracted trace metal amounts vs. pH from soil mixed with amendment whose fixation effect involves a decrease of total extracted amount (at a given pH).

tion about how effective immobilization will be to prevent migration toward waters (groundwaters, but also surface waters).

The aim of the present study was to estimate the trace metal immobilization potential of different amendments, that is to say, the removal of trace metals from the soil solution using a physicochemical approach. Therefore, because soil parameters such as pH are important, we tried to quantitatively determine the effect of several amendments under different physicochemical conditions (i.e., different chemical reagents). Then we compared the amount of trace metals extracted with a given chemical reagent from soils with or without amendment, at the same pH, as is shown on Figure 1.1.

This approach allows us to estimate the true potential of a given amendment and the effect of pH on it to remove trace metals from the solution; it also allows us to compare, from this point of view, different amendments and, thus, gives a fast and reliable way to choose the more appropriate amendment for a given soil.

1.2   Example of Study to Assess the Effectiveness of Several Amendments

In this study, we have investigated the effect of several amendments on two contaminated soils when the samples were subjected to a chemical reagent simulating conditions enhancing exchange phenomena or reducing conditions:

- A solution of $Ba(ClO_4)_2$ (0.5 mol·L^{-1}) was used to assess the reactivity of trace elements when the soil samples were subjected to an exchanger medium simulating natural physicochemical conditions that may occur when a soil receives industrial effluents, fertilizers, or deicing salts from nearby roadways.
- $NH_2OH \cdot HCl$ (0.1 mol·L^{-1}) was used to estimate trace metal behavior when the sample was subjected to reducing conditions that are, for instance, able to occur during flooding.

1.2.1   Methods

1.2.1.1   Soils Characteristics

Two soil samples contaminated by trace metals were used in this study. The first one, named Couhins, is a sewage sludge treated soil; the other, named Evin, is contaminated by industrial atmospheric depositions. Table 1.2 gives their main physicochemical characteristics.

As discussed earlier (see Figure 1.1), several experiments were carried out to plot the amounts of trace elements released as a function of pH at equilibration time. The resulting graphs will be our references to estimate, at a given pH, the effect of the added phase. In a first set of experiments, the soil sample (1 g) and the reagent, $Ba(ClO_4)_2$ or $NH_2OH \cdot HCl$ (50 mL), were shaken with a mechanical stirrer, for different equilibration times determined by kinetic study, i.e., 140 min in barium perchlorate solution or 24 h in reducing solution. After mixing, the pH of the solution, pH_e, was measured; the amounts of extracted trace elements (Zn, Cd, Cu, Pb) or major elements (Ca, Mg, Fe, Mn) were determined to establish "reference curves."

Figure 1.2 shows the amount of Zn and Cd extracted using barium perchlorate solution. In the same way, Figure 1.3 represents the amount of Zn and Cd extracted in reducing conditions, obtained when using hydroxylamine solution.

TABLE 1.2

Physical-Chemical Characteristics of the Studied Soil Samples

	Evin	Couhins
pH	8.56	7.64
Organic matter	1.8	2.2
Clay (%)	16.5	2.9
Zinc ($\mu g \cdot g^{-1}$)	1,415	151
Cadmium ($\mu g \cdot g^{-1}$)	23	94.9
Lead ($\mu g \cdot g^{-1}$)	1,120	44.8
Copper ($\mu g \cdot g^{-1}$)	43.5	45.3
Iron ($\mu g \cdot g^{-1}$)	20,900	3,526
Manganese ($\mu g \cdot g^{-1}$)	411	40.4
Calcium ($\mu g \cdot g^{-1}$)	7,600	1,486
Magnesium ($\mu g \cdot g^{-1}$)	33,000	212

TABLE 1.3

Total Trace Element Contents of the Studied Amendments ($\mu g \cdot g^{-1}$)

Element	HMO	HFO	Clay	O.M.	EDF	Valenton
Zinc	10.7	0	57.5	49.1	849	2,912
Cadmium	9	0	4.7	0	0.71	15.4
Lead	0	0	0	0	323	1,270
Copper	0	4.8	10	0	233	1,900
Iron	55.7	28,120	710	10,400	60,150	20,670
Manganese	622,200	39.2	358	44.4	291	13,441
Calcium	77	0	6,150	10,106	5,233	77,300
Magnesium	6.2	0	6,500	1,097	5,175	4,334

Whatever the trace element is, the importance of solution pH on the quantity of toxic elements in soil solution can be seen: the extracted amount increases as pH decreases. This a well-known effect (Sims, 1986), demonstrating the extracting effect of protons.

1.2.1.2   *Rapid Characterization of the Solid Phases Used as Amendments*

Six solid phases were used:

- Synthetic compounds: hydrous manganese oxide (HMO), a goethite (HFO), a clay (a montmorillonite), and fulvic acids (O.M. [organic matter])
- Wastes: two kinds of fly ashes, one being a coal fly ash (EDF), the other a product of sewage sludge combustion from the Valenton Wastewater Plant, called in this study "Valenton"

The trace metal composition (total concentration) of these phases is given in Table 1.3. Synthetic phases, such as HMO, HFO, clay, or O.M., are weakly polluted by trace metals, as opposed to fly ashes that are strongly contaminated. Nevertheless, these compounds could have a good remediation capacity if trace elements are strongly bound and not easily released under the studied physical chemical conditions.

To assess the ability of these phases to remediate polluted soils and to fix trace elements, a rapid study of their capacity to fix several trace elements was performed. Figures 1.4 and 1.5 and Tables 1.4 and 1.5 give the percentage of cations adsorbed by each phase in each medium [Ba(ClO$_4$)$_2$ or NH$_2$OH·HCl] at different pHs: only the results for trace elements leached from soil samples in different media have been shown here [Cd and Zn in

Evin sample

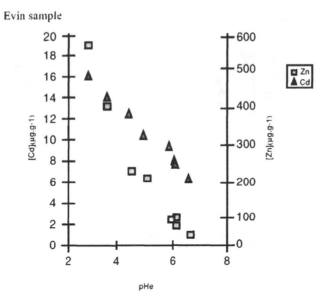

Couhins sample

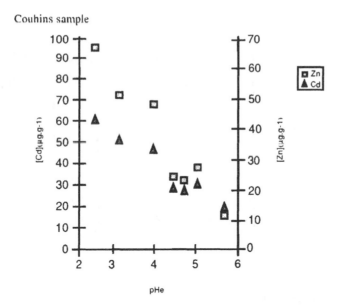

FIGURE 1.2
"Reference curve" in exchanger medium (0.5 M) Ba(ClO$_4$)$_2$. Amount of cations extracted from soil — without amendment — as function of pH at equilibration time (pH$_e$).

Ba(ClO$_4$)$_2$ medium and Cd, Zn, Pb for NH$_2$OH·HCl solution]. The ratio of solid to solution (1 to 2% w/w) and the equilibration times used were equal to those used for experiments carried out with soil samples. The initial concentrations were 0.4 mg·L^{-1} for zinc, 0.2 mg·L^{-1} for cadmium, and 5 mg L^{-1} for lead. In the reducing medium, the number of usable phases was limited as we had to eliminate HMO and HFO because they are easily destroyed by such a reducing agent.

In most cases, the fixation of trace metals increases when pH is increased. Moreover, some results do not show the fixation effect of the phase. For instance, clay minerals do not

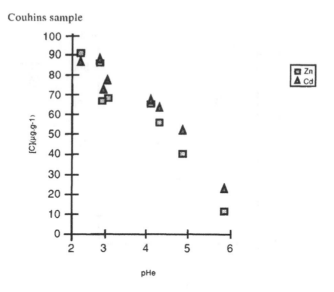

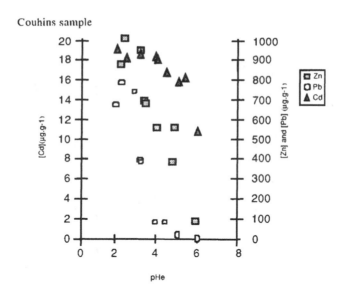

FIGURE 1.3

"Reference curve" for reducing medium (0.1 M) $NH_2OH \cdot HCl$. Amount of cations extracted from soil — without amendments — as function of pH at equilibration time (pH_e).

fix Zn or Cd in the exchanger medium. A competition of these trace elements with Ba^{2+} ions may be involved, which could explain these results.

Zn and Cd exhibit a similar behavior, while more Pb seems to be fixed. Finally, from a general point of view, it could be noticed that even polluted phases are able to diminish total content of trace metals in solution.

1.2.1.3 Calculation of the Amendment Effect

To assess the amendment effect, a second set of experiments was undertaken in each medium for a mixture of soil sample (1 g) plus solid phase in 50 mL of solution. In this case, the amount of solid phase added is 10 mg for EDF, Valenton, O.M., and HMO, 12.5 mg for

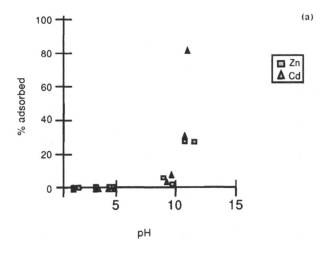

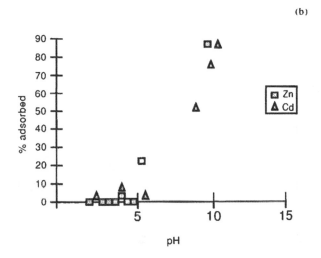

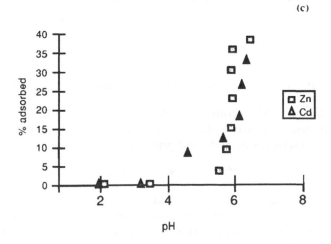

FIGURE 1.4
Amount of cations adsorbed (%) in Ba(ClO$_4$)$_2$ medium for (a) organic matter, (b) HFO, and (c) Valenton.

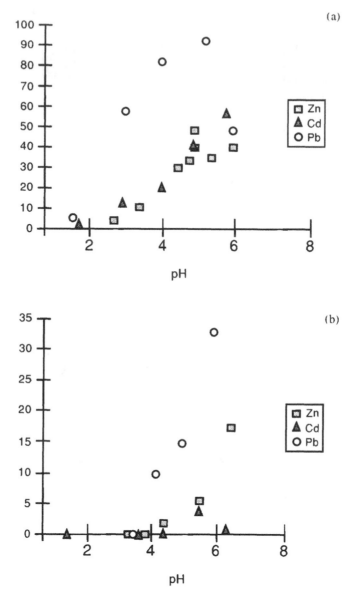

FIGURE 1.5
Amount of cations adsorbed (%) in $NH_2OH \cdot HCl$ medium, for organic matter (a) and clay (b).

HFO, and 20 mg for clay. After equilibration time (i.e., 1440 min or 24 h), the pH_e was recorded and trace element concentrations in solution were determined.

The effect E produced by the added solid phase was then calculated as follows:

$$E\ (\%) = 100 \times [(R - A)\ /\ R] \tag{1}$$

where A is the amount of trace metals released from mixing the soil sample and amendment at a given pH_e, and R is the amount of trace metals released from the soil sample without amendment, presented on the reference curves at the same pH. The input of trace elements was adjusted by the phase, but in fact this correction was not necessary because the amounts of trace elements brought by the phase were always (much) smaller than the amount released by soil.

TABLE 1.4

Percentage of Cations Adsorbed by Amendments in $Ba(Cl)_4)_2$ Medium

	pH	HMO[a]	Clay[b]	EDF[a]
Zn	4	9	0	0
	6	66	0	5
	7	98	0	21
Cd	4	38	0	0
	6	66	0	0
	7	73	0	0

[a] 10 mg has been mixed to 50 mL of solution.
[b] 20 mg has been mixed to 50 mL of solution.

TABLE 1.5

Percentage of Adsorbed Cations by Amendments in $NH_2OH \cdot HCl$ Medium

	pH	EDF[a]	Valenton[a]
Zn	4	0	8
	6	0	22
Cd	4	0	0
	6	0	6
Pb	4	3	32
	6	8	96

[a] 10 mg has been mixed to 50 mL of solution.

TABLE 1.6

Comparison of E Values in Both Media at pH_e about 6 — Evin Sample

	Zinc		Cadmium	
	BA(ClO$_4$)$_2$	NH$_2$OH	Ba(ClO$_4$)$_2$	NH$_2$OH
Valenton	42	20	29	7
EDF	28	16	32	5
Clay	28	23	28	4
HMO	90	[a]	56	[a]
HFO	42	[a]	30	[a]

[a] These phases are destroyed in reducing medium.

Assuming an error on the amount experimentally determined equal to $\pm 10\%$, we have therefore considered that the influence of amendment was significant when the value of E was higher than 20 (Table 1.6).

1.2.1.4 Analytical Methods

Solutions were analyzed with a flame atomic absorption spectrophotometer equipped with an air-acetylene flame. The following wavelengths were used: 213.9 nm for Zn, 228.8 nm for Cd, 324.8 nm for Cu, and 283.3 nm for Pb. The external standards method was chosen, and standards were made with $NH_2OH \cdot HCl$ (0.1 mol·L^{-1}) or $Ba(ClO_4)_2$ (0.5 mol·L^{-1}) solutions obtained from analytical grade salts, titrisol solutions, and pure MilliQ water.

1.3   Results and Discussion

Figure 1.6 shows the values of E and pH_e obtained when the samples are leached with barium perchlorate solution. As $Ba(ClO_4)_2$ is a weak reagent, only Zn and Cd were extracted from our samples and measured.

Whatever the compound added to the Evin sample, the effect of E involved was obvious and higher than 20. In the case of the Couhins sample, the range of pH_e was lower and the influence of solid phase addition was slightly more complicated. It can be said from a general point of view that the effect E depends on the final pH, the amendment, the soil sample, and the studied cation:

- The E value is generally higher for Zn than Cd; for example, the amount of Zn in the soil solution for Evin sample diminishes about 90% by HMO addition, while the amount of Cd in soil solution is simultaneously reduced to 60%.

- The value of E also depends on pH_e and, moreover, its evolution with pH is not necessarily the same as the variation of the percentage of fixed cations by the amendment with pH, as is expected. For instance, the E value increases for Valenton amended soil (Evin) when pH decreases, while the quick phase characterization shows the percentage of adsorbed Zn decreases as pH decreases (see Table 1.4 and Figure 1.4).

- The E value for the same pH_e is also a function of the studied soil sample. The studied amendments seem to give a more effective remediation for the Evin sample. Several explanations could be proposed. The presence of competitive cations in solution or different chemical form of cations (Cd^{2+}, $CdOH^+$, $CdCl^+$...) can affect the potential of trace metal fixation by amendment (Bar-Tal et al., 1988; Fu et al., 1991; Garcia-Miragaya and Page, 1977). In other respects, the nonlinear adsorption isotherm can be mentioned. For the same phase, the amounts released in solution from the Evin and Couhins samples are different (see reference curve, for a given pH_e); the amount of trace metals to be fixed by a phase thus differs.

According to these results, to remediate soils submitted to media enhancing exchange phenomena, HMO seems to be the best amendment. It is able to significantly reduce Zn as well as Cd concentrations in soil solutions.

Figure 1.7 shows the results obtained in the NH_2OH-HCl medium. As previously mentioned for the barium perchlorate medium, the calculated effect of E under reducing conditions depends on the amendment, the soil sample, and the studied cation. O.M. seems to be the best amendment to reduce the concentration of trace metals in the solution. On the other hand, other compounds, as fly ashes, seem ineffective in this condition to significantly influence the amounts of toxic trace elements in solution.

The comparison of results obtained for the same sample in both media (Figures 1.6 and 1.7) points out that the effectiveness of a given phase depends also on the studied medium. Thus, for pH_e about 6, the E values obtained in barium perchlorate medium are different from those obtained under reducing conditions (see Table 1.6). For instance, fly ashes (Valenton and EDF) are able to limit the Cd content in Evin soil solution in the barium perchlorate medium — E values are equal to 29 and 32% — but seem to be largely unable to limit Cd concentration released in the $NH_2OH \cdot HCl$ medium. (E values were equal to 7 and 5%, respectively, and considered non-significant.) Several explanations could be proposed

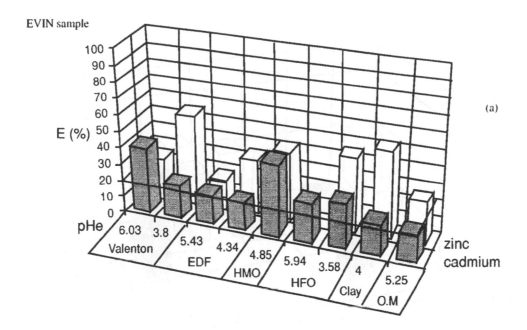

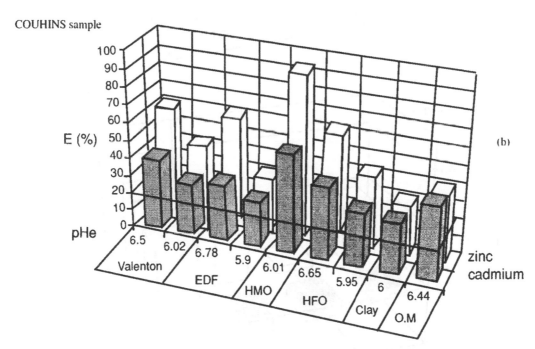

FIGURE 1.6
E values calculated for both (a) Evin and (b) Couhins soil samples in perchlorate barium solution (*E* calculated from Equation 1).

— competition, nonlinear isotherms, etc. However, according to these results, a trace-metal-polluted soil subjected to reducing conditions seems to be more difficult to remediate by immobilization techniques; the amendments seem to involve a lower reduction of the toxic trace element concentration in solution, and the number of phases able to be used in this medium is limited.

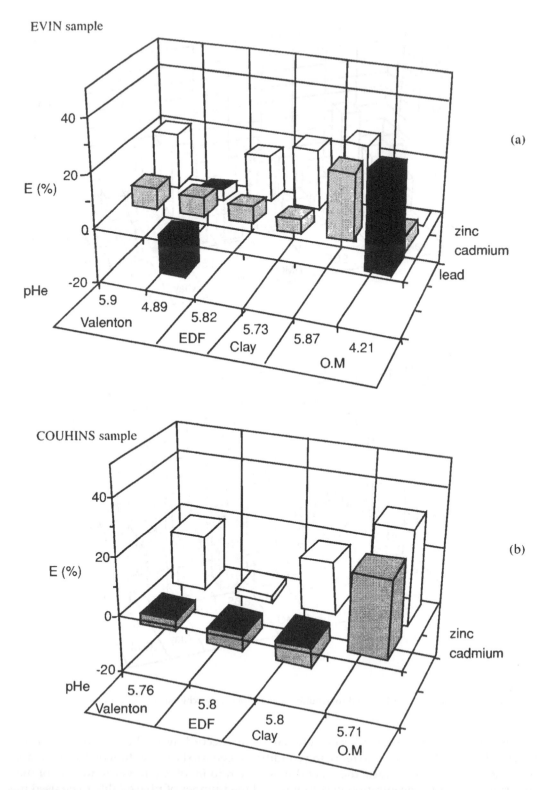

FIGURE 1.7
E values calculated for both (a) Evin and (b) Couhins soil samples under reducing conditions (E calculated from Equation 1).

From a general point of view, whatever the studied medium [$NH_2OH \cdot HCl$ or $Ba(ClO_4)_2$] was, outlining the rapid characterization of the potential of the amendment to fix trace metals (see Tables 1.4 and 1.5) was not enough to predict these results. For instance, we noticed that the percentage of Zn or Cd fixed by clay in barium perchlorate medium is null (see Table 1.4), as it is for EDF in the reducing medium. However, E values for Zn and Cd are significant for both soil samples: about 35% for Zn and Cd for Evin sample, and close to 50% for Zn and 22% for Cd in the case of Couhins sample. This different reactivity of the phase in the presence of a soil sample can be explained by the difference of equilibria involved; among these, competition of Ba for clay sites could be less important in the presence of a soil sample, a part of Ba^{2+} being able to be fixed on the different soil sites. So whatever the assumptions, the simple characterization of adsorption capacity of phase alone is not enough to evaluate the effect of a given phase when it is added to the soil sample.

To improve the effect of amendments, the ratio of soil to the added phase quantity is usually studied. This study was carried out in all cases where the effect was close to 20%. Some of the representative results are shown in Figures 1.8 and 1.9. In most cases, the effect of this ratio is obvious: for the close values of pH_e, E increases slightly when the quantity of phase added increases. For instance, it is the observed effect for either the fixation of Zn or Cd (Figure 1.9b) onto clay when reducing conditions were used.

As shown in Figures 1.8a and 1.9a, no significant effect of the ratio soil/added phase (HFO, Valenton) was observed for Cd. These examples seem to outline a threshold effect for phase fixation. As shown in Figure 1.9b where the E value for Cd decreases slightly, probably due to weak change of pH between experiments, these results seem to present evidence that the fixation effect of phases does not depend directly only on the quantity of added compound when the amount is greater than the minimum added in this study. The phenomena invoked by adding phase are complex and, as we noticed earlier, difficult to predict and extrapolate.

1.4   Conclusion

This study shows that the addition of different solid compounds, such as HMO, clay, and O.M., to a contaminated soil can be a useful way for soil trace metal remediation. Nevertheless, the effect of a given added phase is complex and cannot be easily predicted with classical adsorption experiments. Actually, this study has pointed out that this effect is a complex function of pH, trace element, soil, and medium; consequently, all soil + amendment associations will not lead to the same results from the remediation point of view. At least, it seems necessary to take into account the change of pH involved by adding a solid phase to assess and compare the effectiveness of different amendments.

Finally, the study presented here, in which we have compared the amounts released in the same physicochemical conditions, particularly at the same pH, gives a fast tool to estimate the true ability of amendments to immobilize trace metals. Thus, this physicochemical approach appears to be a tool to use to make a first selection among amendments able to limit trace element concentration in soil solutions (i.e., to limit the risk of the propagation of toxic trace elements to plants as well as toward groundwaters). In other words, this physiochemical approach allows the reduction of the number of subsequent experiments with plants.

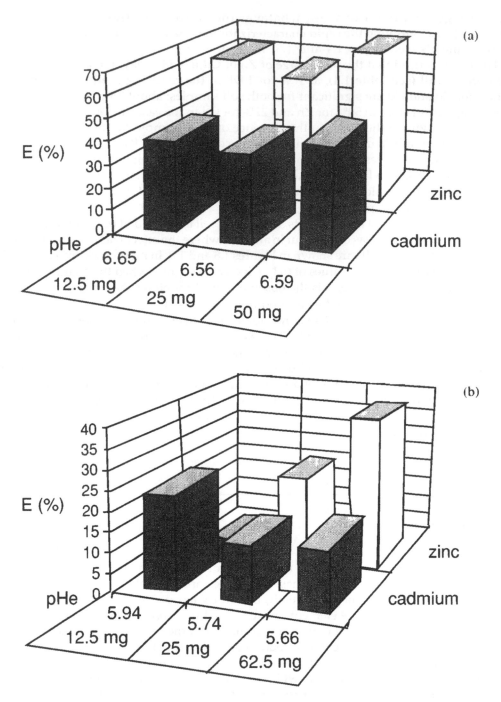

FIGURE 1.8
Effect of soil/added phase ratio in $Ba(ClO_4)_2$ (0.5 *M*) medium: (a) for Evin sample + HFO, and (b) Couhins sample + HFO.

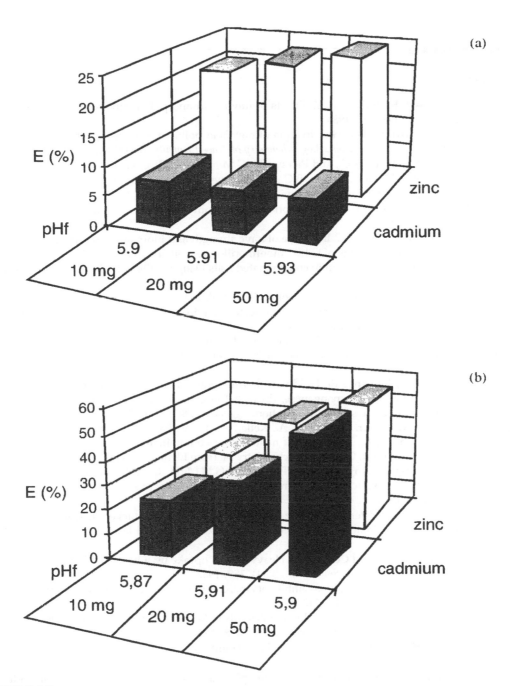

FIGURE 1.9
Effect of soil/added phase ratio in NH$_2$OH·HCl (0.1 M) medium: (a) for Evin sample + Valenton and (b) Evin sample + O.M.

References

Bar-Tal, A., Bar-Yosef, B., and Chen, Y., Effects of fulvic acid and pH on zinc sorption on monmorillonite, *Soil Sci.*, 146, 367, 1988.

Chlopecka, A. and Adriano, D.C., Inactivation of metals in polluted soils using natural zeolite and apatite, in *Proc. Fourth Int. Conf. Biogeochemistry of Trace Elements*, Berkeley, CA, 1997. Iskandar, I.K., Hardy, S.E., Chang, A.C., and Pierzynski, G.M., Eds., CRREL, Hanover, NH, 415.

Davies, B.E., Paveley, C.F., and Wixson, B.G., Use of limestone wastes from metal mining as agricultural lime: potential heavy metal limitations, *Soil Use Manage.*, 9(2), 47, 1993.

Fu, G., Allen, H.E., and Cowan, C.E., Adsorption of cadmium and copper by manganese oxide, *Soil Sci.*, 152(2), 72, 1991.

Garcia-Miragaya, J. and Page, A.L., Influence of ionic strength and inorganic complex formation on the sorption of trace amounts of Cd by montmorillonite, *Soil Sci. Am. J.*, 41, 718, 1977.

Gworek, B., Inactivation of cadmium in contaminated soils using synthetic zeolites, *Environ. Pollut.*, 75, 269, 1992a.

Gworek, B., Lead inactivation in soils by zeolite, *Plant Soil*, 143, 71, 1992b.

Hamby, D.M., Site remediation techniques supporting environmental restoration and activities: review, *Sci. Tot. Environ.*, 191, 203, 1996.

Homer, F.A., Morrison, R.S., Brooks, R.R., Clemens, J., and Reeves, R.D., Comparative studies of nickel, cobalt and copper uptake by some nickel hyperaccumulators of the genus *Alyssum*, *Plant Soil*, 138, 195, 1991.

Juste, C. and Soldâ, P., Influence de l'addition de différentes matières fertilisantes sur la biodisponibilité du cadmium, du manganèse, du nickel et du zinc contenus dans un sol sableux amendé par des boues de station d'épuration, *Agronomie*, 8(10) 897, 1988.

Keller, C., Attinger, W., Furrer, G., Kayser, A., Keller, A., Lothenbach, B., Ludwig, C., Merki, M., Stenz, B., and Schulin, R., Extraction of metals from contaminated agricultural soils by crop plants, in Proc. Fourth Int. Conf. Biogeochemistry of Trace Elements, Berkeley, CA, 1997. Iskandar, I.K., Hardy, S.E., Chang, A.C., and Pierzynski, G.M., Eds., CRREL, Hanover, NH, 473.

Li, Z. and Shuman, L.M., Extractability of zinc, cadmium, and nickel in soils amended with EDTA, *Soil Sci.*, 161, 226, 1996.

Mench, M.J., Didier, V.L., Löffer, M., Gomez, A., and Masson, P., A mimicked in situ remediation study of metal contaminated soils with emphasis on cadmium and lead, *J. Environ. Qual.*, 23, 58, 1994.

Petruzzelli, G., Lubrano, L., and Cervelli, S., Heavy metals uptake by wheat seedlings grown in fly ash-amended soils, *Water Air Soil Pollut.*, 32(4), 389, 1987.

Pierzynski, G.M. and Schwab, A.P., Bioavailibity of zinc, cadmium, and lead in a metal-contaminated alluvial soil, *J. Environ. Qual.*, 22, 247, 1993.

Sappin-Didier, V. and Gomez, A., Réhabilitation des sols pollués par des métaux toxiques. Exemple de l'apport de grenaille d'acier, *Analysis*, 22(2), M28, 1994.

Sims, J.T., Soil pH effects on the distribution and availability of maganese, copper and zinc, *Soil Sci. Soc. Am. J.*, 50, 367, 1986.

Sims, J.T. and Kline, J.S., Chemical fractionation and plant uptake of heavy metals in soils amended, *J. Environ. Qual.*, 152, 72, 1991.

Waren, C.J., Evans, L.J., and Sheard, R.W., Release of some trace elements from sluiced fly ash in acidic soils with particular reference to boron, *Waste Manage. Res.*, 11, 3, 1993.

Wong, J.W.C. and Wong, M.H., Effects of ash on yields and elemental composition of two vegetables, *Brassica parachinensis* and *B. chinensis*, *Agr. Ecosyst. Environ.*, 30, 251, 1990.

2

Remediation of Metal- and Radionuclides-Contaminated Soils by In Situ Stabilization Techniques

A.S. Knox, J.C. Seaman, M.J. Mench, and J. Vangronsveld

CONTENTS .

2.1 Introduction

The extent of metal and radionuclide contamination in the world is immense. In the soil environment, metals and radionuclides can be dissolved in solution, held on inorganic soil constituents through various sorption or ion exchange reactions, complexed with soil organics, or precipitated as pure or mixed solids. Soluble contaminants are subject to migration with soil water, uptake by plants or aquatic organisms, or loss due to volatilization (Smith et al., 1995). Lead (Pb), chromium (Cr), zinc (Zn), arsenic (As), and cadmium (Cd) are the most frequently identified inorganic contaminants in soil and groundwater in the order of their relative occurrence (National Research Council, 1994; Knox et al., 1999). Unlike degradable organic contaminants and even short-lived radionuclides that can become less toxic over time, metals can be considered conservative because they are not decomposed in the environment. However, many metals, especially redox-sensitive elements such as As and Cr, can undergo transformations or sorption reactions that alter both mobility and relative toxicity.

Soil contamination can have dire consequences, such as loss of ecosystem and agricultural productivity, diminished food chain quality, tainted water resources, economic loss, and human and animal illness. Public attention generally focuses on dramatic examples of contamination such as the nuclear accident in Chernobyl, Ukraine, where significant releases of radioactivity occurred (Adriano et al., 1997). The most dramatic ecological effects, however, were confined to a 30-km radius from the reactor. In contrast, extensive areas of eastern and central Europe suffer from diseases associated with elevated levels of Pb in the air, Co in the soil, and a food chain that is contaminated by metals related to heavy industry (Tikhonov, 1996). At present, there is a critical need for the development of cost-effective remediation technologies that reduce such risks.

Attention has focused on the development of *in situ* stabilization methods that are generally less expensive and disruptive to the natural landscape, hydrology, and ecosystem than conventional excavation, treatment, and disposal methods. A major drawback to such an approach, however, is that the metal or radionuclide contaminant remains in place and may, due to various physicochemical and biological processes, become a health or regulatory concern at a later date.

In situ remediation techniques for metals and radionuclides typically rely on a fundamental understanding of the natural geochemical processes governing the speciation, migration, and bioavailability of a given element in the environment. Remediation techniques can be placed in one of three general categories: (1) physical methods that simply restrict access to the contamination through containment or removal; (2) chemical methods that attempt to alter contaminant speciation to either enhance mobility under various extraction scenarios or decrease mobility to reduce potential exposure hazards; and (3) biological methods that attempt to use natural or enhanced biochemical processes to either increase contaminant mobility for extraction (e.g., phytoaccumulation) or reduce mobility by altering metal speciation. For instance, natural geochemical processes such as precipitation, sorption, ion exchange, and even redox manipulation have been employed as remediation methods.

This chapter will focus on the use of natural and synthetic soil amendments to reduce the mobility and bioavailability of contaminant metals and radionuclides without drastically altering the physical or chemical properties of the soil. An overview of the important mechanisms controlling contaminant stabilization will be given for additives such as lime, zeolites, phosphate compounds (e.g., apatite minerals), Fe and Mn oxides and other minerals, biosolids, and industrial by-products. Many of these amendments are inexpensive, readily available, and can be applied to large areas of contaminated soil without the need for costly excavation.

The efficacy of *in situ* stabilization is usually evaluated using one of the following approaches: (1) batch studies evaluating contaminant solubility and migration potential under controlled equilibrium conditions, many of which are designed to identify and characterize resulting mineral phases or specific sorption processes using spectroscopic techniques (e.g., Berti and Cunningham, 1997; Chen et al., 1996, 1997; Lothenbach et al., 1998; Ma et al., 1993; Ruby et al., 1994; Xu et al., 1994; Zhang et al., 1997); (2) dynamic leaching studies (i.e., columns) that simulate kinetically limiting conditions (e.g., Seaman et al., 1995, 1999); and (3) plant growth experiments that indicate bioavailability and long-term stability under variable moisture conditions which are more indicative of the field environment (e.g., Chlopecka and Adriano, 1996; Laperche et al., 1997; Lothenbach et al., 1998). This reflects a general progression from well-defined systems to more complex conditions typical of the real world. Often sequential extractions such as Tessier's method (Tessier et al., 1979) are combined with the above techniques to operationally define contaminant mineral associations, bioavailability and potential mobility, and chemical liability (Arey et al., 1999; Chlopecka and Adriano, 1996, 1997a, b, c, and d, 1999; Ma and Rao, 1997).

Selective extraction techniques successively liberate less-chemically labile phases and, ideally, their associated contaminants. The operational nature of such techniques is generally illustrated, however, by the apparent association of contaminants with specific phases that may not actually be present in the soil or that seem to contradict the likely mechanism of metal immobilization. For example, much of the research on *in situ* stabilization methods has focused on the use of apatite addition to promote the precipitation of sparingly soluble metal phosphates. Although selective extraction methods have been widely applied to such studies, it is still unclear which extract would likely solubilize such precipitates (Arey et al., 1999). In addition, the possible redistribution of contaminant metals during the extraction process cannot be discounted. Numerous studies have demonstrated that various soil amendments such as hydrous ferric oxid (HFO) and lime can result in significant phase redistribution of contaminant metals, as defined by extraction, which is not reflected by plant uptake (Sappin-Didier et al., 1997a and b; Müller and Pluquet, 1997; Chlopecka and Adriano, 1996). Therefore, batch extraction methods provide a means of rapidly screening numerous alternative treatment scenarios, especially when evaluating contaminant mobility, but the limitations of such methods necessitate the use of plant growth experiments and bioassays to assess biological availability.

In a sense the study of soil additives for contaminant metal immobilization parallels earlier research efforts related to soil fertility and the development of chemical fertilizers. Such studies are often empirical in nature because the addition of a reactive component to soils can have numerous predictable and unforeseen consequences. As will be discussed, a reduction in the solubility or plant availability of a given metal can result from several different and distinct chemical mechanisms. Like previous soil fertility studies, factors such as soil heterogeneity (i.e., soil type, mineralogy, contamination level, etc.), climate (i.e., moisture conditions, temperature, etc.), and the concentration and speciation of a given contaminant, as well as the specific plant species and genotype, all play an important role in determining the relative effectiveness of a given amendment under a specific set of conditions.

2.2 Techniques of *In Situ* Stabilization

2.2.1 Lime

Soils differ considerably in their pH and most crops grow best when the soil pH is between 6.5 to 7.0. For centuries, lime in various forms [e.g., $CaCO_3$, $(Ca, Mg)CO_3$, CaO, $Ca(OH)_2$] has been used to increase soil pH and thereby improve soil fertility. Lime is a cheap and effective ameliorant for many metals; however, repeated applications are required to maintain metal immobilization. Lime is applied frequently (2 to 5 years) and in quantities larger than any other inorganic soil amendment (typically 2 to 10 t/ha). Soil pH is an important factor controlling metal mobility and bioavailability. Usually the mobility of many metals increases with decreasing pH. With increasing pH, the solubility of most trace cations will decrease (Kabata-Pendias and Pendias, 1992). Sims and coworkers have demonstrated the effects of pH on micronutrient distribution among soil fractions. They found that more Cu, Fe, Mn, and Zn were in the exchangeable and organic fractions at low pH than at high (Sims and Patrick, 1978). Also, Iyengar et al. (1981) found that exchangeable Zn generally increases in soils with decreasing pH. This observation can be explained by the precipitation of metal hydroxides, changes in the carbonate and phosphate concentrations in the soil water, adsorption and desorption of metals by hydrous oxides and organic matter, and the formation and dissolution of Fe and Mn oxides. For example, the metals Cd and Zn are illustrative of the effect of pH on their mobility. Cadmium exists in the divalent form to pH 7.8 and only 50% is converted to the $Cd(OH)_2$ precipitate at pH 11. In contrast, 50% of

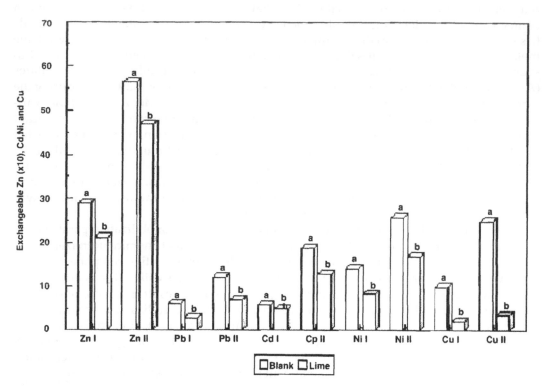

FIGURE 2.1
Exchangeable fractions extracted with 0.5 M MgCl$_2$ (in mg kg^{-1}) of Zn, Pb, Cd, Ni, and Cu in soil without lime (blank treatment) and in soil with lime. There were two following doses of metals (I, II; in mg kg^{-1}) in soil: Zn — 1000, 2000; Pb — 1500, 3000; Cd — 20, 40; Ni — 350, 700; Cu — 500, 1000; means followed by letters a and b are significantly different at $P<0.05$. (Reprinted with permission from Knox, 1998a, unpublished data).

Zn is in the Zn(OH)$_2$ form at pH 7.5, suggesting that at a given pH, Zn will be less mobile than Cd in a soil system.

Presented in this section are greenhouse data demonstrating the effect of lime on metal mobility and plant uptake (Knox, 1998a). In this experiment, an Appling soil was contaminated with the following metals at two treatment levels (mg kg^{-1}): Zn – 1000, 2000; Cd – 20, 40; Pb – 1500, 300; Cu – 500, 1000; Ni – 350, 700. The metals were added to the soil separately as a mixture of various metal sources (40% of sulfate, 25% of carbonate, 20% of oxide, and 15% of chloride). After equilibrating 4 weeks, lime was added at variable rates to adjust the pH from 5.4 to 6.5. In the control soil, native Zn was found mostly in the residual form (71% of total), with lesser amounts in the organic and the iron-manganese oxide fractions. In the contaminated soil (two levels of Zn), Zn increased in all fractions, with the largest increase in the exchangeable fraction (30 to 42% of the total Zn concentration in the soil) (Figure 2.1). Addition of lime significantly decreased the exchangeable fraction of Zn by increasing concentrations in the carbonate, iron-manganese oxide, and residual fractions.

In treatments where Pb was added, the initial soil pH was low (5.1) and the mobility of Pb was high, with 58.8 and 123 mg kg^{-1} found in the exchangeable fraction at low and high treatment levels, respectively. When the soil pH was raised to 6.1 by addition of lime, Pb mobility was significantly reduced by 54 and 45%, respectively, at the first and second level of Pb (Figure 2.2). For the other studied metals (Cd, Cu, and Ni) addition of lime to the soil significantly reduced the exchangeable fraction (Figure 2.1). In Zn, Cu, and Ni treated soil early plant mortality resulted, and yield of rye and maize was not obtained (Figure 2.2). Yield of these plants was obtained only in treatments with both levels of Pb and Cd

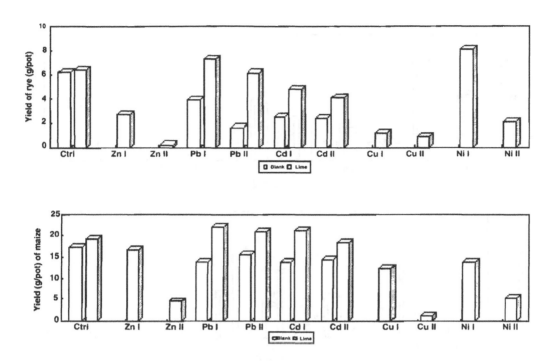

FIGURE 2.2
Yield of rye and maize (g/pot, DW); I and II — levels of metals in soil in mg kg⁻¹: Zn — 1000, 2000; Pb — 1500, 3000; Cd — 20, 40; Ni — 350, 700; Cu — 500, 1000; means followed by letters a and b are significantly different at P<0.05. (Reprinted with permission from Knox, 1998a. Unpublished data.).

(Figure 2.2). When soil pH increased to 6 or 6.5, rye and maize yield was obtained in treatments with Zn, Cu, and Ni. However, at the high metal level (i.e., Zn – 2000, Cu – 1000, Ni – 700 mg kg⁻¹) yield was much lower than in treatments with low metal level (Zn – 1000, Cu – 500, and Ni – 350 mg kg⁻¹) (Figure 2.2). Metal concentrations in plant tissues for treatments without lime were high. For example, Cd and Pb at the high metal application rate were 79.6 and 138 mg kg⁻¹ in rye tissues, respectively. In the control soil, Cd and Pb concentrations were only 0.7 and 1.3 mg kg⁻¹, respectively (Figure 2.3). Lime significantly decreased metal uptake by plants (Figure 2.3).

2.2.2 Zeolites

Zeolites are framework aluminosilicates consisting of extended three-dimensional networks of linked SiO_4 and AlO_4 tetrahedra. They possess interconnected channels or voids that form ideal sorption sites for both water and specific alkali and alkali-earth metals. Nearly 50 natural forms of zeolite have been identified and over 100 forms having no natural analogs have been synthesized in the lab. The unique physical and chemical properties of zeolites combined with their natural abundance in sedimentary deposits and volcanic parent materials have made them useful in many industrial processes.

Zeolites derive cation exchange capacity from Al^{3+} substitution for Si^{4+} with the size of the channel determining the type of exchangeable cation that is preferred (Breck, 1974). Recent greenhouse and field studies have demonstrated the ability of zeolites to reduce the uptake of Cs, Sr, Cu, Cd, Pb, and Zn in plants (Chelishchev, 1995; Leppert, 1990; Mumpton 1984; Mineyev et al., 1990; Rebedea, 1997; Rebedea and Lepp, 1994). However, zeolites are not effective sorbents for transuranic species, such as uranyl (UO_2^{2+}), that are commonly

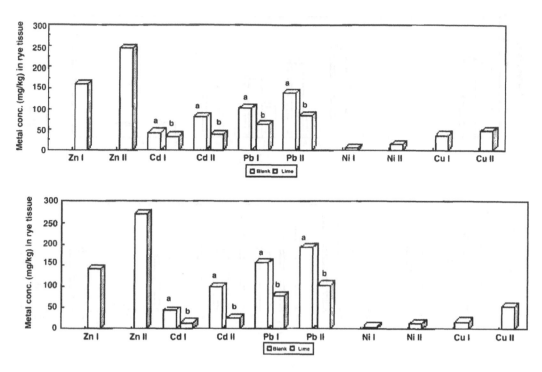

FIGURE 2.3
Concentrations (mg kg⁻¹) of Cd, Cu, Ni, Pb, and Zn in rye and maize tissues (I and II — levels of metals in soil in mg kg⁻¹: Zn — 1000, 2000; Pb — 1500, 3000; Cd — 20, 40; Ni — 350, 700; Cu — 500, 1000; means followed by letters a and b are significantly different at P<0.05 (Reprinted with permission from Knox, 1998a, unpublished data.).

found at sites with elevated Cs and Sr levels (Vaniman and Bish, 1995). To date, results from the use of zeolites as soil amendments to reduce plant uptake of radionuclides and contaminant metals have been mixed (Adriano et al., 1997; Leppert, 1990; Mineyev et al., 1990). For example, Mineyev et al. (1990) observed that clinoptilolite application to an acidic podzolic soil that had been spiked with Zn, Pb, and Cd reduced acid-extractable Zn, but not the concentration of acid-soluble Pb and Cd. Clinoptilolite application also reduced Zn and Pb accumulation in barley grain, but had no effect on yield. However, Chlopecka and Adriano (1996, 1997a) observed that the application of clinoptilolite (15 g kg⁻¹ soil) significantly decreased exchangeable Zn, Cd, and Pb. For example, the exchangeable Zn concentration decreased from 237 to 189 mg kg⁻¹ with zeolite application.

In another set of studies, Chlopecka and Adriano (1997c), Knox (1998b), and Knox and Adriano (1999) applied the natural zeolite, phillipsite, to soils from Canada, Poland, Taiwan, and the Czech Republic, which were contaminated with As, Cd, Pb, and Zn from mining, smelting, and other industrial activities. Zeolite was added at two rates, 25 and 50 g kg⁻¹, both of which significantly enhanced the yield of maize and oats and reduced the plant uptake of Cd, Pb, and Zn (Figure 2.4, Table 2.1). Zeolite application also influenced the plant uptake of both macro- and micronutrients for the Czech soil with an increase in plant tissue concentrations observed for Ca and Mg and a decrease of Mn from 933 to 256 mg kg⁻¹. For the highly polluted Czech soil, zeolite addition reduced the exchangeable Cd, Pb, and Zn by 43, 46, and 29%, respectively, but increased the concentration of each of those metals in the residual fraction (Knox, 1998b; Knox and Adriano, 1999). Other research groups have reported similar reductions in the uptake of Cd by lettuce after application of zeolite material from the foyazite group, type 4A (Rebedea et al., 1994; Rebedea, 1997). Rebedea et al. (1994) investigated the metal-binding capacity of three synthetic zeolites, 4A,

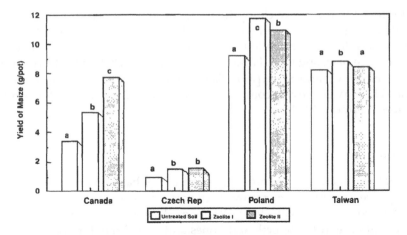

FIGURE 2.4

Yield of 6-week-old maize in the contaminated soils with two levels of zeolite (phillipsite); I — 25 g kg⁻¹, II — 50 g kg⁻¹; means followed by letters a, b, and c are significantly different at P<0.05 (Reprinted with permission from Chlopecka, A. and D.C. Adriano, Inactivation of metals in polluted soils using natural zeolite and apatite, in Proc. Extended Abstracts from the Fourth Int. Conf. on the Biogeochemistry of Trace Elements, Berkeley, CA, 415, 1997c.)

TABLE 2.1

Cadium, Pb, and Zn Concentration (mg kg⁻¹, DW) in Maize in Polluted Soils with Two Levels of Zeolite (Phillipsite)

Country	Treatment	Soil pH	Leaves			Roots		
			Cd	Pb	Zn	Cd	Pb	Zn
Canada	Untreated	7.08	3.30ᵃ	4.25ᵃ	104ᵃ	9.31ᵃ	25.3ᵃ	172ᵃ
	Zeolite I	7.39	1.60ᵇ	2.01ᵇ	101ᵃ	6.87ᵇ	14.0ᵇ	108ᵇ
	Zeolite II	7.55	0.91ᶜ	1.92ᵇ	79.2ᵇ	6.21ᵇ	13.6ᵇ	99.8ᵇ
Czech Rep.	Untreated	5.41	11.8ᵃ	9.50ᵃ	3188ᵃ	36.1ᵃ	121ᵃ	4823ᵃ
	Zeolite I	5.46	6.55ᵇ	7.25ᵇ	2065ᵇ	26.3ᵇ	76.5ᵇ	3558ᵇ
	Zeolite II	5.98	5.93ᶜ	7.20ᵇ	1874ᶜ	15.6ᶜ	73.7ᵇ	3025ᶜ
Poland	Untreated	6.23	2.71ᵃ	3.73ᵃ	149ᵃ	6.47ᵃ	34.5ᵃ	283ᵃ
	Zeolite I	6.39	2.12ᵃ	1.70ᵇ	103ᵇ	5.40ᵇ	9.33ᶜ	148ᵇ
	Zeolite II	6.74	1.82ᵇ	1.35ᶜ	96.7ᵇ	4.43ᶜ	12.2ᵇ	115ᶜ
Taiwan	Untreated	5.32	19.3ᵃ	44.0ᵃ	26.2ᵃ	56.9ᵃ	125ᵃ	39.4ᵃ
	Zeolite I	5.86	17.7ᵇ	12.3ᵇ	22.8ᵇ	43.1ᵇ	40.3ᵇ	32.8ᵇ
	Zeolite II	6.36	14.6ᶜ	8.23ᵈ	22.3ᵇ	30.8ᶜ	31.7ᶜ	29.3ᶜ

Note: Rates of amendments: I — 25 g kg⁻¹, II — 50 g kg⁻¹; means followed by letters a, b, c, d, e are significantly different at P <0.05.

Reprinted with permission from Chlopecka, A. and D.C. Adriano, Inactivation of metals in polluted soils using natural zeolite and apatite, in Proc. Extended Abstracts from the Fourth Int. Conf. on the Biogeochemistry of Trace Metals, Berkeley, CA, 415, 1997c.

P, and Y. Each displayed a high affinity for Cd, Cu, Pb, and Zn at low solution concentrations. Greenhouse pot experiments using contaminated soils demonstrated the ability of these synthetic zeolites to reduce the phytotoxicity for maize (Rebedea et al., 1994).

More than 30 years ago, Ames (1959) demonstrated the high selectivity of clinoptilolite for Cs. Certain zeolites also display high ion exchange selectivity for Sr (Chelishchev, 1973, 1995), even in the presence of Ca and Mg (Tsitsishvili et al., 1992). Chelishchev and others (Chelishev and Chelishcheva, 1980; Vaniman and Bish, 1995) demonstrated that the presence of clinoptilolite can reduce plant uptake of Cs and Sr. The fact that isotopes of Cs and Sr represent much of the radioactivity released at Chernobyl (Krooglov et al., 1990) has heightened interest in the use of clinoptilolite as a soil amendment. Selectivity for Cs and Sr is generally highest under slightly acidic conditions (pH 5.0–6.0), but drops off dramatically at lower pHs due to competition with hydrogen for exchange sites (Tsitsishvili et al., 1992). The presence of strong ligands, such as the synthetic chelate EDTA or citrate and tartate, can also reduce Cs and Sr sorption (Tsitsishvili et al., 1992). Other common metals, such as Ca and K, may effectively compete with the target contaminant for sorption sites in some instances (Leppert, 1990).

2.2.3   Apatite

The use of apatite minerals as a remediation strategy is based on recognized geochemical principles. Apatite minerals form naturally and are stable across a wide range of geologic conditions (Nriagu, 1974; Wright, 1990). Wright et al. (1987) investigated the trace element composition of apatite in fossil teeth and bones and in sedimentary phosphorite deposits through geologic time. They found that apatite deposited in seawater adsorbs metals and radionuclides from the seawater to millions of times the ambient concentration. The metals remain within the apatite structure indefinitely with little subsequent desorption, leaching, or exchange, even in the face of subsequent digenetic changes in the pore water chemistry, pH, or temperatures up to 1000°C. Apatite minerals act as natural collectors for metals and radionuclides. Apatite deposits in Florida, for example, have accumulated large amounts of uranium (U), enough in fact to be considered a commercial source of uranium (Eisenbud, 1987). A younger apatite deposit located in North Carolina is mined primarily for fertilizer. Young deposits have had less exposure to metals in the environment and are, therefore, generally more reactive.

Remediation studies on metal-contaminated wastes and soils using apatite or hydroxyapatite have focused mainly on Pb (Berti and Cunningham, 1997; Chen et al., 1996, 1997; Laperche et al., 1997; Ruby et al., 1994; Zhang et al., 1997). Ma et al. (1993, 1994, 1995) reported that before hydroxyapatite can be successfully used as a Pb-immobilizing material, three factors need to be considered. Hydroxyapatite must immobilize Pb^{2+} in the presence of interfering cations, anions, and dissolved organic matter; the resulting products must be stable in the contaminated environment, and the reaction should be rapid.

Current research demonstrates the successful precipitation not only of Pb, but also other metals when phosphate minerals are added to a contaminated medium (Misra and Bowen, 1981; Ma et al., 1993; Xu and Schwarz, 1994). LeGeros and LeGeros (1984) showed that there are three types of substitutions that can occur in hydroxyapatite or hydroxypymorphite structures. The cations Pb^{2+}, Ba^{2+}, Zn^{2+}, Fe^{3+}, and Mg^{2+} can substitute for Ca^{2+}, while the oxyanions AsO_4^{3-}, VO_4^{3-}, CO_3^{2-}, and SO_4^{2-} can replace structural PO_4^{3-}. Additionally, anions such as F^- and Cl^- can substitute for OH^- in the apatite structure. Ma et al. (1994) showed that hydroxypyromorphite $[Pb_5(PO_4)_3OH]$ precipitated after the reaction of hydroxyapatite with Pb^{2+} in the presence of NO_3^-, SO_4^{2-}, and CO_3^{2-}, while chloropyromorphite $[Pb_5(PO_4)_3Cl]$ and fluoropyromorphite $[Pb_5(PO_4)_3F]$ formed in the presence of Cl^- and F^-, respectively.

The sorption mechanisms are variable in the reaction between the apatite mineral and Pb, Cd, and Zn from contaminated soils. Lead removal results primarily from the dissolution of apatite followed by the precipitation of hydroxyl fluoropyromorphite. Minor otavite precipitation was observed in the interaction of the apatite with aqueous Cd, but other sorption mechanisms, such as surface complexation, ion exchange, and the formation of amorphous solids, are primarily responsible for the removal of aqueous Zn and Cd (Wright et al., 1995).

Other researchers found that the pH under which a reaction between metals and apatite occurs plays an important role. Wright et al. (1995) reported that the immobilization of Pb was primarily through a process of apatite dissolution followed by precipitation of various pyromorphite-type minerals under acidic condition, or the precipitation of hydrocerussite [$Pb_3(CO_3)_2(OH)_2$ or $Pb(OH)_2 2PbCO_3$] and lead oxide fluoride (Pb_2OF_2) under alkaline conditions. Otavite ($CdCO_3$) and cadmium hydroxide [$Cd(OH)_2$], and zincite (ZnO) were formed in the Cd or Zn system, respectively, especially under alkaline conditions; while hopeite [$Zn_3(PO_4)_2 4H_2O$] might only precipitate under alkaline conditions. Alternative sorption mechanisms other than precipitation were important in immobilizing Cd and Zn in the presence of apatite. The selectivity order of heavy metal sorption by apatite depends on pH. The selectivity order at pH below 7 was Pb>Cd>Zn, but at higher pH it was Pb>Zn>Cd. Removal of Pb, however, is less sensitive to pH (Wright et al., 1995).

Saeed and Fox (1979) studied the influence of phosphate fertilization on Zn adsorption by tropical soils and found that this component increased Zn adsorption in variable charge soils, suggesting that phosphate addition increases the negative charge on iron and aluminum oxides. However, for less-weathered soils, prior phosphate applications decreased Zn sorption because the fertilizer contained Zn as an accessory element.

Recent studies on the stabilization of metals by phosphate compounds have focused on the reduction of plant uptake, as well as the reduction in solubility/mobility. Chlopecka and Adriano (1997b, c), Knox (1998b), and Knox and Adriano (1999) found that apatite from North Carolina, consisting mainly of hydroxyapatite or fluorapatite, effectively immobilized Pb, Cd, and Zn in contaminated soils and decreased plant uptake by several crop species. In these studies, apatite (25 and 50 g kg^{-1}) addition to the contaminated soils significantly reduced the potential mobility of Cd, Zn, and Pb, as indicated by sequential extraction (Figure 2.5). For example, the concentration of Zn in the exchangeable fraction was reduced from 272 to 126 mg kg^{-1} at the highest apatite addition rate. The exchangeable fraction of Cd and Pb was reduced by 55 and 60%, respectively, at the highest apatite application rate, 50 g kg^{-1}. Data from this study clearly show partitioning of the contaminant metals to the residual fraction, with the highest increase observed for Pb (Figure 2.5). Apatite significantly improved plant growth and yield on highly contaminated soils (Figure 2.6). Also, concentrations of these metals in plant tissues like leaves or roots significantly decreased. For example, the Cd concentration in maize leaves decreased from 11.8 mg kg^{-1} to 5.8 and 4.8 mg kg^{-1}, respectively, in the control treatment (contaminated soil from Czech Republic) and treatments with 25 and 50 g apatite kg^{-1}. Reduction of Zn concentrations in these tissues was higher than for Cd (50, 59% and 51, 64%, respectively, for first and second dose of apatite for Cd and Zn) (Figure 2.7) (Chlopecka and Adriano, 1997c). In maize leaves the lowest reduction was obtained for Pb, 41 and 44% for first and second doses of apatite. Generally, the reduction in metal concentrations in maize roots was about 10 to 20% lower than for maize leaves (Chlopecka and Adriano, 1998). Apatite affects not only heavy metal concentrations in plants, but also essential micronutrients. Several studies have demonstrated that increasing addition of apatite to the soil decreases the concentration of P, Mn, and Fe in the plant tissues and generally increases the concentration of Ca and Mg (Boisson et al., in press; Grant and Bailey, 1992). Other studies have shown that heavy metal concentrations are not reduced in all plant tissues. Laperche et al. (1997) found the Pb content in shoot tissue decreased with increasing apatite addition. However, Pb and P contents in the plant roots increased as the quantity of added apatite increased. The authors hypothesized that Pb accumulates in/on the roots because it precipitates as lead phosphate. Without phosphate, Pb is readily translocated from roots to shoots, with similar Pb contents observed in both shoot and root tissues in unamended soils. This study also strongly suggests that, in the absence of other phosphate sources, plants can induce the dissolution of pyromorphite to facilitate P uptake. To prevent the release of Pb due to pyromorphite dissolution, soil-P levels for plants must be maintained in excess of that needed to immobilize Pb.

Arey et al. (1999) demonstrated the effectiveness of apatite addition in reducing uranium (U) solubility and TCLP-extractability in contaminated sediments from the Department of

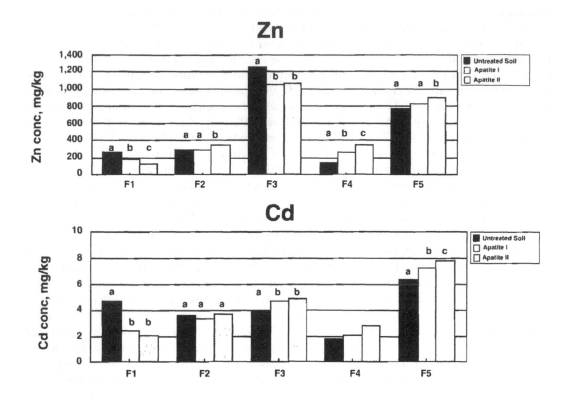

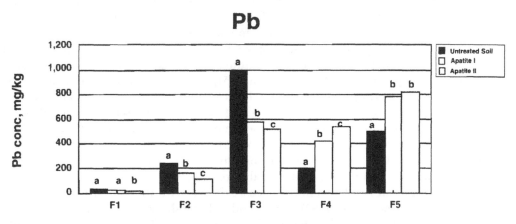

FIGURE 2.5

Zinc, Cd, and Pb fractions in Czech contaminated soil treated with apatite (fractions: F1 — exchangeable, F2 — carbonate, F3 — Fe-Mn oxides, F4 — organic, F5 — residual; sequential extraction by Tessier et al., 1979); means followed by letters a, b, and c are significantly different at P<0.05. (Reprinted with permission from Knox, 1998b, unpublished data.)

Energy's Savannah River Site, Aiken, SC, as well as several other contaminant metals, including Pb and Cd (Figure 2.8). However, complexation of U by dissolved organic carbon (DOC) in materials containing higher organic matter slightly decreased the effectiveness of apatite addition, presumably by lowering its free-ion activity in solution and thereby increasing U solubility for a given phosphate level.

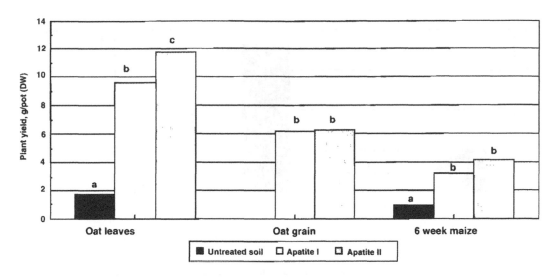

FIGURE 2.6

Yield of 6-week-old maize and oat (doses of apatite I — 25 g kg^{-1}, II — 50 g kg^{-1}; Cd, Pb, and Zn contaminated soil from Czech Republic); means followed by letters a, b, and c are significantly different at P<0.05. (Reprinted with permission from Chlopecka, A. and D.C. Adriano, Inactivation of metals in polluted soils using natural zeolite and apatite, in Proc. Extended Abstracts from the Fourth Int. Conf. on the Biogeochemistry of Trace Metals, Berkeley, CA, 415, 1997c.

2.2.4 Fe and Mn Oxides, Fe- and Mn-Bearing Amendments

Iron and Mn oxides as well as Fe- and Mn-bearing amendments have been investigated with respect to the reduction of metal mobility and plant and microorganism uptake (Juste and Solda, 1988; Czupyrna et al., 1989; Didier et al., 1992; Mench et al., 1994 a, b, c; Sappin-Didier, 1995; Boularbah et al., 1996; Müller and Pluquet, 1997; Verkleij et al., 1998; Mench et al., 1998 a, b). Iron oxides (hematite, maghemite, magnetite), oxyhydroxides (ferrihydrite, goethite, akaganeite, lepidocrocite, feroxyhite), and Mn oxides (phyllomanganates, birnessite group of minerals) are common and occur naturally in soils. The OH-OH distance in Fe, Mn, and Al oxides matches well with the coordination polyhedra of many trace metals; therefore, such hydroxyl groups form an ideal template for bridging trace metals. Consequently, hydrous oxides of Fe, Mn, and Al are highly reactive for many trace metals (Manceau et al., 1992a; Charlet and Manceau, 1993; Spadini et al., 1994; Hargé, 1997). Reactions between iron oxides and trace elements are well documented (Gerth and Brümmer, 1983; Kabata-Pendias and Pendias, 1992; Manceau et al., 1992 a, b; Spadini et al., 1994). They have been related to adsorbent aging, coagulation and rearrangement processes, and penetration of ions into the crystal lattice by exchange with lattice constituents near the mineral surface. Electron-microprobe studies confirm that metals in contaminated soils can accumulate in iron oxides (Hiller and Brümmer, 1995).

Trace elements form different types of surface complexes with the hydrous oxides. AsO_4^{3-} and Pb^{2+} form isolated inner-sphere surface complexes with hydrous ferric oxide (HFO), often used synonymously for ferrihydrite. Mononuclear complexes are formed by Zn^{2+}, Cd^{2+}, and Pb^{2+} on goethite and ferrihydrite surfaces (Manceau et al., 1992a; Spadini et al., 1994; Hargé, 1997). Most Cd ions sorb onto HFO at the termination of chains by sharing edges and corners with adjacent chains (Spadini et al., 1994). At high surface coverages, arsenate forms binuclear bidentate complexes on HFO, whereas at low surface coverages, mononuclear monodentate complexes are formed. Anions sorb mainly to goethite and ferrihydrite by two

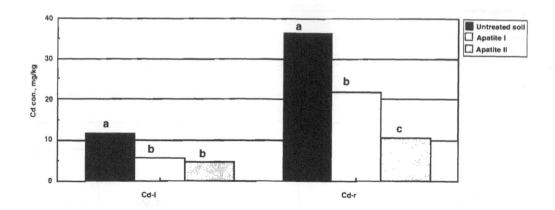

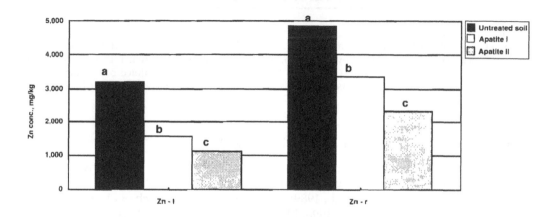

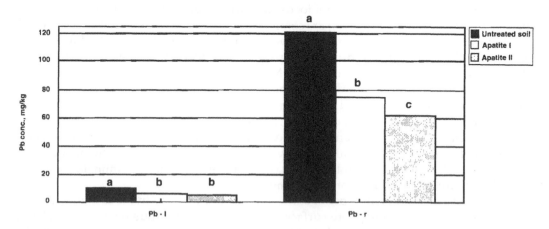

FIGURE 2.7
Cadmium, Zn, and Pb concentrations (mg kg^{-1}) in maize (l — leaves, r — roots; untreated soil — Cd, Pb, and Zn contaminated soil from Czech Republic); means followed by letters a, b, and c are significantly different at P<0.05. (Reprinted with permission from Chlopecka, A. and D.C. Adriano, Inactivation of metals in polluted soils using natural zeolite and apatite, in Proc. Extended Abstracts from the Fourth Int. Conf. on the Biogeochemistry of Trace Elements, Berkeley, CA, 415, 1997c.)

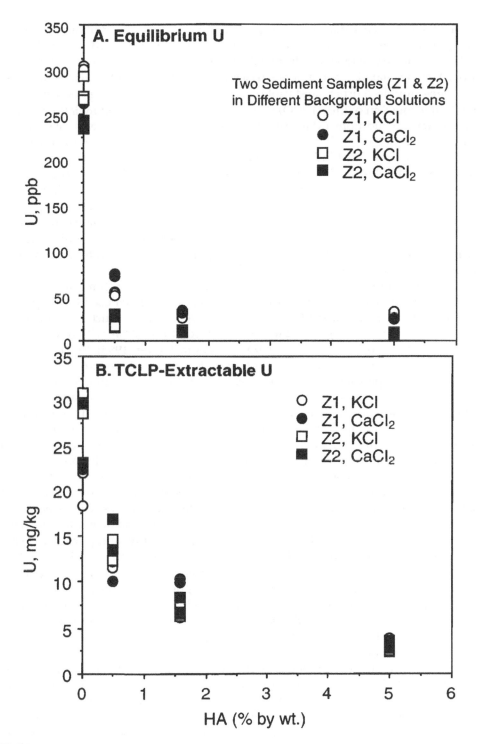

FIGURE 2.8
Effect of apatite amendment on equilibrium (A) and TCLP (Toxicity Characteristic Leaching Procedure) extractable U. (Reprinted with permission from Arey, J.S., J.C. Seaman, and P.M. Bertsch, Immobilization of U(VI) in contaminated sediments by apatite addition, *Environ. Sci. Technol.*, 33, 337, 1999.)

single coordinated groups attaching through double corner links. In addition, Cr, Cd, Pb, and other trace metals can substitute to varying degrees for Fe within the hydrous oxide structure depending on the degree to which the metal distorts the crystal lattice (Gerth, 1990).

Copper forms isolated inner-sphere surface complexes on MnO_2. Various surface sites such as edge-, double-, corner-, and triple corner-polyhedra linkages are observed for Pb^{2+} binding to hydrous Mn oxides (HMO) (Hargé, 1997). HMO can bind metals such as Pb or Cd even in acidic conditions (Manceau et al., 1992a, b). Permanent reactive sites are displayed on its surface layer. With a zero point charge for HMO ranging from 1.5 to 2.0, variable negative sites may bind cations under fairly acidic conditions. For example, more Cd may be bound to Mn oxides than to Fe oxides between pH 4.5 and 6.5 (Fu et al., 1991). Na-birnessite in neutral to acidic conditions can sorb large amounts of metals. Three Me-O-Mn bonds are formed by metals at the birnessite surface, and this mechanism accounts for the high binding affinity and low reversibility of metal sorption (Manceau et al., 1997; Sylvester et al., 1997). Zinc, Pb, and Cu form inner-sphere complexes with birnessite or birnessite-like structures such as the phyllomanganate chalcophanite ($ZnMn_3O_7 \cdot 3H_2O$) (Manceau et al., 1997; Hargé, 1997).

Fe- and Mn-bearing amendments (steel shot, Fe-rich™, and red mud) are waste by-products that contain high amounts of Fe. Steel shot is an industrial material used for shaping metal surfaces. It contains mainly iron (97% α-Fe) and native impurities such as Mn (0.6 to 1%), Si (0.8 to 1.2%), C (0.8 to 1.2%), Cr (0.2 to 0.5%), and Cu (0.1 to 0.3%), with trace amounts of Cd (3.6×10^{-6}%), Zn (0.01%), and Ni (0.074%) (Sappin-Didier, 1995). Steel shots readily corrode and oxidize to form several Fe oxides and Mn oxides depending on the environmental condition (Sappin-Didier, 1995; Hargé, 1997). In water, oxidation is rapid and visible after 15 min. In the same time, solution pH increases from 6 to 8, and maghemite, magnetite, and lepidocrocite are formed. A bag method was used to study the oxidation of steel shots in the soil (Sappin-Didier et al., 1997). All nonbiodegradable filter membranes recovered from soils after 9 months showed extensive oxide deposition and a high increase in Fe and Mn concentrations. Iron and Mn are likely released into solution, subsequently forming oxides in the soil. These may coat soil particles, developing a large surface for reaction with trace elements. X-ray diffraction analysis indicated iron oxides such as maghemite, lepidocrocite, and goethite formed on filter membranes when steel shot was confined to a bag in sandy soil, whereas lepidocrocite, goethite, and hematite were detected when steel shot was scattered into the soil. Hematite, magnetite, and pure iron were also detected following native steel shot amendment in German soil samples (Müller and Pluquet, personal communication, 1998). The increase in cation exchange capacity in a sandy soil with steel shot amendment (1% by soil weight, w/w) appeared marginal, from 8.4 to 9.3 cmol kg^{-1} (Mench et al., 1999b). Extended X-ray Absorption Fine Structure (EXAFS) indicates some α-Mn from the steel shot was transformed to a birnessite-like phyllomanganate compound (Manceau et al., 1997; Hargé, 1997).

Fe-rich, a byproduct from the processing of TiO_2 (Chlopecka and Adriano, 1996, 1997a), has a pH of 8.5 and a calcium carbonate equivalence of 33.5%. It contains poorly crystalline ferrihydrite (31.7% Fe), some Mn (1.76%), and Ca (10.3%), and trace metals (mg kg^{-1}): Cd – 20, Cr – 1272, Pb – 655, Ni – 104, and Zn – 260. Fe-rich™ was tested using a silt loam soil spiked with increasing quantities of flue dust, resulting in elevated Zn, Cd, and Pb (Chlopecka and Adriano, 1996, 1997a). Fe-rich (5% w/w) decreased the exchangeable Zn, Pb, and Cd fraction at each level of flue dust. The greatest decrease (>80%) occurred for Zn with the lowest flue dust dose, but the ameliorative effect was evident even at the highest flue dust application rates.

Red mud from the aluminum industry, sludge from drinking water treatment, bog iron ore, and steel shot waste from descaling of untreated steel plate have been used (Müller and Pluquet, 1997). Precipitated sludge from drinking water may consist mainly of ferrihydrite,

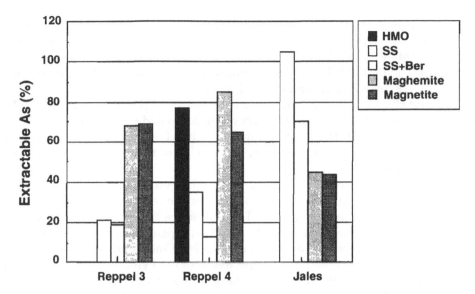

FIGURE 2.9

Arsenic concentration in soil extracts using 0.1 M calcium nitrate solution after soil treatment with various iron and manganese oxides and iron- and/or manganese bearing materials (concentration expressed in °₀ based on concentration in the untreated soil). (Reprinted with permission from Boisson, J., A. Ruttens, M. Mench, and J. Vangronsveld, Evaluation of hydroxyapatite as a metal immobilizing soil additive for the remediation of polluted soils, *Environ. Pollut.*, 104, 225, 1999.)

whereas red mud contains hematite, quartz, boehmite, gibbsite, and Ca silicates. Changes in the surface area of the soil occurred, from 10 to 15 m^2 g^{-1} after single application of either sludge from drinking water or native steel shot (Müller and Pluquet, 1997).

Early studies were concerned with contaminated soils treated with either iron sulfate or iron oxides (Förster et al., 1983; Juste and Solda, 1988; Czupyrna et al., 1989; Didier et al., 1992). In some initial tests, iron oxide addition to the soil appeared to be promising for trace metal immobilization (Didier et al., 1992; Sappin-Didier et al., 1997b). For example, calcium nitrate-exchangeable fractions of Cd and Zn decreased after a single application of HFO (1% w/w) in the Louis Fargue and Evin soils, but to a lesser extent than with HMO. Reduction of both extractable Cd and Zn by either maghemite and magnetite amendment in the soils was not as effective. The effectiveness of direct application of crystallized iron oxides into the soil was highest for As, with the water-soluble fraction reduced from 40 to 60°₀ depending on the soil sample (Figure 2.9).

Mn oxides and oxyhydroxides (birnessite and HMO) have been evaluated as additives in various metal contaminated soils, especially to reduce plant exposure to Cd and Pb (Didier et al., 1992; Mench et al., 1994a, b, c; Sappin-Didier, 1995). In coarse-textured, sludge-amended soils such as Ambares, addition of Na-birnessite (1% w/w) reduced extractable Cd and Zn. The effectiveness of HMO was similar to that of maghemite and magnetite for decreasing the water-soluble As fraction (Figure 2.9).

Single applications of steel shot (average size 0.35 mm), separately and in combination with beringite, have been tested in more than ten contaminated soils (Mench et al., 1994a, 1998, 1999a, b; Sappin-Didier, 1995; Gomez et al., 1997; Boisson in Verkleij et al., 1999). In all soils, the addition of steel shot reduced Cd (from 20 to 70%) and especially Zn mobility (from 10 to 90%), but the extent of immobilization likely depends on the specific soil conditions and metal speciation. Remarkable success was obtained in the reduction of water-soluble As in the Reppel soil (Figure 2.9). The combination of beringite (5% w/w) with steel shot was generally more effective in decreasing extractable metals and As (Figure 2.9).

Ameliorants such as HMO, with or without lime, beringite (5% w/w), and birnessite (1% w/w) were also effective in immobilizing Cd and Zn, and sometimes more effective than steel shot (Mench et al., 1999b). However, Mn oxides are not readily available and are difficult to apply in the field. To be effective, the soil application rates for beringite are generally three to five times higher than for steel shots.

Red mud and sludge from drinking water have been found to reduce the 1 M NH$_4$NO$_3$ extractable fractions of Cd and Zn in several contaminated German soils by over 50% (Müller and Pluquet, 1997). The other treatments such as bog iron ore and native steel shot showed less impact on extractable metals, and steel shot waste actually caused a small increase in extractable Zn.

The discussed amendments significantly reduce the potential mobility of contaminant metals as determined by sequential extraction. This section will provide examples of the biological evaluation of metal immobilization. The addition of HMO (0.5 to 1% w/w) in metalpolluted soils generally reduces the level of Cd and Zn in plant tissues, but the relative effectiveness depends both on soil type and plant species. HMO addition decreased the shoot Cd uptake by ryegrass by 76, 14, and 57%, respectively, in the Evin, Seclin, and Louis Fargue soils. Among the tested ameliorants, HMO was found to display the highest efficiency in reducing Cd availability to ryegrass shoots, irrespective of soil type or time of harvest (Mench et al., 1999b). The transfer factors for Cd in ryegrass, expressed as shoot dry weight Cd level vs. soil Cd level, were 0.15 and 0.27 for the untreated Louis Fargue and Evin soils. The addition of the HMO decreased these values to 0.07 (Evin) and 0.065 (Louis Fargue). The application of HMO was also effective at reducing Cd uptake by dwarf bean, e.g., Ambares soil; however, this plant species is susceptible to Mn toxicity. When Mn sensitive plant species are used, it is recommended that the application rate be reduced and combined with alkaline material such as lime.

Birnessite as HMO combined with lime was effective in reducing Zn and Cd accumulation in aerial bean parts. But the effect of birnessite addition on Zn availability for ryegrass did not persist beyond the third harvest. Excess Zn sorption may reflect the roots being potted in a small soil volume. Indeed, roots of some plant species are able to release Mn from Mn-oxides in the rhizosphere, and inorganic elements may also be recycled from root decomposition. Subsequent ryegrass cultures in pot experiments suggest the alteration of birnessite, which precludes its effectiveness, but further information must be gained in field trials.

Magnetite, maghemite, and hematite have been investigated in several experiments with mixed results depending on the target contaminant and soil type. Hematite was found to decrease leaf Cd uptake by 50% in dwarf bean, whereas maghemite was less effective and magnetite had no effect. Iron oxide treatment of As-contaminated garden soils resulted in a 50% reduction of water-soluble As and a similar decrease of As accumulation in dwarf bean leaves (Vangronsveld and Cunningham, 1998). However, use of maghemite and magnetite in the Reppel R3 soil led to an important decrease in shoot As uptake by maize (Figure 2.10).

Native steel shot (0.35 mm) was efficient in decreasing shoot Cd and Zn in ryegrass foliage cultivated in the Louis Fargues soil, especially for the first two harvests (Sappin-Didier, 1995). In contrast, Zn and Cd concentrations in ryegrass shoots were not affected by the addition of either lepidocrocite or stainless steel shot. This demonstrates the importance of Fe and Mn release in the soil solution and perhaps the *in situ* crystallization of some Fe and Mn oxides for immobilizing toxic metals in contaminated soils. Since the addition of crystalline iron oxides such as lepidocrocite, maghemite, and magnetite has been proven to be more or less unsuccessful in decreasing Cd and Zn mobility to a significant degree, changes induced by steel shot might be attributed to the presence of manganese. EXAFS indicates some Mn from steel shot was transformed to a birnessite-like phyllomanganate compound (Manceau et al., 1997), and birnessite addition to soil changes Cd and Zn mobility and plant

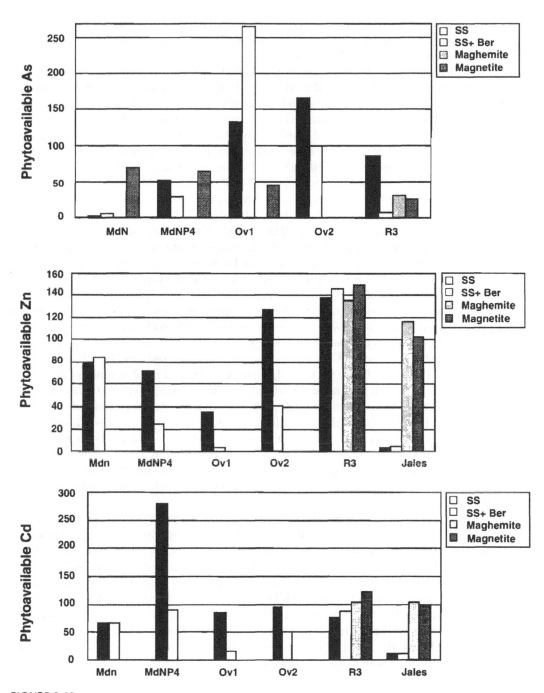

FIGURE 2.10

Shoot-Cd, Zn, and As uptake by maize (expressed in % based on plant uptake in untreated soil). (Reprinted with permission from Boisson, J., A. Ruttens, M. Mench, and J. Vangronsveld, Evaluation of hydroxyapatite as a metal immobilizing soil additive for the remediation of polluted soils, *Environ. Pollut.*, 104, 225, 1999.)

uptake (Mench et al., 1998 a, b). Thus, metal mobility and plant availability in steel shot-treated soils may be controlled by Mn-oxides as in the HMO- and birnessite-treated soils. The particle size of steel shot and their addition rate to soil are significant factors affecting plant availability of Cd and Zn in contaminated soils (Sappin-Didier, 1995). Steel shot with larger particle size was less effective in reducing shoot Cd and Zn uptake compared to the finest ones, despite a similar chemical composition. Indeed, 2% steel shot was more efficient than 1% for decreasing Cd and Zn uptake by ryegrass shoots, but this must be balanced against the additional cost of treating the contaminated soils. In addition, shoot phosphorus uptake by plants can be reduced by the addition of steel shot to the soil. Consequently, a decrease in ryegrass dry matter yield (20%) was found with 2% steel shot compared with the untreated soil, whereas the 1% rate had no effect (Sappin-Didier, 1995).

Other Fe-bearing materials like red mud, sludge from drinking-water treatment, and steel shot waste were added (1% pure Fe in soil) to a mud dredged from the harbor of Bremen (Germany) and deposited in a settling basin (Müller and Pluquet, 1997). This mud was polluted with Cd (4.2–7.1 mg kg^{-1}) and Zn (453–790 mg kg^{-1}). All treatments caused a reduction of the Cd concentration in both wheat grain and straw by over 30%. The relative order of effectiveness of the treatments in decreasing grain Cd content was red mud>steel shot waste> sludge from drinking water. All treatments lowered Zn uptake (e.g., 10 to 15% in wheat grain). Red mud, sludge from drinking-water treatment, and steel shot waste showed the best results for reducing Cd uptake by spinach (20 to 50%) and ryegrass (25 to 30%). Again, Zn uptake was influenced less by the treatments than Cd. Similar results on the effectiveness of Fe-bearing amendments were obtained with two other German soils contaminated by either mining effluents transported by a river or fallout from a former Pb/Zn smelter. Red mud was also tested using French soils. In all soils studied, a decrease in shoot Cd and Zn uptake by both ryegrass and beans was observed. After seven subsequent harvests, total Cd and Zn uptake by ryegrass was reduced by 60 and 30% in Evin soil, and by 51 and 18% in the Seclin soil, amended with red mud (Gomez et al., 1997).

Compared to lime, natural zeolite, and hydroxyapatite, Fe-rich™ was the most effective ameliorant in reducing the availability of Zn, Cd, and Pb from flue dust amended soils to maize, barley, and radish (Chlopecka and Adriano, 1996, 1997a, b). Only Fe-rich enhanced radish growth at all flue dust rates. The effectiveness of Fe-rich could have been partly due to its creation of alkaline conditions as well as its Fe-Mn fraction. Concomitant with the largest decrease of exchangeable Zn, Cd, and Pb by Fe-rich were substantial increases in these metals associated with the Fe-Mn oxide and carbonate fractions that may be indicative of their role as sorbents.

The overall effect of ameliorants and the persistence (durability) of trace metal or As immobilization in contaminated soils must be evaluated using various living organisms from different trophic levels because of the diversity of exposure pathways. Ecotoxicity tests can be used as a first approach. For example, a microbial assay (i.e., solid-phase MetPLATE™) has been used to study the effectiveness of steel shot (1% w/w), compared to that of basic slags (0.28% w/w) and HMO (1% w/w) for reducing metal toxicity in the Evin soil (Boularbah et al., 1996). This soil was highly toxic to bacteria (72% inhibition using MetPLATE); however, all three ameliorants were somewhat effective in reducing metal toxicity, with HMO being the most effective (44% inhibition) compared with basic slags (55% inhibition) and steel shot (62% inhibition).

2.2.5 Alkaline Composted Biosolids

Soil humic substances consist of a heterogeneous mixture of interacting functional groups, which include strong acidic (sulfonic acid -SO$_2$OH), weak acidic (carboxyl -COOH; hydroxyl-OH) as well as carbonyl (-C=O) and amine (-NH$_2$) groups (Schnitzer and Khan, 1972). Humic substances may interact in several ways with metal ions including ion exchange, complexation, adsorption and desorption, precipitation, and dissolution, thus affecting several physical and chemical properties of metals including their oxidation state and chemical form, apparent solubility, phase distribution, and speciation. This, in turn, will influence the mobility and transport, immobilization and geoaccumulation, bioavailability and bioaccumulation, and toxicity to organisms of environmental metals (Senesi and Sakellariadou, 1997).

Biosolids contain considerable amounts of organic solids and plant-essential nutrients like N or P. Biosolids can be applied to soils to enhance crop production; however, they may also contain high concentrations of trace elements like Cd, Cu, or Zn which may accumulate with repeated applications. In biosolids-amended soils, the pH may increase or decrease, depending upon the type and amount of biosolids added and the length of time since incorporation. Increasing pH generally enhances the sorption of metals by soils (Kuo and Baker, 1980; Soon, 1981). To minimize metal mobility and bioavailability in biosolids-amended soils, the U.S. EPA recommends the application of alkaline-stabilized biosolids to increase the soil pH to 6.5 or greater. Chaney (1997) considered that high Fe, lime-rich composted sludges have potential for the inactivation of metals in soil. This was demonstrated in remediation of Zn- and Cd-contaminated soils at Palmerton, PA. Chaney also found that limestone reacted very slowly in high Zn soils. In contrast, a mixture of limestone in biodegradable organic amendments such as biosolids or compost can raise soil pH to a depth of at least 1 m in coarse-textured soils. In France, the application of lime combined with organic matter has been used for more than 30 years to reduce Cu phytotoxicity in vineyards (Mench et al., 1999a). Pierzynski and Schwab (1993) showed that N-Viro (a heat-treated mixture of cement kiln dust and sewage sludge; Nviro Resources, Inc., Sioux City, IA) significantly increased soybean yields, and significantly decreased tissue Zn concentrations as compared to the control. Knox (1998c) also evaluated the efficacy of N-Viro in reducing the bioavailability of metals from a sandy loam soil as indicated by chemical extraction, crop growth, and tissue metal content. Metals were added to soil at two levels (in mg kg^{-1}): Cd–20, 40; Cu–500, 1000; Ni–350, 700; Pb–1500, 3000; Zn–1000, 2000, from various sources (40% sulfate, 25% carbonate, 20% oxide, and 15% chloride). After equilibration, addition of N-Viro (25g kg^{-1}) increased soil pH in all treatments. Early plant mortality resulted from Cu, Ni, and Zn-treated soil without N-Viro. In contrast, N-Viro application reduced the mobility and plant uptake of metals and significantly increased yield of rye and maize (Figures 2.11, 2.12; Table 2.2). Reduction of metal content in both plants was substantial for Cd, Cu, Pb, and Zn, with the highest reduction obtained for Cu and Ni (Table 2.2). The exchangeable fraction of these metals and metal contents in plant tissues were further reduced by N-Viro, in comparison with lime (Figure 2.11, Table 2.2).

Previous studies indicate that alkaline biosolids could possibly immobilize metals in contaminated soils, but further research is needed to determine the persistence of this effect in this area. However, humic substances are ineffective at stabilizing some radionuclides. Adriano et al. (1997) showed that ^{137}Cs bioavailability as indicated by plant uptake increased with increasing soil organic matter content in the soil.

2.2.6 Other Minerals and Industrial By-Products

Micas (illites) and vermiculites, 2:1 phyllosilicate clays possessing relatively high layer charge, have the ability to irreversibly fix or "sorb" certain elements and molecules within the clay interlayer region that are similar in size and hydration energy to K$^+$, such as NH$_4^+$ and Cs$^+$, significantly reducing their mobility and bioavailability in the soil environment. This ability to irreversibly fix such elements increases dramatically with pH from 2.5 to 5.5 and to a lesser degree from 5.5 to 7.0 (Sparks and Huang, 1985). Illitic materials may be more effective long-term stabilizing agents because of their ability to fix Cs$^+$ under a greater range of moisture conditions compared to vermiculites. Such fixation processes

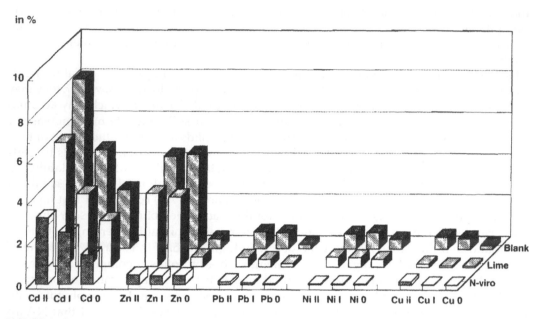

FIGURE 2.11

Exchangeable form of metals (extracted with 0.5 M MgCl$_2$) as % of their total amount in soil (0 — control soil, I and II levels of metals; Zn — 1000, 2000; Pb — 1500, 3000; Cd — 20, 40; Ni — 350, 700; Cu — 500, 1000). (Reprinted with permission from Knox, unpublished data, 1998c.)

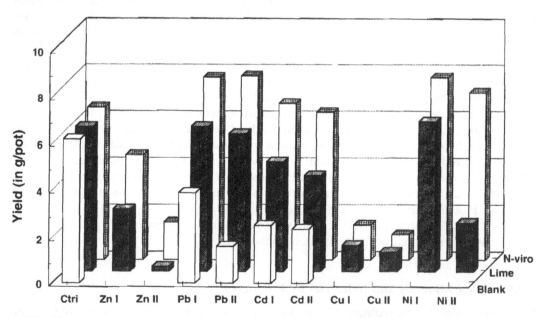

FIGURE 2.12

Yield of rye (g/pot, DW) (I and II levels of metals (mg kg^{-1}); Zn — 1000, 2000; Pb — 1500, 3000; Cd — 20, 40; Ni — 350, 700; Cu — 500, 1000). (Reprinted with permission from Knox, unpublished data, 1998c.)

are limited by the presence of Al-hydroxy interlayer polymers, such as those found in hydroxy-interlayered vermiculite/smectite (HIV/HIS), which block both general cation exchange and specific fixation sites, and inhibit interlayer collapse (Elprince et al., 1977; Maes et al., 1998).

TABLE 2.2

Reduction (%) of Metals by N-Viro in Rye and Maize;
Comparison with Blank and Lime Treatments

Metal	Blank		Lime	
	Rye	Maize	Rye	Maize
Zn 1000	—[a]	—	40.6	84.4
		—	15.9	71.7
Cu 500	—	—	33.9	28.3
Cu 1000	—	—	42.2	73.4
Ni 350	—	—	81.2	54.2
Ni 700	—	—	89.6	84.7
Pb 1500	63.7	68.6	39.3	36.4
Pb 300	51.7	65.6	20.3	36.5
Cd 20	52.4	85.7	38.5	54.9
Cd 40	65.3	89.5	28.1	55.4

[a] Yield was not obtained due to metal toxicity.

Reprinted with permission from Knox, unpublished data, 1998c.

In comparison, zeolites can initially be more effective at reducing Cs^+ uptake by plants than 2:1 clays, yet the effects may be less persistent than fixation within clay interlayers because of the dynamics of cation exchange on such minerals. The reversibility of such reactions reflects the subsequent release of the sorbed metal to the aqueous phase in response to changing geochemical conditions, a critical factor in the development of an effective remediation scheme. To a degree, this reflects the kinetics and reversibility of the two sorption processes with Cs exchange on zeolite and even external surfaces of phyllosilicate clays occurring more rapidly (Comans and Hockley, 1992; Komarneni, 1978; Sawhney, 1966). Exchange with interlayer cations, however, can be a much slower process, which may not be readily reversible (Comans et al., 1991; Comans and Hockley, 1992; Komarneni, 1978; Sawhney, 1966). Other factors such as the soil moisture content, and the presence of organic acids and cations of similar ionic radius, such as K^+ and NH_4^+, can affect Cs^+ sorption and fixation by soil clays (Comans et al., 1991; Hsu and Chang, 1994; McLean and Watson, 1985; Staunton and Roubaud, 1997). This discussion suggests that the relative effectiveness of interlayer fixation as a stabilization method depends largely on factors such as background solution composition (i.e., predominant cation), soil pH, clay mineralogy, and organic matter content (Staunton and Roubaud, 1997), indicating the necessity for site-specific data to predict the relative effectiveness of such a remediation scheme.

The ability of phyllosilicate clays to sorb heavy metals such as Pb, Cd, Cu, and Zn suggests their possible use as an amendment for contaminated soils, especially when added to extremely coarse-textured materials. The specific adsorption capacity of clay minerals for heavy metals, however, is relatively small compared to the nonspecific adsorption capacity. Distinguishing between adsorption and metal precipitation may be difficult under some circumstances, with sorption on clay minerals generally increasing with pH until the threshold for precipitation of the hydroxyl species is exceeded. Adsorption to minerals such as illite or kaolinite, which possess little external permanent negative charge or cation exchange capacity, is often more sensitive to changes in pH because the amphoteric clay edges play the dominant role in the sorption process. For clay minerals such as montmorillonite, sorption to basal surfaces that derive surface charge from mineral lattice imperfections and substitutions is less sensitive to external solution conditions.

Prior to precipitation, the mechanisms of sorption may be largely electrostatic in nature and, thus, nonspecific and subject to exchange with other common solution cations (Farrah and Pickering, 1976a, b, 1977; Lothenbach et al., 1997, 1998). Generally, the presence of the clay minerals can somewhat lower the pH threshold for metal precipitation because the clay can act as a nucleation center (Farrah and Pickering, 1976b, 1977). Since most contaminated soils already contain clay minerals, such benefits may be achieved by simply modifying the soil pH. As one might expect with any stabilization mechanism that depends to some degree on metal speciation and free-ion activity, the threshold for precipitation can shift to higher pH values in the presence of ligands that form stable anionic ligand-metal complexes (Farrah and Pickering, 1976b, 1977).

The chemical modification of 2:1 phyllosilicate clays through the precipitation of Al-hydroxy interlayers has shown promise as a means of increasing their ability to specifically bind heavy and transition metal cations (Keizer and Bruggenwert, 1991; Lothenbach et al., 1997, 1998). Research interest in this area originated to some degree because such materials are thought to be synthetic analogs for natural hydroxy-interlayer vermiculites and smectites which are widely distributed in highly weathered, acidic soils throughout the world (Harsh and Doner, 1984; Keren et al., 1977). As discussed above for contaminant metals, the clay mineral acts as a nucleation site for the precipitation and stabilization of a highly reactive, amphoteric, Al-hydroxy polymer. Batch sorption studies have revealed that such a surface modification enhances the ability to specifically bind metals such as Cu better than either the clay or similar hydrous aluminum oxide phases separately. This is presumably due to the higher specific surface area ($m^2 g^{-1}$) of the Al precipitate formed within the clay interlayer (Harsh and Doner, 1984; Keizer and Bruggenwert, 1991; Lothenbach et al., 1998). The ability to easily exchange sorbed cations from the modified clays is somewhat specific for a given metal. Such modification appears to be more effective at increasing the sorption affinity for Zn, Ni, Cu, and to a lesser degree Cd and Pb (Keizer and Bruggenwert, 1991). The sorption of Ni, Cu, and Zn on Al-hydroxy polymer modified clays appears to be specific and, therefore, less subject to competition from other cations such as Ca and Mg which are commonly present in the soil solution.

Specific adsorption by the Al-modified clays is highly dependent on the pH. At pH values below the threshold for specific sorption of a given metal on the Al-hydroxy interlayer, nonspecific cation exchange sites control metal sorption and the presence of the interlayer actually reduces the sorption capacity of the clay mineral (Keizer and Bruggenwert, 1991). At pH values above the threshold, the negative effect on CEC is compensated by specific sorption on the Al-hydroxy interlayer phase. The addition of exchangeable Ba does little to remobilize Zn, Ni, and Cu, while Cd was readily exchanged (Lothenbach et al., 1997). In addition, the affinity for some metals (e.g., Cu and Ni) appears to increase with prolonged sample aging (Harsh and Doner, 1984; Lothenbach et al., 1997).

Batch and greenhouse results look promising although the production costs may restrict widespread usage of Al-modified clays. However, as with any engineered or surface-modified material, the relative effectiveness may depend largely on the exact method of synthesis and aging conditions (Keizer and Bruggenwert, 1991; Keren et al., 1977). The addition of Al polymers reduces the available permanent negative charge, as indicated by a decrease in CEC, but actually increases the mineral surface area determined by BET gas sorption because the polymers act to prop open clay interlayers. However, the presence of such an interlayer can actually inhibit the fixation of poorly hydrated alkali cations such as K^+ and Cs^+ (Maes et al., 1998). Soil pH is a critical factor controlling the effectiveness of Al-modified clays, with the greatest reduction in metal availability occurring at pH conditions above 5.0. Under acidic conditions (pH < 5.0), however, the nontreated clay minerals may be more effective at reducing metal availability due to the higher permanent negative surface charge (i.e., CEC) (Keizer and Bruggenwert, 1991; Lothenbach et al., 1997).

Lothenbach et al. (1998) found that Al-montmorillonite was effective in specifically sorbing Zn and, thereby, reducing its exchangeability. They contend that such metals may eventually become incorporated or encased within the hydroxy Al polymer. In pot experiments

TABLE 2.3

Element Contents (Mn and Fe in mg kg^{-1} fresh weight; Mg, Ca, K, P, and S in mg g^{-1} fresh weight) in the First Leaf Pair of 6-Week-Old Tomato Plants Grown on the Untreated and Treated Zn, Pb, Cd, and Cu Contaminated Soil 3 Months after Application of the Amendments

Treatment	Mn	Fe	Mg	Ca	K	P	S
Untreated	1.23	4.39	1.24	1.90	2.10	0.47	0.44
CA	1.21	6.16	1.32	3.10	2.10	0.44	0.64
BE	2.80	3.87	0.98	1.00	1.03	0.22	0.32

Note: Given are means of 3 measurements. CA = treated with 5% of cyclonic ashes; BE = treated with 1% of bentonite.

Reprinted with permission from Vangronsveld, J., J. Colpaert, and K. Van Tichelen, Reclamation of a bare industrial area contaminated by non-ferrous metals: physico-chemical and biological evaluation of the durability of soil treatment and revegetation, *Environ. Pollut.*, 94, 131, 1996.

using contaminated soils, the Al-montmorillonite was far more effective at reducing plant uptake and Na$_2$NO$_3^-$ extractable Zn on an equivalent clay mass basis than montmorillonite. In batch sorption studies, the Al-montmorillonite was somewhat less effective at sorbing Cd and more dependent on the exact solution pH. Below pH 6.0, the unmodified montmorillonite was at least as effective, if not more, than the Al-montmorillonite in sorbing Cd, suggesting that permanent CEC was the important factor under acidic conditions. Cadmium sorption changed little with long-term aging and was subject to greater displacement than Zn in the presence of exchanging cations, suggesting an electrostatic sorption mechanism. For both Cd and Zn, greater remobilization after acidification was observed for the montmorillonite compared to the Al-treated montmorillonite.

The use of the natural untreated clay bentonite as a barrier lining at dumping grounds is well known. When tested as a soil additive to reduce metal mobility in contaminated soils, possible disadvantages of the use of this product were encountered (Spelmans et al., unpublished results). Although the addition of bentonite resulted in a strong reduction of the exchangeable fraction of metals, bacterial toxicity and phytotoxicity seemed to remain at increased levels. The adverse effects of bentonite could be explained by changes in the physicochemical properties of the soil, such as waterlogging, compaction, and reduced phytoavailability of Ca, K, and P (Table 2.3).

In Belgium different products originating from coal mine waste material (i.e., Beringite, Elutrilite, Metir) have been evaluated as "soil ameliorants." The main "active component" in this material are different clay minerals. The products called Elutrilite and Metir can be classified as "cold treated mine waste materials." The main treatments consist of grinding and wet classification (using flotation techniques) of the schists. Metir can be considered a more purified version of the Elutrilite. Both products were evaluated in terms of their metal immobilizing capacity after mixing them in contaminated soils with only limited success (Vangronsveld et al., 1992).

Beringite is the name given to cyclonic ashes originating from the fluidized bed burning of coal refuse (mine pile material) from the former coal mine in Beringen (northeast of Belgium). The primary purpose of the fluidized bed burning of this material was the recuperation of energy left in the coal refuse. The burnt material contains about 30% coal; the remaining fraction is inorganic and mainly consists of schists containing quartz, illite, kaolinite, chlorite, calcite, dolomite, anhydrite, siderite, and pyrite (De Boodt, 1991). Illite is the dominant clay present. The schists are burned by heating in an electronically guided fluidized bed oven at temperatures between 800 and 825°C. During the heating process, the schists undergo partial breakdown and recrystallization. Kaolinite, chlorite, and pyrite are transformed during the heating process. A new crystalline mineral called ettringite (6CaO Al$_2$O$_3$·3SO$_4$ 31H$_2$O) appears and the formation of minerals of the pyroaurite and hydrotalcite families is postulated (De Boodt, 1991). An important difference in the products Elutrilite and Metir mentioned above is that by heating the clay minerals at this temperature, the swelling and shrinking characteristics (in presence or absence of water) are almost eliminated, although the lamellar structure is conserved and the inner surfaces become more accessible for metal immobilization processes.

Due to the use of air suction (air current) most of the particles with a mean diameter of less than 0.002 mm (clay fraction) are separated in a cyclone (about 25% of the total ash fraction, mainly the modified clay fraction), which were shown to possess a very high metal immobilizing capacity (De Boodt, 1991; Vangronsveld et al., 1990, 1991, 1993, 1995 a and b, 1996; Vangronsveld and Clijsters, 1992; Mench et al., 1994, 1998). This capacity to immobilize metals is not surprising since the minerals mentioned above are known to possess high sorption capacities. In terms of chemical composition, the product contains the same elements as the original schists; SiO_2 and Al_2O_3 represent 52 and 30%, respectively. The pH of the product is strongly alkaline (about 11) due to the presence of MgO and CaO, which are formed during the heating of $CaCO_3$ and $(Ca, Mg)CO_3$ minerals present in the schists. The oxides form hydroxides ($Ca(OH)_2$ and $Mg(OH)_2$) when they come in contact with water, that are responsible for the high buffering capacity of the cyclonic ashes.

De Boodt (1991) suggests the high metal immobilizing capacity of the product to be based on a combination of chemical precipitation, ion exchange, and crystal growth. Based on changes in pH after application, results from selective and sequential extractions and plant availability of the metals in long-term growth and field experiments suggest a three-step sorption process (Vangronsveld et al., 1998): (1) an initial rapid first step (hours) representing adsorption onto highly accessible sites on the surface of the modified clay and on binding sites of the original soil components resulting from the "liming effect" (presence of $Ca(OH)_2$ and $Mg(OH)_2$); followed by (2) a slower type (days) of sorption characteristic for modified surfaces (i.e., coprecipitation associated with Al, Fe, and Mn oxides); and (3) on the longer term (years) crystal growth and metal diffusion into the mineral surface. This last step should be responsible for the permanent decrease observed for the chemical extractability and availability to plants (Vangronsveld, 1998). With aging metal diffusion into the mineral structure can occur. Gerth and Brummer (1983) reported the diffusion of Zn, Ni, and Cd into goethite with the rates of these three trace elements in the order Ni<Zn<Cd, paralleling their ionic radii of 0.35, 0.37, and 0.49 Å, respectively. The same authors suggest that metal adsorption is determined by three different steps that parallel the mechanisms described above: surface adsorption, diffusion into the mineral, and fixation at positions within the mineral. Similar metal diffusion processes have been observed for manganese oxides, illite (one of the predominant clays in the Beringite), and smectite clays (Gerth, 1985). Based on EXAFS observations, the formation of metal silicates after Beringite addition to soils also is postulated (Hargé, 1997). Experiments are in progress to further elucidate the working mechanism of the product in the field.

Improved metal affinity can be obtained by addition of aluminum salts during the heating process. This results in surface coating by aluminum hydroxides which increases chemical adsorption of the metals (De Boodt, 1991). This "ameliorated" product which is similar to the Al-treated montmorillonite described earlier has not been extensively tested in soils since (1) the efficiency of the original cyclonic ashes was satisfactory, and (2) this treatment significantly increased the cost of the final product.

The optimal amount of cyclonic ashes to be added to metal-contaminated soils can vary as a function of the soil type and the degree of contamination. Application rates must be determined based on both the efficiency of the additive to immobilize metals and its effect on the plant availability of essential elements. For a soil contaminated with Zn, the optimal concentration was found to be 5% (Vangronsveld et al., 1990, 1995a).

In sandy soils, the addition of cyclonic ashes also enhances the water holding capacity. Increased amounts of available Ca, Mg, and SO_4, which are present in the cyclonic ashes, are reflected in a better growth of plants compared to the untreated soils, even for noncontaminated soils (Table 2.3). The increase in nutrients is due to the presence of rather high amounts of $CaSO_4$ and $MgSO_4$ in the mine pile material.

It could be argued that reduction of metal availability observed in cyclonic ash amended soils results solely from an increase in soil pH, with sorption mechanisms described above being of lesser importance. The effect of pH was tested in a series of experiments that compared the addition of cyclonic ashes or lime to an acidic sandy soil that was contaminated with Zn, Cd, and Pb. The cyclonic ashes were applied at a 5% rate; lime was added at such a concentration that the same soil pH was obtained. A significant difference was found between both treatments for all the parameters that were investigated (selective extractions, phytotoxicity tests; Vangronsveld et al., unpublished results). Metal immobilization efficacy was much lower after liming. Zinc uptake, for instance, was 74 to 126% higher for plants grown on the limed soils compared to the Beringite-treated soils 6 weeks after the treatment (unpublished results). At longer time intervals, the efficacy of lime further decreases (Chlopecka and Adriano, 1996).

Results indicate that the cyclonic ashes have a long-lasting residual effect on the bioavailability and leaching of metals from both heavily contaminated industrial soils and garden soils (Vangronsveld et al., 1995b, 1996; Vangronsveld, 1998).

In 1990, 3 ha of a highly metal polluted acidic sandy soil at the site of a former pyrometallurgical Zn smelter were treated with a combination of cyclonic ashes and compost (Vangronsveld et al., 1995). After soil treatment and sowing of a mixture of metal-tolerant *Agrostis capillaris* and *Festuca rubra*, a healthy vegetation cover developed. Metal assimilation by the plant cover was minimal with the effect of soil treatment clearly reflected in the metal contents in the grasses (Figure 2.13). Five years after application, the soil physico-chemical parameters, potential phytotoxicity, floristic and fungal diversity, and mycorrhizal infection of the plant community were evaluated (Vangronsveld et al., 1996). Phytotoxicity still was at the low level observed immediately after soil treatment. The water-extractable metal fraction of the treated soil was up to 70 times lower than the nontreated soil (Table 2.4). The vegetation was still healthy and regenerating by vegetative means and by seed. Species diversity of higher plants and saprophytic fungi remained extremely low in the untreated area due to the high soil toxicity. On the treated soil, the diversity of higher plants was much higher. Several perennial forbs which are not noted to be metal tolerant had colonized the amended area. In 1997, a total of 32 higher plant species, including seedlings of two woody species, were observed.

Kitchen gardens with sandy soils contaminated by aerial deposition from the same pyrometallurgical Zn smelter mentioned above were treated with cyclonic ashes (1 ton/acre, incorporated to a depth of 25 cm). Soils were contaminated with Zn (92 to 983 mg kg^{-1}), Cd (3.1 to 9.4 mg kg^{-1}), Pb (170 to 682 mg kg^{-1}), and Cu (31 to 107 mg kg^{-1}). Total Cd contents in the top 25 cm of soil are presented in Table 2.3. The same vegetables were grown on untreated and treated plots in each garden. A comparison of Cd contents in edible parts of these plants at harvest time (after normal washing) showed strong (factor 2 to 4) reductions in total Cd content (Table 2.5).

The persistence of metal immobilization in these garden soils treated by cyclonic ashes (5% by wt) was also tested in a simulation experiment. Column tests were performed using a soil containing Zn (730 mg kg^{-1}), Cd (8 mg kg^{-1}), and Pb (300 mg kg^{-1}). The effect of natural rainfall (600 mm/year) for 30 years was simulated using a slightly acid rain water as the percolating fluid with the metal concentrations determined in the column effluent. The evolution of the plant available fraction of metals was evaluated using both chemical

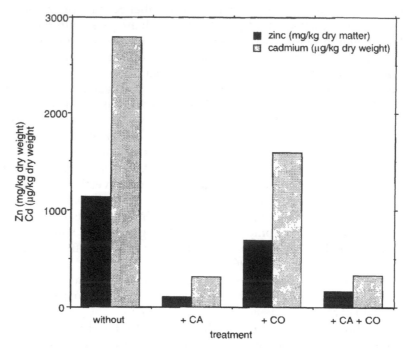

FIGURE 2.13

Metal concentration (Zn in mg kg⁻¹, DW; Cd in mg kg⁻¹, DW) in the aerial parts of the grasses grown in the field experiment after soil treatment with cyclonic ashes (CA) and/or compost (CO). (Reprinted with permission from Vangronsveld, J., J. Colpaert, and K. Van Tichelen, Reclamation of a bare industrial area contaminated by non-ferrous metals: physico-chemical and biological evaluation of the durability of soil treatment and revegetation, *Environ. Pollut.*, 94, 131, 1996.)

extraction and biological methods (phytotoxicity test, metal accumulation in bean, spinach, and lettuce). The total leachable Zn and Cd decreased significantly for the treated soil (Vangronsveld, 1998). At the end of the 30-year simulation, the mean total amount of percolated Cd and Zn from the untreated columns was 2371 and 296 mg, respectively, compared to 383 mg (= 16% of the untreated) and 27 mg (9% of the untreated) for the Beringite-treated columns. The cumulative percolation of Cd is given as an example (Figure 2.14).

For the treated soils, the bioavailability of metals decreased (Table 2.6). Zinc and Cd uptake in the leaves of lettuce plants grown on cyclonic ash treated soils gradually decreased to the level observed in control plants (Table 2.4). For the untreated soil, the bioavailability of Cd showed a very drastic increase at the end of the 30-year simulation period. For Zn, a slight decrease was observed. In the leaves of bean and spinach similar results were observed.

The field and simulation studies mentioned above confirm the long-term sustainability of metal immobilization after soil treatment with cyclonic ashes. These results were confirmed by Wessolek and Fahrenhorst (1994) who studied the efficacy of cyclonic ashes to immobilize heavy metals in soils from a former sewage disposal site with high levels of Zn and Cd. The immobilization effects for different additions of cyclonic ashes were described in laboratory studies based on the cation exchange capacity and a two-component Freundlich sorption isotherm. Long term displacement of Zn and Cd was then simulated with a numerical model. The simulation studies show a high immobilization of Zn and Cd in the solid phase for at least the next 80 years. Higher rates of cyclonic ash addition are required for the immobilization of Cd than for Zn.

TABLE 2.4

Total Zn Concentration (mg kg⁻¹ soil), Water-Extractable Zn (mg kg⁻¹ soil), and Ratio of
Water-Extractable Zn on Total Zn Concentration Measured 5 Years after the Soil Treatment
(given are mean ± SD of three extractions)

Treatment	Total Zn	Water-Extractable Zn	Ratio
Uncontaminated garden soil	106 ± 7	0.7 ± 0.06	0.660
Contaminated without vegetation	114 ± 506	141.0 ± 111.2	1.234
Contaminated with natural vegetation	960 ± 51	10.4 ± 0.81	1.085
Contaminated after soil treatment	12075 ± 701	2.1 ± 0.23	0.017

Reprinted with permission from Vangronsveld, J., J. Colpaert, and K. Van Tichelen, Reclamation of a bare
industrial area contaminated by non-ferrous metals: physico-chemical and biological evaluation of the durability
of soil treatment and revegetation, *Environ. Pollut.*, 94, 131, 1996.

TABLE 2.5

Cadmium Contents (mg kg⁻¹) in Soil of 10 Kitchen Gardens and in Spinach, Lettuce, and Celery
(mg kg⁻¹, FW) Cultivated *in situ* with and without 5% Cyclonic Ashes Treatments

Garden	Soil Cd	Cd in Spinach		Cd in Lettuce		Cd in Celery	
		Original	Treated	Original	Treated	Original	Treated
Control	0.6	0.11	a	0.06	a	0.08	a
1	4.1	0.96	0.32	0.18	0.06	0.39	0.18
2	5.1	0.46	0.23	0.14	0.05	0.26	0.13
3	9.4	0.77	0.47	0.25	0.09	1.28	0.42
5	7.4	1.48	0.57	0.78	0.18	1.37	0.81
6	6.0	0.68	0.49	0.13	0.07	0.55	0.30
7	6.1	1.85	0.71	0.15	0.08	0.43	0.24
8	8.5	a	0.53	0.63	0.15	0.68	0.32
9	8.8	0.87	0.42	0.33	0.12	0.46	0.25
10	3.1	1.23	0.42	a	a	a	a

a No plant material available. Given are means of 3 measurements.

Reprinted with permission from Vangronsveld, J., J. Colpaert, and K. Van Tichelen, Reclamation of a bare
industrial area contaminated by non-ferrous metals: physico-chemical and biological evaluation of the durability
of soil treatment and revegetation, *Environ. Pollut.*, 94, 131, 1996.

2.2.7 *In Situ* Redox Manipulation

The manipulation of redox status offers a promising treatment alternative for several common soil and groundwater contaminants that display distinctively different toxicity (e.g., Cr) and/or mobility (e.g., Cr, U, Tc) characteristics depending on redox speciation. For example, chromium (Cr), a common waste product resulting from various industrial plating processes, exists in the soil environment in one of two redox states, the relatively benign and immobile Cr^{3+}, and the toxic hexavalent forms, chromate (CrO_4^{2-}) or dichromate ($Cr_2O_7^{2-}$), which are generally considered to be more mobile in the environment (Bartlett, 1991; Davis and Olsen, 1995; James, 1996; Kent et al., 1995). In fact, reduction of Cr(VI) within the subsurface environment is often cited as the major process limiting its migration (Anderson et al., 1994; Henderson, 1994; Kent et al., 1995; Puls, 1994). The difference in Cr mobility and toxicity as a function of redox state, however, suggests two disparate approaches to remediation: (1) enhanced recovery through the use of oxidizing agents (Davis and Olsen, 1995) and (2) *in situ* stabilization through chemical reduction of Cr(VI) to form Cr(III) (Davis and Olsen, 1995; James, 1996; Saleh et al., 1989).

Redox manipulation provides a potential strategy for addressing mixed-metal/organic waste sites that are typical of many contaminated areas, especially those contaminated with

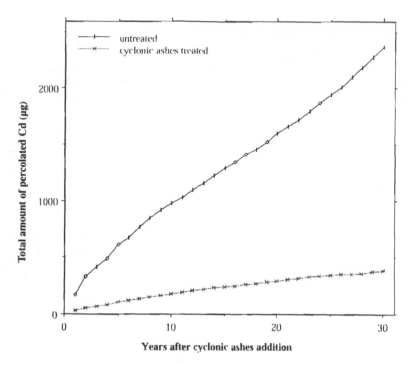

FIGURE 2.14

Total amount of Cd (µg) percolated from untreated and cyclonic ash treated soil columns during the simulated 30 year leaching period. Given are means of 3 untreated and 4 treated columns. (Reprinted with permission from Vangronsveld, J., J. Colpaert, and K. Van Tichelen, Reclamation of a bare industrial area contaminated by non-ferrous metals: physico-chemical and biological evaluation of the durability of soil treatment and revegetation, *Environ. Pollut.*, 94, 131, 1996).

TABLE 2.6

Zinc and Cd Contents (mg kg⁻¹ FW) in Lettuce Grown on Soils Originating from the Percolation Experiment

	Metal Content Leaves	
Soil	Zn (mg kg⁻¹ FW)	Cd (mg kg⁻¹ FW)
Control	4.82	0.08
Untreated	12.01	0.31
1 year	6.99	0.18
3 years	6.49	0.12
6 years	6.67	0.15
10 years	7.84	0.16
15 years	8.47	0.18
20 years	1.98	0.09
30 years	2.48	0.09
Untreated	5.82	0.55

Note: Given are means of 3 measurements.

Reprinted with permission from Vangronsveld, J., J. Colpaert, and K. Van Tichelen, Reclamation of a bare industrial area contaminated by non-ferrous metals: physico-chemical and biological evaluation of the durability of soil treatment and revegetation, *Environ. Pollut.*, 94, 131, 1996.

halogenated aliphatic organics, such as TCE, which tend to resist aerobic degradation. The complexities associated with redox processes, however, which are often kinetically controlled and difficult to predict based on chemical equilibria relations (Cui and Eriksen, 1996a; Cui and Eriksen, 1996b; Lindberg, 1984; White and Yee, 1985), may limit widespread application of such techniques in the field.

This section will deal mainly with the potential obstacles encountered during *in situ* reduction of Cr(VI), a major inorganic groundwater contaminant that is second only to Pb in terms of the number of impacted sites in the United States (National Research Council, 1994). Effective reductants of Cr(VI) include organic wastes such as animal manure, organic acids, sulfides, elemental Fe, and Fe(II) salt solutions (Davis and Olsen, 1995; Eary and Rai, 1988; Elovitz, 1994; 1995; James, 1996; Patterson et al., 1997; Saleh et al., 1989; Suciu et al., 1991; Wittbrodt and Palmer, 1995, 1996). At pH values above 5, Cr(III)aq readily precipitates as insoluble Cr-hydroxides (Gauglhofer and Bianchi, 1991). Unfortunately, Cr(III) may be reoxidized in the presence of Mn oxides (Bartlett, 1979; Saleh et al., 1989).

The inherent ability of Fe(II)-bearing minerals to reduce Cr(VI) has also been demonstrated (Bidoglio et al., 1993; Patterson et al., 1997); however, their effectiveness in reducing Cr(VI) may depend in part on the solubility of the solid-phase Fe(II) source, indicating that reduction in some instances may actually occur in solution rather than at the mineral surface (Eary and Rai, 1991; Patterson et al., 1997). Such an attenuation mechanism, rather than sorption of the chromate species itself, has often been implicated in Cr retardation (Kent, 1995; Anderson, 1994; Henderson, 1994). Similar surface-mediated reactions may also be operable for Se(VI), As(V) (Myneni et al., 1997), and technetium (^{99}Tc), a long-lived ($t_{1/2} = 2.13 \times 10^5$ yr) fission product commonly found in nuclear waste. Technetium generally persists as the anionic pertechnetate species, TcO_4^-, under oxic conditions which is only weakly sorbed and generally thought to be quite mobile. Recent studies have demonstrated the surface-mediated reduction of Tc to the sparingly soluble tetravalent form, $TcO_2 \cdot nH_2O_{(s)}$, and demonstrated the effectiveness of sorbed Fe(II) and various Fe(II) bearing solids as reductants (Cui and Eriksen, 1996a; Cui and Eriksen, 1996b), illustrating the importance of surface mediated redox reactions in natural or intrinsic remediation processes.

The use of various solid-phase reduced Fe sources, such as iron filings, chemically reduced aquifer sediments, and even pyritic materials has been proposed as engineered reactive, flow-through barriers to the migration of redox sensitive contaminants, such as Cr(VI) and chlorinated hydrocarbons (Amonette et al., 1994; Blowes et al., 1997; Gillham et al., 1994; Johnson and Tratnyek, 1994; Kaplan, 1994; Powell et al., 1995; Sayles et al., 1997). A discussion of engineered reductive barriers is beyond the scope of this review; however, the use of such barriers may be impractical at many sites due to the depth and extent of contamination. The general effectiveness of Fe(II) solutions in reducing Cr(VI), even under fairly aerobic conditions (Eary and Rai, 1988), suggests their surface application or subsurface injection may be an effective means of dealing with extensive Cr(VI) contamination.

In the presence of Fe(II), Cr(VI) reduction proceeds rapidly as:

$$Cr(VI)_{(aq)} + 3Fe(II)_{(aq)} \rightarrow Cr(III)_{(aq)} + 3Fe(III)_{(aq)} \tag{1}$$

with the reduction of 1 mol of Cr(VI) requiring 3 mol of Fe(II) (Buerge and Huh, 1997; Eary and Rai, 1988; Fendorf and Li, 1996; Saleh et al., 1989; Sedlak and Chan, 1997). In the presence of Fe(III), Cr(III) readily precipitates as a mixed Cr(III)/Fe(III) phase (Eary and Rai, 1988; Fendorf and Li, 1996):

$$xCr(III) + (1 - x)Fe(III) + 2H_2O = (Cr_xFe_{1-x})(OH)_{3(s)} + 3H^+ \tag{2}$$

for which the solubility of Cr(III) is less than that of the pure $Cr(OH)_{3(am)}$, and thus more resistant to reoxidation by Mn oxides (James and Bartlett, 1983). Such a technique is commonly used in the wastewater treatment industry, but poses certain obstacles to adoption as an *in situ* method, many of which have not been adequately investigated or discussed.

Seaman et al. (1999) demonstrated the efficacy of such an approach in batch systems where the Cr(VI) concentration decreased with increasing Fe(II) addition in a similar fashion for both buffered and unbuffered treatment solutions (Figure 2.15). However, the $Cr_{dissolved}$ [i.e., Cr(III)] in unbuffered, filtered samples leveled off at ≈ 2–3 mg L^{-1} despite the presence of residual unreacted Fe(II) (Figure 2.15). The high levels of Cr(III), well above the current drinking water standards (DWS), 0.1 mg L^{-1}, resulted from the lower pH, ≈ 3.3, induced by Fe(III) and Cr(III) hydrolysis (Figure 2.15). Despite having a minimum rate at pH 4.0 (Buerge and Huh, 1997; Sedlak and Chan, 1997), oxidation of Fe(II) by Cr(VI) in the presence of dissolved oxygen is kinetically favored under acidic conditions and influenced by the presence of reactive surfaces and the identity of the predominate counter anion (i.e., SO_4^{2-} vs. Cl^-) (Langmuir, 1997; Stumm and Morgan, 1981; Sung and Morgan, 1980). In buffered systems (pH ≈ 5.6), the level of $Cr_{dissolved}$ was below the DWS despite even greater Fe(II) oxidation than required for Cr(VI) reduction because of the higher pH which remained above 5.4 at all Fe(II) treatment levels (Figure 2.15).

Similar results have been reported for batch studies evaluating the intrinsic reduction potential of subsurface soils under acidic conditions (Eary and Rai, 1991), with the total $Cr_{dissolved}$ in this and previous studies being slightly less than the predicted value based on the solubility of a mixed Fe/Cr hydroxide, $Fe_{0.75}Cr_{0.25}(OH)_3(am)$, which begins to exceed current DWS at a pH of ≈ 4.7. In addition to the increase in solubility, the resulting acidic pH is below the adsorption edge for Cr(III) on hydrous ferric oxides (Dzombak and Morel, 1990), another potential mechanism retarding Cr migration. A decrease in pH has also been observed after the application of $FeCl_2 \cdot 4H_2O$ to Cr(VI) contaminated soils by Davis and Olsen (1995), but the initial pH and buffering capacity of the soil was adequate to induce Cr(III) precipitation.

Although certain problems have been identified, batch studies clearly suggest that *in situ* reduction may be an appealing treatment alternative if the solution pH can be controlled to ensure Cr(III) precipitation. Under aerobic conditions where Cr(VI) is likely to persist, however, a greater percentage of the added Fe(II) may be consumed by O_2 in buffered systems presumably due to the faster oxidation kinetics at higher pH. These concerns are not a problem in the water-treatment industry because of the ability to both add additional Fe(II) reactants and adjust the solution pH with base prior to final disposal. Any attempt to buffer or control the pH in the field is likely to reduce the efficiency of Cr(VI) reduction by increasing competition with O_2.

The application of such an *in situ* reduction method is most likely warranted in highly weathered, aerobic, oxide-coated systems where Cr persists in the oxidized state, i.e., Cr(VI), and can be significantly retarded by anion sorption to Fe and Al oxides (Seaman et al., 1999; Stollenwerk and Grove, 1985). This type of system poses both the greatest health risks due to Cr(VI) and difficulty in remediation because of the impact of sorption on contaminant extraction techniques, such as pump-and-treat. However, dynamic column experiments used to mimic the batch studies described above confirmed the importance of pH and identified additional obstacles to the use of Fe(II)-containing solutions as an *in situ* reductant for Cr(VI). The *in situ* fixation of Cr(VI) as Cr(III) was difficult in column systems (i.e., columns), not because of the inability to reduce Cr using Fe(II), but because the acidity generated by such a process inhibits Cr(III) precipitation, consistent with the batch results. Since Cr(VI) contamination is likely to persist in oxic environments, one must buffer such a reaction in favor of the formation of a mixed Fe(III)/Cr(III) precipitate (i.e., pH > 4.7)

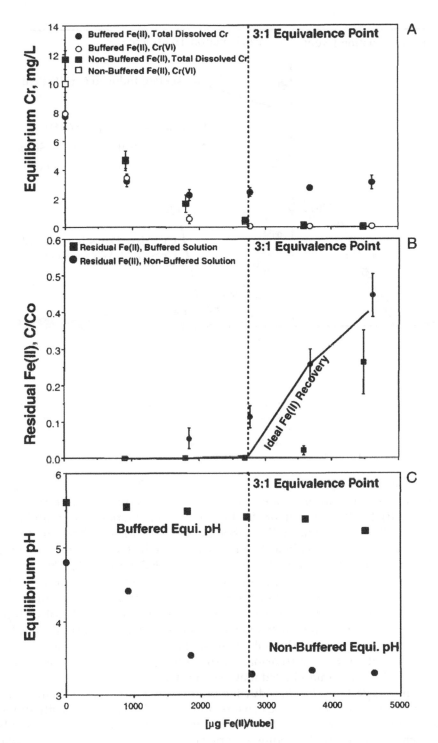

FIGURE 2.15

In situ reduction of Cr(VI) in batch using buffered and unbuffered Fe(II) containing solutions: (A) equilibrium Cr, (B) Fe(II), and (C) pH. (Reprinted with permission from Seaman, J.C., P.M. Bertsch, and L. Schwallie, In situ reduction of Cr(VI) within coarse-textured, oxide-coated soil and aquifer system using Fe(II) containing solutions, *Environ. Sci. Technol.*, 33, 938, 1999.)

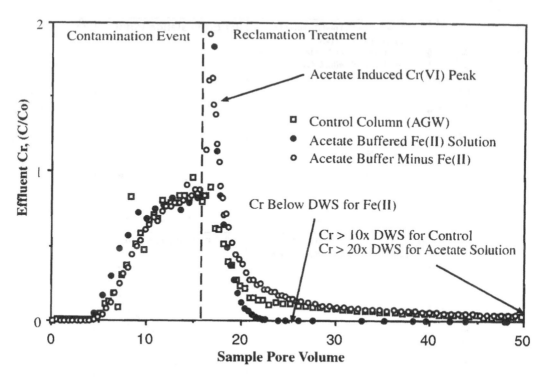

FIGURE 2.16

Chromium breakthrough behavior for columns leach for 16 PV with Cr (VI) (52 mg/L) followed by one of three remediation treatments: (1) artificial ground water (i.e., control), (B) acetate buffer solution, and (C) buffered Fe(II)–sulfate solution. (Reprinted with permission from Seaman, J.C., P.M. Bertsch, and L. Schwallie, In situ reduction of Cr(VI) within coarse-textured, oxide-coated soil and aquifer system using Fe(II) containing solutions, *Environ. Sci. Technol.*, 33, 938, 1999.)

without greatly increasing Fe(II) losses due to dissolved O_2. Acetate was chosen to buffer the system because of its appropriate pKa, 4.75; its relative inability to form strong complexes with Fe(II), Fe(III), or Cr(III) or affect the reduction kinetics (Buerge and Huh, 1997); compatibility with the ferrozine Fe(II) method (Stookey, 1970); and the desire to reduce the potential for other unforeseen photochemical reactions which can occur in the presence of ligands such as oxalate and citrate (Buerge and Huh, 1997; Hug et al., 1997). As observed for SO_4^{2-}-containing treatments, the competition with acetate for anion sorption sites enhanced Cr(VI) migration ahead of the Fe(II) reactant front, significantly enhancing mobility and leaving less Cr(VI) available for reduction (Figure 2.16). Until such problems have been solved, the application of *in situ* reduction using soluble reactants may be limited to surficial contamination where both the reductant and possibly a liming material to neutralize acidity resulting from Fe(II) oxidation and subsequent hydrolysis can be effectively incorporated within the contaminated material of interest.

These results suggest that *in situ* reduction or any *in situ* treatment scheme that utilizes soluble reactive constituents may be considerably less effective in field situations due to the imperfect mixing associated with advective transport, a process that cannot be mimicked in batch experiments. Amendment application may actually increase contaminant mobility if any constituent of the treatment solution competes with the contaminant of concern for sorption sites. This suggests that such a system must only be used under circumstances where hydraulic control of both the contaminant and the reactant plume can be maintained.

2.3 Summary and Conclusions

In situ stabilization techniques are relatively simple, inexpensive, and dramatically less disruptive to the natural landscape, hydrology, and ecosystem than conventional excavation, extraction, and disposal methods. The studies discussed in this review highlight the potential application and limitations of several natural and industry-derived materials as soil ameliorants. Both the immediate effectiveness and long-term sustainability, however, of such stabilization methods must be demonstrated to achieve widespread public and regulatory acceptance. With the exception of the use of Fe(II)-containing solutions to reduce and stabilize Cr(VI), the techniques discussed in this review generally rely on insoluble or sparingly soluble materials such as hydroxyapatite, lime, clay minerals, and zeolites. Their use, however, physically limits these techniques to areas of shallow contamination where such materials can be effectively applied and intimately incorporated within the zone of interest. This restriction to the surficial environment also dictates that plant uptake be evaluated as a potential vector for transfer to animals and humans, assuming the soil amendment was effective at restricting contaminant migration to groundwater. Simple batch (i.e., sorption and extractability) and column techniques can be used to easily estimate the potential mobility of a given contaminant under various leaching scenarios, but the evaluation of bioavailability may be somewhat more difficult and subjective. A high rate of amendment application not only increases reclamation costs, but also increases the chance of detrimentally altering soil properties. For example, the same sorption mechanisms that act to limit toxic metal availability can reduce the availability of certain plant essential micronutrients as well. In many instances, periodic application of additional stabilizing/immobilizing agents, such as lime or hydroxyapatite, may be necessary to ensure continued effectiveness. As the previous studies demonstrate, the complexities of the soil system often make it difficult to predict the consequences of such innovative methods without site-specific data.

Acknowledgments

The authors would like to acknowledge the support of the U.S. Department of Energy through Financial Assistance Award Number DE-FC09-96SR18546 to the University of Georgia Research Foundation. The authors would also like to acknowledge the thoughtful comments of Dr. C. Strojan on an early version of the manuscript.

References

Adriano, D.C., J. Albright, F.W. Whicker, I.K. Iskandar, and C. Sherony, Remediation of soil contaminated with metals and radionuclide-contaminated soils, in *Remediation of Soils Contaminated with Metals*, I.K. Iskandar and D.C. Adriano, Eds., Science Reviews, Northwood, 1997.

Ames, L.L., Zeolitic Extraction of Cesium Aqueous Solutions, U.S. Atomic Energy Comm. Unclass. Rep. HY-62607, 23, 1959.

Amonette, J.E., J.E. Szecsody, H.T. Schaef, J.C. Templeton, Y.A. Gorby, and J.S. Fruchter, Abiotic reduction of aquifer materials by dithionite: a promising in-situ remediation technology, in *In-Situ Remediation: Scientific Basis for Current and Future Technologies*, G.W. Gee and N.R. Wing, Eds., Vol. 2., Battelle Press, Pasco, WA, 1994.

Anderson, L.C.D., D.B. Kent, and J.A. Davis, Batch experiments characterizing the reduction of Cr(VI) using suboxic material from a mildly reducing sand and gravel aquifer, *Environ. Sci. Technol.*, 28, 178, 1994.

Arey, J.S., J.C. Seaman, and P.M. Bertsch, Immobilization of U(VI) in contaminated sediments by apatite addition, *Environ. Sci. Technol.*, 33, 337, 1999.

Bartlett, R. and B. James, Behavior of chromium in soils: III. Oxidation, *J. Environ. Quality*, 8, 31, 1979.

Bartlett, R.J., Chromium cycling in soils and water: links, gaps, and methods, *Environ. Health Perspectives*, 92, 17, 1991.

Berti, W.R. and S.D. Cunningham, In-place immobilization of Pb in Pb-contaminated soils, *Environ. Sci. Technol.*, 31, 1359, 1997.

Bidoglio, G., P.N. Gibson, M. O'Gorman, and K.J. Roberts, X-ray absorption spectroscopy investigation of surface redox transformations of thallium and chromium on colloidal mineral oxides, *Geochim. Cosmochim. Acta*, 57, 2389, 1993.

Blowes, D.W., C.J. Ptacek, and J.L. Jambor, In-situ remediation of Cr(VI)-contaminated groundwater using permeable reactive walls: laboratory studies, *Environ. Sci. Technol.*, 31, 3348, 1997.

Boisson, J., A. Ruttens, M. Mench, and J. Vangronsveld, Evaluation of hydroxyapatite as a metal immobilizing soil additive for the remediation of polluted soils, *Environ. Pollut.*, 104, 225, 1999.

Boularbah, A., J.L. Morel, G. Bitton, and M. Mench, A direct solid-phase assay specific for heavy-metal toxicity. II. Assessment of heavy-metal immobilization in soils and bioavailability to plants, *J. Soil Contam.*, 5, 395, 1996.

Breck, D.W., *Zeolite Molecular Sieves*, Krieger Publishing Co., Malabar, FL, 1974, 771.

Buerge, I.J. and S.J. Huh, Kinetics and pH dependence of chromium(VI) reduction by iron(II), *Environ. Sci. Technol.*, 31, 1426, 1997.

Chaney, R., personal communication, 1997.

Charlet, L. and A. Manceau, Structure, formation and reactivity of hydrous oxide particles: insights from X-ray absorption spectroscopy, in *Environmental Particles*. Environmental Analytical and Physical Chemistry Series, Vol. 2, Buffle, J. and H.P. van Leeuwen, Eds., Lewis Publishers, Boca Raton, FL, 1993.

Chelishchev, N.F., *Ion-Exchange Properties of Minerals*, Nauka, Moscow, 1973, 202 (in Russian).

Chelishchev, N.F. and R.V. Chelishcheva, Importance of ion-exchange properties of natural zeolites for removal of toxic metals from the digestive canal, in *Natural Zeolites in Agriculture*, Krupennikova, A.J., Ed., Metsniereba Press, Tbilisi, 1980 (in Russian).

Chelishchev, N.F., Use of natural zeolites at Chernobyl, in *Natural Zeolites '93 Occurrence, Properties, Use*, D.W. Ming and F.A. Mumpton, Eds., Int. Comm. Natural Zeolites, Brockport, NY, 1995.

Chen, X., L.M. Peurrung, J.V. Wright, and J.L. Conca, In-situ immobilization of lead from aqueous solutions and contaminated soils by phosphate rocks, in *Proc. Spectrum 96 Int. Conf. on Nuclear and Hazardous Waste Manage.*, 1996, 1.

Chen, X., J.V. Wright, J.L. Conca, and L.M. Peurrung, Effect of pH on heavy meals sorption on mineral apatite, *Environ. Sci. Technol.*, 31, 624, 1997.

Chlopecka, A. and D.C. Adriano, Mimicked in-situ stabilization of metals in a cropped boil: bioavailability and chemical form of zinc, *Environ. Sci. Technol.*, 30, 3294, 1996.

Chlopecka, A. and D.C. Adriano, Influence of zeolite, apatite, and Fe-oxide on Cd and Pb uptake by crops, *Sci. Total Environ.*, 207, 195, 1997a.

Chlopecka, A. and D.C. Adriano, Zinc uptake by plants on amended polluted soils, *Soil Sci. Plant Nutr.*, 43, 1031, 1997b.

Chlopecka, A. and D.C. Adriano, Inactivation of metals in polluted soils using natural zeolite and apatite, in Proc. Extended Abstracts from the Fourth Int. Conf. on the Biogeochemistry of Trace Elements, Berkeley, CA, 1997c, 415.

Chlopecka, A. and D.C. Adriano, Mimicked in-situ stabilization of lead-polluted soils, in Proc. Extended Abstracts from the Fourth Int. Conf. Biogeochemistry of Trace Elements, Berkeley, CA, 1997d, 423.

Comans, R.N., M. Haller, and P.D. Pretter, Sorption of cesium on illite: non-equilibrium behaviour and reversibility, *Geochim. Cosmochim. Acta*, 55, 433, 1991.

Comans, R.N.J. and D.E. Hockley, Kinetics of cesium sorption on illite, *Geochim. Cosmochim. Acta*, 56, 1157, 1992.

Cui, D. and T.E. Eriksen, Reduction of pertechnetate by ferrous iron in solution: influence of sorbed and precipitated Fe(II), *Environ. Sci. Technol.*, 30, 2259, 1996a.

Cui, D. and T.E. Eriksen, Reduction of pertechnetate in solution by heterogeneous electron transfer from Fe(II)-containing material, *Environ. Sci. Technol.*, 30, 2263, 1996b.

Czupyrna, G., R.D. Levy, A.I. MacLean, and H. Gold, *In-Situ Immobilization of Heavy-Metal-Contaminated Soils*, Noyes Data Corp., Park Ridge, NJ, 1989, 155.

Davis, A. and R.L. Olsen, The geochemistry of chromium migration and remediation in the subsurface, *Groundwater*, 33, 759, 1995.

De Boodt, M.F., Application of the sorption theory to eliminate heavy metals from waste waters and contaminated soils, in *Interactions at the Soil Colloid – Soil Solution Interface*, G.H. Bolt, M.F. De Boodt, M.H.B. Hayes, and M.B. McBride, Eds., NATO ASI Series, Series E: Applied Sciences, 1991, 190.

Didier, V., M. Mench, A. Gomez, A. Manceau, D. Tinet, and C. Juste, Rehabilitation of cadmium-contaminated soils: efficiency of some inorganic amendments for reducing Cd-bioavailability, *C. R. Acad. Sci. Paris*, 316 (série III), 83, 1992.

Dzombak, D.A. and F.M.M. Morel, *Surface Complexation Modelling: Hydrous Ferric Oxide*, John Wiley & Sons, New York, 1990, 393.

Eary, L.E. and D. Rai, Chromate removal from aqueous wastes by reduction with ferrous ion, *Environ. Sci. Technol.*, 22, 972, 1988.

Eary, L.E. and D. Rai, Chromate reduction by subsurface soils under acidic conditions, *Soil Sci. Soc. Am. J.*, 55, 676, 1991.

Eisenbud, M., *Environmental Radioactivity*, 3rd ed., Academic Press, San Diego, CA, 1987, 125.

Elovitz, M.S. and W. Fish, Redox interactions of Cr(VI) and substituted phenols: products and mechanisms, *Environ. Sci. Technol.*, 29, 1933, 1995.

Elovitz, M.S. a. W.F., Redox interactions of Cr(VI) and substituted phenols: kinetic investigation, *Environ. Sci. Technol.*, 28, 2161, 1994.

Elprince, A.M., C.I. Rich, and D.C. Martens, Effect of temperature and hydroxy aluminum interlayers on the adsorption of trace radioactive cesium by sediments near water-cooled nuclear reactors, *Water Resour. Res.*, 13, 375, 1977.

Farrah, H. and W.F. Pickering, The sorption of copper species by clays. II. Illite and montmorillonite, *Aust. J. Chem.*, 29, 1976a.

Farrah, H. and W.F. Pickering, The sorption of zinc species by clay minerals, *Aust. J. Chem.*, 29, 1649, 1976b.

Farrah, H. and W.F. Pickering, The sorption of lead and cadmium species by clay minerals, *Aust. J. Chem.*, 30, 1417, 1977.

Fendorf, S.E. and G. Li, Kinetics of chromate reduction by ferrous iron, *Environ. Sci. Technol.*, 30, 1614, 1996.

Förster, C., H. Kuntze, and E. Pluquet, Influence of iron in soils on the cadmium uptake of plants, in *Proc. 3rd Int. Symp. on Sewage Sludge, Brighton*, P. L'Hermite, Ed., Reidel Publishing Co., Dordrecht, 1983, 426.

Fu, G., H.E. Allen, and C.E. Cowan, Adsorption of cadmium and copper by manganese oxide, *Soil Sci.*, 152, 72, 1991.

Gauglhofer, J. and V. Bianchi, Chromium, in *Metals and Their Compounds in the Environment*, E. Merian, Ed., VCH Publishers, New York, 1991.

Gerth J. and G. Brümmer, Adsorption und Festlegung von Nickel, Zink und Cadmium durch Goethit (-FeOOH), *Fresenius Zeitschrift für Anal. Chem.*, 316, 616, 1983.

Gerth, J., Untersuchungen zur Adsorption van Nickel, Zink und Cadmium durch Bodentonfractionene Unterschiedlichen Stoffbestandes und Verschiedene Bodenkomponenten, dissertation, University of Kiel, 1985.

Gerth, J., Unit-cell dimensions of pure and trace metal associated goethites, *Geochim. Cosmochim. Acta*, 54, 363, 1990.

Gillham, R.W., D.W. Blowes, C.J. Ptacek, and S.F. O'Hannesin, Use of zero-valent metals in in-situ remediation of contaminated groundwater, in *In-Situ Remediation: Scientific Basis for Current and Future Technologies*, G.L. Gee and N.R. Wing, Eds., Vol. 2, Battelle Press, Pasco, WA, 1994.

Gomez, A., A. Vives, V. Didier-Sappin, T. Prunet, P. Soulet, and R. Chignon, Immobilization In Situ des Mètaux Lourds par Ajout de Composés Minéraux en Sols Pollués. Final report 92056, Ministère de l'Environnement, Paris, 1997.

Grant, C.A. and L.D. Bailley, Interactions of zinc with banded broadcast phosphorus fertilizer on the concentration and uptake of P, Zn, Ca, and Mg in plant tissue of oilseed flax, *Can.J. Plant Sci.*, 73, 17, 1993.

Hargé, J.C., Speciation comparée du zinc, du plomb et du manganese dans des sols contaminés. Ph.D. thesis, Université Joseph Fourier, Grenoble, France, 1997.

Harsh, J.B. and H.E. Doner, Specific adsorption of copper on an hydroxy-aluminum-montmorillonite complex, *Soil Sci. Soc. Am. J.*, 48, 1034, 1984.

Henderson, T., Geochemical reduction of hexavalent chromium in the trinity sand aquifer, *Ground Water*, 32, 477, 1994.

Hiller, D.A. and G.W. Brümmer, Mikrosondenuntersuchungen an unterschiedlich Stark mit Schwermetallen belasteten Böden. 1. Methodische Grundlagen und Elementanalysen an pedogenen Oxiden, *Z. Pflanzenernählr. Bodenk.*, 158, 147, 1995.

Hsu, C.-N. and K.P. Chang, Sorption and desorption behavior of cesium on soil components, *Int. J. App. Radiat. Isotopes*, 45, 433, 1994.

Hug, S.J., H.U. Laubscher, and B.R. James, Iron(III) catalyzed photochemical reduction of chromium(VI) by oxalate and citrate in aqueous solutions, *Environ. Sci. Technol.*, 31, 160, 1997.

Iyengar, S.S., D.C. Martens, and W.P. Miller, Distribution and plant availability of soil zinc fractions, *Soil Sci. Soc. Am. J.*, 45, 735, 1981.

James, B.R., The challenge of remediating chromium-contaminated soil, *Environ. Sci. Technol.*, 30, 248, 1996.

James, B.R. and R.J. Bartlett, Behavior of chromium in soils. VI. Interactions between oxidation-reduction and organic complexation, *J. Environ. Qual.*, 12, 173, 1983.

Johnson, T.L. and P.G. Tratnyek, A column study of geochemical factors affecting reductive dechlorination of chlorinated solvents by zero-valent iron, in *In-Situ Remediation: Scientific Basis for Current and Future Technologies*, G.L. Gee and N.R. Wing, Eds., Vol. 2., Battelle Press, Pasco, WA, 1994.

Juste, C. and P. Solda, Influence de l'addition de différentes matières fertilisantes sur la biodisponibilité du cadmium, du manganèse, du nickel, et du zinc contenus dans un sol sableux amendés par des boues de station d'epuration, *Agronomie*, 8, 897, 1988.

Kabata-Pendias, A. and H. Pendias, *Trace Elements in Soils and Plants*, CRC Press, Boca Raton, FL, 1992, 365.

Kaplan, D.I., K.J. Cantrell, and T.W. Wietsma, Formation of a barrier to groundwater contaminants by the injection of zero-valent iron colloids: Suspension Properties, in *In-Situ Remediation: Scientific Basis for Current and Future Technologies*, G.L. Gee and N.R. Wing, Eds., Vol. 2, Battelle Press, Pasco, WA, 1994.

Keizer, P. and M.G.M. Bruggenwert, Adsorption of heavy metals by clay-aluminum hydroxide complexes, in *Interactions at the Soil Colloid-Soil Solution Interface*, chap. 6, G.H. Bolt, M.F.D. Boodt, M.H.B. Hayes, and M.B. McBride, Eds., Kluwer Academic Publishers, Boston, MA, 1991.

Kent, D.B., J.A. Davis, L.C.D. Anderson, and B.A. Rea, Transport of chromium and selenium in a pristine sand and gravel aquifer: role of adsorption processes, *Water Resour. Res.*, 31, 1041, 1995.

Keren, R., R.G. Gast, and R.I. Barnhisel, Ion exchange reactions in nondried chambers montmorillonite hydroxy-aluminum complexes, *Soil Sci. Soc. Am. J.*, 41, 34, 1977.

Knox, A.S., unpublished data, 1998a.

Knox, A.S., unpublished data, 1998b.

Knox, A.S., unpublished data, 1998c.

Knox, A.S. and D.C. Adriano, Environmental availability of metals in remediated soil, *Environ. Sci. Technol.*, submitted.

Knox, A.S., A.P. Gamerdinger, D.C. Adriano, R.K. Kolka, and D.I. Kaplan, Sources and practices contributing to soil contamination, in *Bioremediation of Contaminated Soils*, D.C. Adraino, J.-M. Bollag, W.T. Frankenberger, Jr., and R.C. Sims, Eds., ASA 37 Monograph, ASA, Madison, WI, 1999, 53.

Komarneni, S., Cesium sorption and desorption behavior of kaolinites, *Soil Sci. Soc. Am. J.*, 42, 531, 1978.

Krooglov, S.V., R.M. Aleksahin, and N.A. Vasileva, Forming of radionuclide composition of soils on Chernobyl area, *Pochvovedenie*, 10, 26, 1990 (in Russian).

Kuo, S. and A.S. Baker, Sorption of copper, zinc, and cadmium by some acid soils, *Soil Sci. Soc. Am. J.*, 44, 969, 1980.

Langmuir, D., *Aqueous Environmental Geochemistry*, Prentice Hall, Upper Saddle River, NJ, 1997, 600.

Laperche, V., T.J. Logan, P. Gaddan, and S.J. Traina, Effect of apatite amendments on the plant uptake of lead from contaminated soil, *Environ. Sci. Technol.*, 31, 2745, 1997.

Leppert, D., Heavy metal sorption with clinoptilolite zeolite. Alternatives for treating contaminated soil and water, *Min. Eng.*, 42, 604, 1990.

LeGeros, R.Z. and J.P. LeGeros, Phosphate minerals in human tissues, in *Phosphate Minerals*, Nriagu, J.O. and Moore, P.B., Eds., Springer-Verlag, Berlin, 1984.

Lindberg, R.D. and D.D. Runnels, Ground water redox reactions: an analysis of equilibrium state applied to Eh measurements and geochemical modeling, *Science*, 225, 925, 1984.

Lothenbach, B., G. Furrer, and R. Schulin, Immobilization of heavy metals by polynuclear aluminum and montmorillonite compounds, *Environ. Sci. Technol.*, 31, 1452, 1997.

Lothenbach, B., R. Krebs, G. Furrer, S.K. Gupta, and R. Schulin, Immobilization of cadmium and zinc in soil by Al-montmorillonite and gravel sludge, *Eur. J. Soil Sci.*, 49, 141, 1998.

Ma, L.Q., T.J. Logan, and S.J. Traina, Pb immobilization from aqueous solutions and contaminated soils using phosphate rocks, *Environ. Sci. Technol.*, 29, 1118, 1995.

Ma, L.Q., T.J. Logan, S.J. Traina, and J.A. Ryan, Effect of NO_3^-, Cl^-, F^-, SO_4^{-2}, and CO_3^{-2} on Pb^{2+} immobilization by hydroxyapatite, *Environ. Sci. Technol.*, 28, 408, 1994.

Ma, L.Q. and G.N. Rao, Effects of phosphate rock on sequential chemical extraction of lead in contaminated soils, *J. Environ. Qual.*, 26, 788, 1997.

Ma, L.Q., S.J. Traina, and T.J. Logan, In situ lead immobilization by apatite, *Environ. Sci. Technol.*, 27, 1803, 1993.

Maes, E., B. Delvaux, and Y. Thiry, Fixation of radiocaesium in an acid brown forest soil, *Eur. J. Soil Sci.*, 49, 133, 1998.

McBean, E.A., J. Balek, and B. Clegg, *Remediation of Soil and Groundwater Opportunities in Eastern Europe*, Kluwer Academic Publishers, Dordrecht, The Netherlands, 1996, 458.

Manceau, A., L. Charlet, M.C. Boisset, B. Didier, and L. Spadini, Sorption and speciation of heavy metals on hydrous Fe and Mn oxides. From microscopic to macroscopic, *Appl. Clay Sci.*, 7, 201, 1992a.

Manceau, A., A.I. Gorshkov, and V.A. Drits, Structural chemistry of Mn, Fe, Co, and Ni in Mn hydrous oxides: II. Information from EXAFS spectroscopy, electron and X-ray diffraction, *Am. Mineral*, 77, 1144, 1992b.

Manceau, A., J.C. Hargé, C. Bartoli, E. Sylvester, J.L. Hazemann, M. Mench, and D. Baize, Sorption mechanism of Zn and Pb on birnessite. Application to their speciation in contaminated soils, in *Extended Abstract Proc. 4th Int. Conf. Biogeochemistry of Trace Elements*, I.K. Iskandar, S. E. Hardy, A.C. Chang, and G.M. Pierzynski, Eds., Berkeley, CA, 1997, 403.

Mench, M., V. Didier, M. Löffler, A. Gomez, and P. Masson, A mimicked in-situ remediation study of metal contaminated soils with emphasis on cadmium and lead, *J. Env. Qual.*, 23, 58, 1994a.

Mench, M., V. Didier, M. Löffler, A. Gomez, and P. Masson, A mimicked in situ remediation study of metal-contaminated soils, in *Workshop 92 Soil Remediation*, C. Avril and R. Impens, Eds., Faculté des Sciences Agronomiques, Gembloux (B), 1994b, 48.

Mench, M., J.Vangronsveld, V. Didier, and H. Clijsters, Evaluation of metal mobility, plant availability and immobilization by chemical agents in a limed-silty soil, *Environ. Pollut.*, 86, 279, 1994c.

Mench, M., J. Vangronsveld, H. Clijsters, N.W. Lepp, and R. Edwards, In-situ metal immobilisation and phytostabilisation of contaminated soils, in *Phytoremediation of Contaminated Soils and Water*, T. Logan, G. Banuelos, J. Vangronsveld, and N. Terry, Eds., CRC Press, Boca Raton, FL, 1999a, 325.

Mench, M., J. Vangronsveld, N. Lepp and R. Edwards, Physico-chemical aspect and efficiency of trace element immobilisation by soil amendments, in *In Situ Inactivation and Phytorestoration of Metals-Contaminated Soils*, J. Vangronsveld and S. Cunningham, Eds., Landes Bioscience, 1998, 151.

Mench, M., A. Manceau, J. Vangronsveld, H. Clijsters, and B. Mocquot, Capacity of soil amendments in lowering the plant availability of sludge-borne zinc, *Comm. Soil Sci. Plant Anal.*, submitted.

McKenzie, R.M., The adsorption of lead and other heavy metals on oxides of manganese and iron, *Australian J. Soil Res.*, 18, 61, 1980.

McLean, E.O. and M.E. Watson, Soil measurement of plant-available potassium, in *Potassium in Agriculture*, 277, ASA, Madison, WI, 1985.

Mineyev, V.G., A.V. Kochetavkin, and N. Van Bo, Use of natural zeolites to prevent heavy-metal pollution of soils and plants, *Soviet Soil Sci.*, 22, 72, 1990.

Misra, D.N. and R.L. Bowen, Interaction of zinc ions with hydroxyapatite, in *Adsorption from Aqueous Solutions*, P.H. Tewari, Ed., Plenum, New York, 1981.

Müller, I. and E. Pluquet, Immobilization of heavy metals in mud dredged from a seaport, in *Int. Conf. Contaminated Sediments*, 09/7–11/1997, Rotterdam, in press.

Mumpton, F.A., Natural zeolites, in *Zeo-Agriculture: Use of Natural Zeolites in Agriculture and Aquaculture*, W.G. Pond and F.A. Mumpton, Eds., Westview Press, Boulder, CO, 1984.

Myneni, S.C. B., T.K. Tokunaga, and G.E. Brown, Jr., Abiotic selenium redox transformations in the presence of Fe(II,III) oxides, *Science*, 278, 1106, 1997.

National Research Council, *Alternatives for Groundwater Cleanup*, National Academy Press, Washington, D.C., 1994.

Nriagu, J.O., Lead orthophosphate. IV. Formation and stability in the environment, *Geochim. Cosmochim. Acta*, 38, 887, 1974.

Patterson, R.R., S. Fendorf, and M. Fendorf, Reduction of hexavalent chromium by amorphous iron sulfide, *Environ. Sci. Technol.*, 31, 2039, 1997.

Pierzynski, G.M. and A.P. Schwab, Bioavailability of zinc, cadmium, and lead in a metal-contaminated alluvial soil, *J. Environ. Qual.*, 22, 247, 1993.

Powell, R.M., R.S. Puls, S.K. Hightower, and D. Sabatini, Coupled iron corrosion and chromate reduction: mechanisms for subsurface remediation, *Environ. Sci. Technol.*, 29, 1913, 1995.

Puls, R. W., D.A. Clark, C.J. Paul, and J. Vardy, Transport and transformation of hexavalent chromium in soils and into ground water, *J. Soil Contam.*, 3, 203, 1994.

Rebedea, I., An Investigation into the Use of Synthetic Zeolites for in situ Land Reclamation, Ph.D. thesis, John Moores University, Liverpool, 1997.

Rebedea, I. and N.W. Lepp, The use of synthetic zeolite to reduce plant metal uptake and phytotoxicity in two polluted soils, in *Biogeochemistry of Trace Elements*, D.C. Adriano et al., Eds., Sci. Technol. Letters, Northwood, 1994, 81.

Ruby, M.V., A. Davis, and A. Nicholson, In situ formation of lead phosphates in soils as a method to immobilize lead, *Environ. Sci. Technol.*, 28, 646, 1994.

Saeed, M. and R.L. Fox, Influence of phosphate fertilization on zinc adsorption by tropical soils, *Soil Sci. Soc. Am. J.*, 43, 683, 1979.

Saleh, F. Y., T.E. Parkerton, R.V. Lewis, J.H. Huang, and K.L. Dickson, Kinetics of chromium transformations in the environment, *Sci. Total Environ.*, 86, 25, 1989.

Sappin-Didier, V., Utilisation de Composés Inorganiques pour Diminuer les Flux de Métaux dans Deux Agrosystèmes Pollués: Etude des Mécanismes Impliqués par l'Emploi d'un Composé du Fer, Ph.D. thesis, Analytical Chemistry and Environment, Bordeaux University, France, 1995.

Sappin-Didier, V., A. Tremel, and M. Mench, A bag method for studying the influence of steel shots on cadmium mobility in a field experiment, in *Extended Abstract Proc. 4th Int. Conf. Biogeochemistry of Trace Elements*, I.K. Iskandar, S.E. Hardy, A.C. Chang, and G.M. Pierzynski, Eds., Berkeley, CA, 1997a, 569.

Sappin-Didier, V., M. Mench, A. Gomez, and C. Lambrot, Use of inorganic amendments for reducing metal bioavailability to ryegrass and tobacco in contaminated soils, in *Remediation of Soil Contaminated with Metals*, I.K. Iskandar, Ed., Science & Technology Letters, Northwood, 1997b, 79.

Sayles, G.D., G. You, M. Wang, and M.J. Kupferle, DDT, DDD and DDE dechlorination by zero-valent iron, *Environ. Sci. Technol.*, 31, 3488, 1997.

Sawhney, B.L., Kinetics of cesium sorption by clay minerals, *Soil Sci. Soc. Am. J.*, 30, 565, 1966.

Schnitzer, M. and S.U. Khan, Reaction of humic substance with metal ions and hydrous oxide, in *Humic Substances in the Environment*, Marcel Dekker, New York, 1972.

Seaman, J.C., P.M. Bertsch, and L. Schwallie, In situ reduction of Cr(VI) within coarse-textured, oxide-coated soil and aquifer system using Fe(II) containing solutions, *Environ. Sci. Technol.*, 33, 938, 1999.

Seaman, J.C., L. Schwallie, P.M. Bertsch, and W.J. Walker, In-situ remediation of Cr(VI) contaminated soils and aquifer sediments, in *1995 Fall Meeting, AGU*, San Franciso, CA, 1995.

Sedlak, D.L. and P.G. Chan, Reduction of hexavalent chromium by ferrous iron, *Geochim. Cosmochim. Acta*, 61, 2185, 1997.

Senesi, N. and F. Sakellariadou, Metal ion complexes in untreated and copper-reacted humic acids from marine and coastal sediments, in *Biogeochemistry of Trace Metals*, D.C. Adriano, Z.S. Chen, S.S. Yang, and I.K. Iskandar, Eds., Science Reviews, Northwood, 1997, 327.

Sims, J.L. and W.H. Patrick, Jr., The distribution of micronutrient cations in soil under conditions of varying redox potential and pH, *Soil Sci. Soc. Am. J.*, 42, 258, 1978.

Smith, L.A., J.L. Means, A. Chen, B. Alleman, C.C. Chapman, J.S. Tixier, Jr., S.E. Brauning, A.R. Gavaskar, and M.D. Royer, *Remedial Options for Metals-Contaminated Sites*, CRC Press, Boca Raton, FL, 1995, 221.

Soon, Y.K., Solubility and sorption of cadmium in soils amended with sewage sludge, *J. Soil Sci.*, 32, 85, 1981.

Spadini, L., A. Manceau, P.W. Schindler, and L. Charlet, Structure and stability of Cd^{2+} surface complexes on ferric oxides. 1. Results from EXAFS spectroscopy, *J. Colloid Interface Sci.*, 168, 73, 1994.

Sparks, D.L. and P.M. Huang, Physical chemistry of soil potassium, in *Potassium in Agriculture*, R.D. Munson, Ed., ASA, Madison, WI, 1985.

Spelman et al., unpublished results.

Staunton, S. and M. Roubaud, Adsorption of Cs-137 on montmorillonite and illite: effect of charge compensating cation, ionic strength, concentration of Cs, K, and fulvic acid, *Clays Clay Min.*, 45, 251, 1997.

Stollenwerk, K.G. and D.B. Grove, Adsorption and desorption of hexavalent chromium in alluvial aquifer near Telluride, Colorado, *J. Environ. Qual.*, 14, 150, 1985.

Stookey, L.L., Ferrozine — A new spectroscopic reagent for iron, *Analytical Chem.*, 42, 779, 1970.

Stumm, W. and J.J. Morgan, *Aquatic Chemistry: An Introduction Emphasizing Chemical Equilibria in Natural Waters*, 2nd, Wiley-Interscience, New York, 1981, 780.

Suciu, D.F., P.M. Wikoff, J.M. Beller, and C.J. Carpenter, *Process for Sodium Sulfide/Ferrous Sulfate Treatment of Hexavalent Chromium and Other Heavy Metals*, United States Air Force, 1991.

Sung, W. and J.J. Morgan, Kinetics and product of ferrous iron oxygenation in aqueous systems, *Environ. Sci. Technol.*, 14, 561, 1980.

Sylvester, E., A. Manceau, and V.A. Drits, The structure of monoclinic Na-birnessite and hexagonal birnessite. 2. Results from chemical studies and EXAFS spectroscopy, *Amer. Mineral.*, 82, 962, 1997.

Tessier, A., P.G.C. Campbell, and M. Bisson, Sequential extraction procedure for the speciation of particulate trace metals, *Analytical Chem.*, 51, 844, 1979.

Tikhonov, S., The Center for International Projects (CIP): opportunities in Russia, in *Remediation of Soil and Groundwater*, E.A. McBean, J. Balek, and B. Clegg, Eds., Kluwer Academic Publishers, Dordrecht, The Netherlands, 1996.

Tsitsishvili, G.V., T.G. Andorikashvili, G.N. Kirov, and L.D. Filizova, *Natural Zeolites*, Ellis Horwood, New York, 1992, 295.

Vangronsveld, J., Case studies in the field: Zn, Cd, Pb contaminated kitchen gardens, in *Metal-Contaminated Soils: In Situ Inactivation and Phytorestoration*, J. Vangronsveld and S. Cunningham, Eds., Springer-Verlag, Landes, in press.

Vangronsveld, J. and H. Clijsters, A biological test system for the evaluation of metal phytotoxicity and immobilization by additives in metal contaminated soils, in *Metal Compounds in Environment and Life*, 4 (*Interrelation between Chemistry and Biology*, E. Merian and W. Haerdi, Eds., Science Reviews, Wilmington, 1992.

Vangronsveld, J., J. Colpaert, and K. Van Tichelen, Reclamation of a bare industrial area contaminated by non-ferrous metals: physico-chemical and biological evaluation of the durability of soil treatment and revegetation, *Environ. Pollut.*, 94, 131, 1996.

Vangronsveld, J. and S. Cunningham, *In-Situ Inactivation and Phytorestoration of Metal-Contaminated Soils*, Landes Bioscience, Springer-Verlag, New York, in press.

Vangronsveld, J., A. Ruttens, and H. Clijsters, The use of cyclonic ashes of fluidized bed burning of coal mine refuse for immobilization of metals in soils, in *Trace Elements in Coal and Coal Combustion Residues*, K.S. Sajwan, R.F. Keefer, and A.K. Alva, Eds., Lewis Publishers, Boca Raton, FL, submitted.

Vangronsveld, J., J. Sterckx, F. Van Assche, and H. Clijsters, Rehabilitation studies on an old non-ferrous waste dumping ground: effects of revegetation and metal immobilization by Beringite, *J. Geochem. Expl.*, 52, 221, 1995b.

Vangronsveld, J., F. Van Assche, and H. Clijsters, Immobilization of heavy metals in polluted soils by application of a modified alumino-silicate: biological evaluation, in *Proc. Int. Conf. Environmental Contamination*, J. Barcelo, Ed., CEP Consultants, Edinburgh, 1990.

Vangronsveld, J., F. Van Assche, and H. Clijsters, Reclamation of a bare industrial area contaminated by non-ferrous metals: in situ metal immobilization and revegetation, *Environ. Pollut.*, 7, 51, 1995a.

Vangronsveld, J., F. Van Assche, J. Sterckx, and H. Clijsgters, Rehabilitation studies on an old non-ferrous waste dumping ground: effects of metal immobilization and revegetation, in *Proc. Int. Conf. Heavy Metals in the Environment*, R.J. Allen and J.O. Nriagu, Eds., CEP Consultants, Edinburgh, 1993.

Vangronsveld, J., F. Van Assche, and H. Clijsters, Reclamation of a desert-like site in the north east of belgium: evolution of the metal pollution and experiments in-situ, in *Proc. Int. Conf. Heavy Metals in the Environment*, J.G. Farmer, Ed., CEP Consultants, Edinburgh, 1991.

Vangronsveld, J., A. Ruttens, and H. Clijsters, Study of the Efficiency and Durability of Metal Immobilization in Soils, Report of a study performed for Union Miniére, Brussels, Belgium, 1996.

Vaniman, D.T. and D.L. Bish, The importance of zeolite in the potential high-level radioactive waste repository at Yucca Mountain, Nevada, in *Natural Zeolites, '93 Occurrence, Properties, Use*, D.W. Ming and F. A. Mumptom, Eds., Int. Comm. on Natural Zeolites, Brockport, NY, 1995.

Verkleij, J., J. Vangronsveld, M. Mench, H. Schat, H. Clijsters, B. Mocquot, M. Margeay, N. Van Der Lelie, S. Kärenlampi, A. Tervahauta, and T. De Koe, *Strategies for Rehabilitation of Metal Polluted Soils: In Situ Phytoremediation, Immobilisation, and Revegetation, a Comparative Study* (PHYTOREHAB). Final Report, European Union, Environment & Climate Programme, No. 4, ENV4-CT95-0083, 1999, 121.

Wessolek, G. and C. Fahrenhorst, Immobilization of heavy metals in a polluted soil of a sewage farm by application of a modified aluminosilicate: a laboratory and numerical displacement study, *Soil Technol.*, 7, 221, 1994.

White, A.F. and A. Yee, Aqueous oxidation-reduction kinetics associated with coupled electron-cation transfer from iron-containing silicates, *Geochim. Cosmochim. Acta*, 49, 1263, 1985.

Wittbrodt, P.R. and C.D. Palmer, Reduction of Cr(VI) in the presence of excess soil fulvic acid, *Environ. Sci. Technol.*, 29, 255, 1995.

Wittbrodt, P.R. and C.D. Palmer, Effect of temperature, ionic strength, background electrolytes, and Fe(III) on the reduction of hexavalent chromium by soil humic substances, *Environ. Sci. Technol.*, 30, 2470, 1996.

Wright, J., Conodont apatite: structure and geochemistry, in *Skeletal Biomineralization: Patterns, Processes and Evolutionary Trends*, J. Carter, Ed., Van Nostrand Reinhold, New York, 1990.

Wright, J.V., H. Schrader, and W.T. Holser, Paleoredox variations in ancient oceans recorded by rare earth elements in fossil apatite, *Geochim. Cosmochim. Acta*, 51, 631, 1987.

Wright, J.V., L.M. Peurrung, T.E. Moody, J.L. Conca, X. Chen, P.D. Didzerekis, and E. Wyse, In situ immobilization of heavy metals in apatite mineral formulation. Milestone Five Report: September 1995, Apatite Mineral Formulations and Emplacement Options, Pacific Northwest Laboratory.

Xu, Y. and F.W. Schwartz, Lead immobilization by hydroxyapatite in aqueous solutions, *J. Contam. Hydrol.*, 15, 187, 1994.

Xu, Y., F.W. Schwartz, and S.J. Traina, Sorption of Zn^{2+} and Cd^{2+} on hydroxyapatite surfaces, *Environ. Sci. Technol.*, 28, 1472, 1994.

Zhang, P., J.A. Ryan, and L.T. Bryndzia, Pyromorphite formation from goethite adsorbed lead, *Environ. Sci. Technol.*, 2673, 1997.

3

Immobilization of Lead by In Situ Formation of Lead Phosphates in Soils

Valérie Laperche

CONTENTS

3.1 Introduction

3.1.1 History of Lead Use

Lead metallurgy began at approximately 5000 B.C. (Settle and Paterson, 1980). This use during the Antiquity was largely a result of the relative abundance of Pb ores, the ease of refining the metal, and the malleability of the finished product. Lead production during the Roman Empire rose to ~80,000 tons per year, declined during medieval times, and subsequently rose again during the 19th century, to 1 million tons per year, with the onset of the industrial revolution. In 1980, about 3 million tons were produced in the world (Settle and Paterson, 1980).

Since 1970, advances in techniques and materials, as well as environmental and health concerns and their inclusion in regulations, have led to the decline of Pb use in certain applications (e.g., piping for drinking water, solder in canning, pigment for certain paints, additives for gases, sheaths for cables, and pesticides). For example, the use of lead in residential paints was not phased out in the United States until the late 1960s and early 1970s; however, Pb-based paints are used in the United States today. Lead in gasoline was eliminated in 1991 only. Lead arsenate ($PbAsHO_4$) also has been banned for use as a pesticide.

Today in France, Pb is mainly used in batteries and storage cells (64%), pigments and stabilizers (10%), pipes and sheets (7%), sheathing (6%), lead shot and fishing sinkers (3%), gasoline (3%), alloys (3%), and glassware and others (4%). North America accounts for 26% of world Pb consumption. The world consumption of lead has increased by nearly 25% in the past two decades. This growth is especially pronounced in Asia and primarily takes place in the battery sector. This is due to the low cost of lead acid batteries for automobiles and the lack of a commercially viable substitute. It is not anticipated that this will diminish in the near future.

3.1.2   Sources of Lead in Soils

Lead is present in uncontaminated soils at concentrations < 20 mg/kg (Davies, 1995), but higher concentrations have been reported (Holmgren et al., 1993). Colbourn and Thornton (1978) reported that lead concentration, apparently in uncontaminated UK soils, ranges from 10 to 150 mg/kg. In polluted areas, concentrations of 100 to 1000 times that of the normal level can be found (Davies, 1995; Adriano, 1986; Peterson, 1978). Sites of this type are common in soils contaminated by petroleum and paint residues, particulate lead from shot at private and military rifle ranges, lead batteries at dump sites, tailings from the mining of lead ores, and soils in orchards and vineyards in which $PbAsHO_4$ was applied. Additionally, other industrial activities can result in lead pollution, some of which are listed in Table 3.1.

Much of the lead in urban soils is thought to have been derived from vehicle exhausts as well as abraded tire material. Soils in close proximity to building walls and foundations can also contain lead from leached and/or exfoliated paints. This multiplicity of potential contamination sources results in a wide range of lead concentrations in urban soils, with values in some instances approaching those found in sites adjacent to mining and smelting activities.

In agricultural soils the repeated application of sludge and pesticides can increase concentrations of soil Pb. Page (1974) reported that the lead in sewage sludge ranged from 100 to 1000 mg/kg, but now the lead concentrations in sewage sludge are lower and range from 200 to 500 mg/kg (Epstein et al., 1992). Chaney and Ryan (1994) suggested a limited high

TABLE 3.1

Identification of Lead Oxides, Salts and Organic, and Their Uses in Industry

Lead Compound	Application
Litharge (PbO)	Composition of the paste used to fill the cells of battery plates, and
Minium (Pb_3O_4)	minium is also used to protect iron and steel from corrosion
Lead naphthenate and octonate	Drying agents for oil paints and alkyd (glyceryl phthalate) paints
Crocoite ($PbCrO_4$)	Colored pigments (from yellow to red)
Wufenite ($PbMoO_4$)	
"Lead sulfate" ($2PbSO_4 \cdot PbO$)	White based pigments
"Lead carbonate" ($2PbCO_3 \cdot Pb(OH)_2$)	Pesticides
Lead stearate	Stabilizers for plastics
Lead arsenate ($PbHAsO_4$)	Pesticides
Organic salts	Lubricants
Tetramethyl ($Pb(CH_3)_4$)	Additives for gasoline as anti-shocks
Tetraethyl ($Pb(C_2H_5)_4$)	

lead concentration for sludge at 300 mg/kg. For some sludges there will need to be an improvement to permit their marketing.

Another situation where lead contamination can occur is in land heavily used for clay pigeon shooting. Where millions of cartridges are fired each year, soils can accumulate several grams of lead per kilogram (Rooney et al., 1997).

In mining and smelting sites, lead concentrations can reach extremely high levels such as 30 g/kg next to an old smelter (Colbourn and Thornton, 1978). This type of pollution is usually referred to as "historical pollution" as it is related to past activities of lead production and processing plants at a time (nineteenth century) when operating techniques and knowledge of the problems linked to lead pollution were not what they are today.

Whereas these levels of lead pollution are large, they are generally restricted in geographic extent and may not represent a mobile form of lead.

3.1.3   Health Hazards of Lead

Numerous human health problems are associated with exposure to Pb. The effects of Pb poisoning occur when Pb is present in the bloodstream. This is typically the result of ingestion and/or inhalation of Pb-containing dust, particulates, fluids, or fumes. In 1979, the effect on neurologic development and IQ was found (Needleman et al., 1979). Other health effects can occur at lower levels (US-DHEW, 1991). The Centers for Disease Control (CDC, 1991) recommended that the blood lead level of concern from the standpoint of protecting the health of sensitive populations was to 10 µg/dl whole blood.

The greatest contributors to human exposure to lead are lead-based paints, urban soils and dust (Chaney and Ryan, 1994), and in isolated cases, drinking water (Adriano, 1986).

Lead rarely occurs naturally in drinking water. Instead, lead contamination usually originates from some point in residential and industrial water delivery systems. It is most commonly caused by the corrosion of lead service connections, pipes, or lead solder used to join copper pipes in municipal water systems and private residences. In 1986, Pb was banned from use in pipes and solder in public water systems within the United States. In 1991, the U.S. EPA set a new nationwide standard of 15 µg/L for Pb in drinking water.

Young children are considered to be at the highest risk of getting lead poisoning. Before 1978, lead was used profusely in interior and exterior paints. Children can ingest paint chips and become exposed to lead. Also, lead paint over time turns to dust and falls to the floor within structures or onto soil surrounding building exteriors. In residential settings, this dust can be tracked into or around the house and crawling children become exposed through hand-to-mouth action or inhalation of lead-based dust. Paint lead, soil lead, and exterior dust lead influence the concentration of lead in house dust. The house dust lead impacts the amount of lead present on a child's hands. The house dust lead and hand lead can be directly correlated with blood levels (Clark et al., 1991). Research by Charney et al. (1983) elucidated the key role that house dust plays in influencing blood lead values. This study compared blood lead concentrations of children from families that did not control household dust (control group) and children from families that control household dust (experimental group). The household dust was controlled by wet-mopping twice monthly (two or three times per week for the "hot spots"), washing children's hands prior to consuming meals and being put in bed, as well as restricting access to areas of high lead levels. During the study year, the children in the control group did not show any significant change in blood lead levels; in contrast, there was a significant fall in the mean blood lead level in the experimental group after 1 year (from 38.6 to 31.7 µg/L).

Another potential problem associated with contaminated urban soils is their use in vegetable gardens without knowledge that trace elements can be taken up by plants and transferred to the food chain. While Pb contamination in urban soils is quite variable, the occurrence of concentrations comparable to those found in mining and smelting environments is of great concern. Preer et al. (1984) reported a lead range from 44 to 5300 mg Pb/kg of soil in gardens in Washington, D.C. Sterrett et al. (1996) studied lead accumulation by lettuce grown in contaminated urban soils (392 to 5210 mg/kg) and found significant increases in lettuce leaf lead concentration compared to the control soil. Lead uptake by lettuce was not significantly higher than the control unless lead levels in the soil exceeded 500 mg/kg soil. In the latter case, lead levels in lettuce reached 37 mg Pb/kg dry matter.

However, extensive uptake studies and comparative risk evaluations suggest that the greatest risk for human exposure to lead in urban soils comes from direct particulate ingestion and not food chain transfer (Chaney and Ryan, 1994; Chaney et al., 1989).

The OSWER (Office of Solid Waste and Emergency Response) guidance sets a residential screening level of soil Pb at 40 mg/kg; in the range 400 to 5000 mg/kg, limited interim controls are recommended depending on conditions at the sites, and soil remediation is recommended above 5000 mg/kg (U.S. EPA, 1994).

As a result of widespread lead contamination in soil, considerable attention and resources are being focused on remediating lead-contaminated sites.

3.1.4 *In Situ* Treatments of Lead in Contaminated Sites

In general, lead accumulates in the topsoil, usually within the top few centimeters (Swaine and Mitchell, 1960), but may also migrate to deeper layers in some cases (Fisenne et al., 1978; Kotuby-Amacher et al., 1992). The development of *in situ* remediation treatments that can be used to reduce Pb bioavailability and transport in soils is desirable.

As described by the Environmental Protection Agency (U.S. EPA, 1990):

> An *in situ* treatment technology is defined as one that can be applied to treat the hazardous constituents of a waste or contaminated environmental medium where they are located and its capability of reducing the risk posed by these contaminants to an acceptable level or completely eliminating that risk. *In situ* treatment implies that the waste materials are treated without being physically removed from the ground.

For inorganic contaminants, the treatments generally applied include soil flushing, solidification/stabilization (pozzolan Portland cement, lime fly ash pozzolan, thermoplastic encapsulation, sorption, vitrification), and chemical and physical separation techniques (permeable barriers, electrokinetics).

The mechanism of *in situ* treatments may be physical, chemical, thermal, biological, or a combination of these. The choice of remediation treatment depends mostly on the type of pollution, the soil properties, and the level of effectiveness desired. Also, amendments should be easily available, relatively low in cost, easy to apply and incorporate, benign to the people using them, and not cause further environmental degradation.

3.1.5 Choice of Phosphate Amendment Treatment: Mechanisms of Lead Immobilization by Apatites

In 1990, Ohio State University and U.S. EPA researchers began an effort to develop a suitable *in situ* treatment technology for lead-contaminated sites with the following objectives:

- To transform all "labile" phases of lead in a chemical form that would not be released under normal environmental weathering conditions

- To use inexpensive, readily available, easy-to-manipulate, nontoxic reagents
- To identify a process that could work for a wide range of lead sources (soil, waste, water treatment)

An examination of existing chemical and geochemical knowledge about lead in the environment suggested that lead phosphates were among the most stable lead compounds in nature, especially under acidic conditions, at surficial temperatures and pressures. In calcareous soils, the solubility of lead is regulated by lead carbonate ($PbCO_3$). In noncalcareous soils, the solubility of lead appears to be regulated by $Pb(OH)_2$, $Pb_3(PO_4)_2$, $Pb_4O(PO_4)_2$, or $Pb_{10}(PO_4)_6(OH)_2$, depending on the pH (Santillan-Medrano and Jurinak, 1975). These results tend to agree with Nriagu's suggestion (1972, 1973) that lead phosphate formation could serve as a significant sink for lead in the environment.

An obvious choice of a phosphate reagent to test for lead treatment was commercial fertilizers that contain phosphate, commonly known as triple superphosphate (produce by reacting phosphoric acid with phosphate rock). This source of phosphate reagent was considered unsuitable because of its relatively high cost, its high water solubility (phosphate in surface waters causes eutrophication, the excessive growth of algae), and its reaction in water producing very acidic phosphoric acid. The latter can actually cause lead to mobilize and would counteract any immobilization effects of the added phosphate (Sterrett et al., 1996).

Ma et al. (1993, 1994a, and 1994b) used a less acid-forming calcium phosphate, hydroxylapatite ($Ca_{10}(PO_4)_6(OH)_2$), which is essentially the same mineral contained in bones, and phosphate rock, primarily fluorapatite ($Ca_{10}(PO_4)_6F_2$), a very insoluble calcium phosphate.

Hydroxylapatite can be acquired commercially in pure form, but commercial use of the technology, if successful, would have to be based on phosphate rock (phosphate rock is cheaper than hydroxylapatite) or bone meal. The principal commercial deposits of phosphate rocks in the United States exist in Florida, North Carolina, and Idaho, and to a lesser degree in Montana and Utah. Prices for phosphate in Idaho, Montana, and Utah averaged \$17.5/tonne, and in Florida and North Carolina \$26.1/tonne (Gurr, 1995). Phosphate rock has long been used as a fertilizer (mostly by organic farmers) since, if finely ground, phosphate rock will partially dissolve in the soil for uptake by plants.

Preliminary research by the Ohio State University and U.S. EPA researchers (Ma et al., 1993, 1994a, and 1994b) showed that lead could be precipitated instantaneously from solution by apatite (hydroxylapatite, fluorapatite, or chlorapatite) and that the solid formed was pyromorphite (hydroxypyromorphite, fluoropyromorphite, or chloropyromorphite).

$$Ca_{10}(PO_4)_6X_2(s) + 6H^+ (aq) \Rightarrow 10Ca^{2+} (aq) + 6HPO_4^{2+} (aq) + 2X^- (aq) \qquad (1)$$

$$10Pb^{2+} (aq) + 6HPO_4^{2+} (aq) + 2X^- (aq) \Rightarrow Pb_{10}(PO_4)_6X_2(s) + 6H^+ (aq) \qquad (2)$$

with X = OH, F, or Cl.

A very important finding of this early work was that pyromorphite could easily be distinguished from apatite by X-ray diffraction (XRD) and by scanning electron microscopy (SEM). Hydroxylapatite is composed of large rectangular crystals, while hydroxypyromorphite exists as needles (Figure 3.1). The ability to identify the mineral forms of the immobilized lead allows one to predict the long-term stability of immobilized lead in the environment.

Ma et al. (1994a) showed that apatite could also be sorbed by other metals (Al, Fe, Cu, Cd, Ni, and Zn), but less effectively than lead. The mechanism of immobilization for these metals was not identified, but it can be adsorption or precipitation (Xu et al., 1994). Another study by Ma et al. (1994b) showed that anions (nitrate, sulfate, and carbonate) had minimal effect on the amount of aqueous lead immobilized at low concentrations.

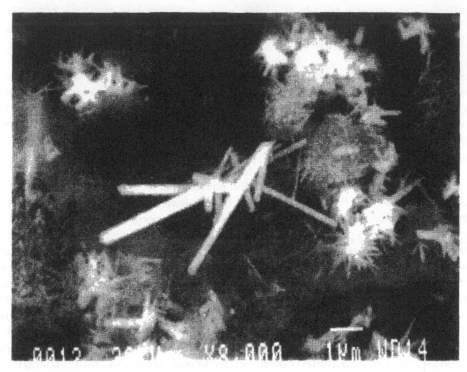

FIGURE 3.1
SEM micrograph of hydroxylapatite (Bio-Rad Laboratory, Rochester, NY) reacted with lead nitrate at pH 5 for 2 h. Hydroxylapatite is composed of large rectangular crystals while hydroxypyromorphite exists as needles. (Micrograph courtesy of John Mitchell, MARC, Geology Department, OSU.)

Zhang et al. (1997) showed that lead sorbed on the surface of oxides (e.g., goethite) was rapidly desorbed in the presence of NaH_2PO_4 to precipitate as chloropyromorphite. By contrast, when hydroxylapatite was used instead of NaH_2PO_4, the formation of chloropyromorphite was slower and appeared to be controlled by the rate of dissolution of the hydroxylapatite.

Nevertheless, the effectiveness of lead removal from solution or desorbed from solids resulted in the formation of pyromorphite in equilibrium with an aqueous lead concentration equal to or below the drinking water limits proscribed by U.S. EPA (15 µg/L).

3.2 Estimation of Lead Bioavailability Using Chemical Extractants

It is known that the total amount of trace elements in soil is an overestimation of the real danger they actually represent. The transfer of trace elements through soil profiles to aquifers or their uptake by plants is related to their physicochemical forms in soil, their mobility, and their bioavailability (Colbourn and Thornton, 1978). Understanding lead mobility and bioavailability in contaminated soils is important for evaluating the potential environmental effects of lead.

Lead compounds entering soil become partitioned among several soil compartments: soil solution, exchangeable, adsorbed, or complexed by organic or inorganic compounds, oxides and carbonates, and primary minerals. Only a small portion of the lead in soil is available for plant uptake. All but the most insoluble surface and solid-phase species are thought to be phytoavailable.

Extractable lead is generally used as the indicator of amount available for plant uptake. Various chemical extractants (single and sequential) have been used to assess bioavailability of lead (Tessier et al., 1979; Mench et al., 1994; Berti and Cunningham, 1997). Prediction of lead availability seems to depend on several factors, such as the type of extractant, the molarity of the extractants, soil properties (such as pH), as well as the actual forms of lead present within a given sample.

Despite the difficulties to find the "perfect extraction method" for measurement of the bioavailability of trace elements in soil, these procedures (Tessier et al., 1979; Mench et al., 1994; Berti and Cunningham, 1997) provide indirect information on the reactivity of the trace elements and their potential mobility, in particular about the residual fraction (nonextractable). The residual fraction should contain mainly primary and secondary minerals which may have trace elements within crystal structure. These trace elements are not expected to be released into solution over a reasonable period of time under normal weathering conditions.

Ma and Rao (1997) studied the effects of phosphate rock on sequential chemical extraction of lead in different contaminated soils using the procedure developed by Tessier et al. (1979). Sequential extraction was used to determine the lead distribution in the soil and evaluate the effectiveness of using phosphate rock to immobilize lead in soil. They showed that most of the lead in the soils was concentrated in the potential available fractions (79 to 96%), sum of the water soluble, exchangeable, carbonate-bound, Fe-Mn oxides-bound and organic bound fractions. The presence of phosphate rock reduced the extractable lead, and the percentage of lead reduction ranged from 10 to 96% (Table 3.2). The effectiveness of phosphate rock treatments in converting lead from available to an unavailable fraction was greater with increasing amounts of phosphate rock added to the soil. Lead precipitation as fluoropyromorphite ($Pb_{10}(PO_4)_6F_2$) was suggested as the reason for reducing lead solubility (Ma and Rao, 1997).

The determination of nonresidual and residual fractions can be the first step in evaluating the real risk of the presence of lead in contaminated soil.

3.3 Effects of Apatite Amendments on Lead-Contaminated Soils

3.3.1 Lead Phosphates in Contaminated Soils

Lead phosphates form naturally in contaminated soils (Cotter-Howells and Thornton, 1991; Cotter-Howells et al., 1994; Ruby et al., 1994). In a historic lead mining village in the U.K., Cotter-Howells and Thornton (1991) identified substantial amounts of lead chlorophosphate in the topsoil (0 to 5 cm). In a more recent study, Cotter-Howells et al. (1994) determined the composition of the lead phosphate as a calcium-rich pyromorphite with 27% of the lead replaced by calcium.

Ruby et al. (1994) showed that the formation of lead phosphate can occur naturally in a short period of time (less than 13 years) in a soil contaminated by smelting activities when phosphorus is not a limited factor. In this site, the phosphorus content ranged from 1400 to 17,700 mg/kg, considerably above the average phosphorus content in the United States (between 200 to 5000 mg/kg; Lindsay, 1979).

TABLE 3.2

Lead Distribution in Eight Contaminated Soils in the Presence of Different Amounts of OC Phosphate Rock (OCPR)

	Burch, Washington (PbAsHO$_4$)			East Field 1, Montana (smelter)		
OCPR added (g)	0	0.125	0.25	0	0.125	0.25
Residual Pb, %	13	56	73	12	32	57
Nonresidual Pb, %	87	44	27	88	68	42
Decrease in nonresidual Pb, %		50	69		23	52
Sum of all fractions	2,920 ± 58	2,850 ± 44	2,660 ± 87	6,310 ± 190	6,190 ± 421	6,380 ± 438
Total via single digestion	2,680 ± 80			5,970 ± 226		

	BPS, Pennsylvania (Battery breaking site)			PTC, Oklahoma (smelter)		
OCPR added (g)	0	0.125	0.25	0	0.125	0.25
Residual Pb, %	8	18	32	9	83	95
Nonresidual Pb, %	92	82	68	91	17	5
Decrease in nonresidual Pb, %		10	26		82	95
Sum of all fractions	41,100 ± 669	47,700 ± 1050	37,900 ± 730	662 ± 37	847 ± 72	736 ± 66
Total via single digestion	40,100 ± 800			1,300 ± 85		

	Twin, Washington (PbAsHO$_4$)		East Field 2, Montana (smelter)		AEC 1-1, Connecticut (incineration ash)		Area 40, Washington (building demolition)	
OCPR added (g)	0	0.125	0	0.125	0	0.125	0	0.125
Residual Pb, %	4	96	6	43	4	20	21	66
Nonresidual PB, %	96	4	94	57	96	78	79	35
Decrease in nonresidual Pb, %		96		40		19		56
Sum of all fractions	768 ± 48	874 ± 62	4,920 ± 170	2,985 ± 310	10,600 ± 790	9,550 ± 740	8,330 ± 550	7,800 ± 474
Total via single digestion	705 ± 50		4,480 ± 250		12,500 ± 710		7,640 ± 510	

Ma, Q.Y. and G.N. Rao, Effects of phosphate rocks on sequential chemical extraction of lead in contaminated soils, *J. Environ. Qual.*, 26, 788, 1997.

Note: Explanation of the abbreviation used in the table — first the abbreviation, second the name of the soil, third the location, and fourth the source of the contamination: BU: Burch, Washington (PbAsHO$_4$); EF1: East Field 1, Montana (smelter); BP: BPS, Pennsylvania (Battery breaking site); PT: PTC, Oklahoma (smelter); TW: Twin, Washington (PbAsHO$_4$); EF2: East Field 2, Montana (smelter); DA: AEC 1-1, Connecticut (incineration ash); DU: Area 40, Washington (building demolition).

The formation of lead phosphates in nature is difficult to predict. Pyromorphite precipitates from aqueous solution and lead immobilization is near completion in less than 30 min for a range of pH from 3 to 7 (Ma et al., 1993). Also, Laperche et al. (1996) provided direct physical evidence of the formation of pyromorphite after a few days at pH 5, subsequent to amendment of an enriched soil fraction with synthetic hydroxylapatite; however, in natural field settings, mineral formation rates are likely to be slower than in laboratory experiments.

Ruby et al. (1994) demonstrated that the weathering of galena (PbS) to anglesite (PbSO$_4$) followed by alteration to insoluble lead phosphate in soil is due to the presence of adequate phosphate content. Forty-six percent of the original galena had been altered to lead phosphates under uncontrolled environmental conditions. Ruby et al. (1994) suggested that the rate of transformation from galena to pyromorphite could be increased by optimizing conditions such as phosphate and chlorine reactivities, pH, water content, and mixing.

3.3.2 Phosphate Amendment to Induce the Formation of Lead Phosphate to Reduce Plant Uptake of Lead

Different types of phosphates have been used to induce the formation of lead phosphates: NPK fertilizers, Ca(H$_2$PO$_4$)$_2$ (Sterrett et al., 1996), Na$_2$HPO$_4$ (Cotter-Howells and Caporn, 1996), and apatite (Laperche et al., 1997; Chlopecka and Adriano, 1997a and b). Cotter-Howells and Caporn (1996) also used peat as phosphate amendment and showed that root exudates containing phosphate enzymes could convert organic-P to phosphate.

Some of the phosphate amendments are more effective than others in reducing plant uptake of lead. Application of NPK fertilizer alone or with Ca(H$_2$PO$_4$)$_2$ showed little or no effect on lead uptake (Sterrett et al., 1996). Sterrett et al. (1996) suggested that the phosphate treatment gave higher metal concentrations in plant tissues probably because of the soil acidification due to the application of the acidic phosphate fertilizer salt.

In the case of the apatite amendment, the pH of the treated soils increased from 0.3 to 0.9 units compared to the untreated soils as a function of the quantity of apatite added to the soils (Chlopecka and Adriano, 1997a). In all cases (Laperche et al., 1997; Chlopecka and Adriano, 1997a and b), apatite amendments reduced lead uptake as a function of the quantity of apatite added to the contaminated soil: 15 to 60% and 12 to 41% in maize and barley tissues, respectively (Chlopecka and Adriano, 1997b), and 45 to 87% in sudax grass shoots for different contaminated soils (Laperche et al., 1997). Laperche et al. (1997) studied the effect of synthetic and natural apatite (phosphate rock) on lead uptake for a soil contaminated by smelting activities: soil A from Butte, Montana (2400 mg Pb/kg) and a soil heavily contaminated by paint; soil B from Oakland, California (37,000 mg Pb/kg). The quantities of phosphate materials added were calculated as a function of the lead content of the soil corresponding at 0.33 (treatment 1) and 1.50 (treatment 4) times the mass of phosphate material necessary for stoichiometric conversion (on a molar basis) of all of the soil lead to pyromorphite. The same production of shoot tissue was measured in both soils even though the lead content of the plants grown on soil B (106 mg Pb/kg of dried shoot) was much higher than that of the plants grown on soil A (6.7 mg Pb/kg of dried shoot). In both soils, Pb uptake by shoots in phosphate-treated soils decreased compared to the untreated soils (Figure 3.2). The greatest reduction of lead uptake was obtained with the largest quantity of apatite amendment. In all treatments, lead showed greater accumulations in root tissues than in shoot tissues. Chlopecka and Adriano (1997a) showed similar results of apatite treatments on Pb uptake by maize in four different contaminated soils. Laperche et al. (1997) showed that, in the presence of a large quantity of phosphate, lead accumulated in root tissues more than in an untreated contaminated soil. Previous research has suggested

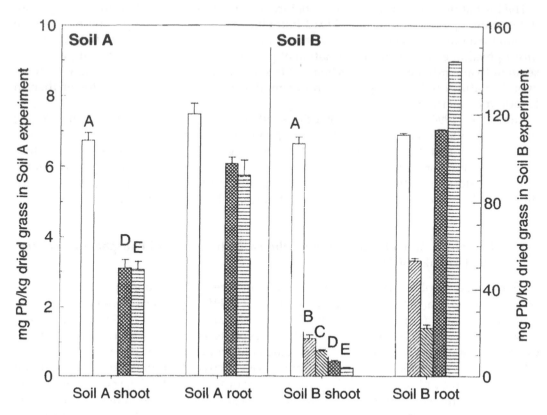

FIGURE 3.2
Shoot (average of three cuts) and root lead concentration in soils A and B. A: untreated soil; B: phosphate rock treatment (0.33); C: hydroxylapatite treatment (0.33); D: phosphate rock treatment (1.50); E: hydroxylapatite treatment (1.50).

that some plants use an exclusion mechanism to accumulate lead in roots and limit transport to shoots. Koeppe (1977) found that lead precipitated on root cell walls in an insoluble, amorphous form which, in maize, has been identified as a lead phosphate.

After only 4 months of incubation, lead phosphate particles were found on the surface of the roots but no lead phosphate particle was identified by X-ray diffraction in the bulk soil. Cotter-Howells et al. (1994) showed that EXAFS (Extended X-ray Absorption Fine Structure) might be more suitable than X-ray diffraction for the identification of compounds at low abundance. Cotter-Howells et al. (1994) identified by EXAFS in a mine waste soil "calcium-rich pyromorphite." Cotter-Howells and Caporn (1996) showed that it is still possible to find and to identify by X-ray diffraction the calcium-rich pyromorphite, but after concentration of the heavy fraction by high-density separation (Figure 3.3).

3.3.3 Stability of Pyromorphite in Soils and under Simulated Gastric Conditions

Laperche et al. (1997) showed that when hydroxypyromorphite is the only source of phosphate, sudax grass can induce its dissolution (Figure 3.4). In the presence of hydroxypyromorphite and apatite, the lead content in shoots decreased an average 10 to 100 times as function of the quantity of apatite and the type of apatite used. A fine ground phosphate rock (<250 µm) was more efficient than unground phosphate rock (2 to 0.5 mm). At a hydroxypyromorphite/apatite ratio of 0.5, fine ground phosphate rock had the same efficiency as synthetic hydroxylapatite. Thus it is likely that the dissolution rates of both

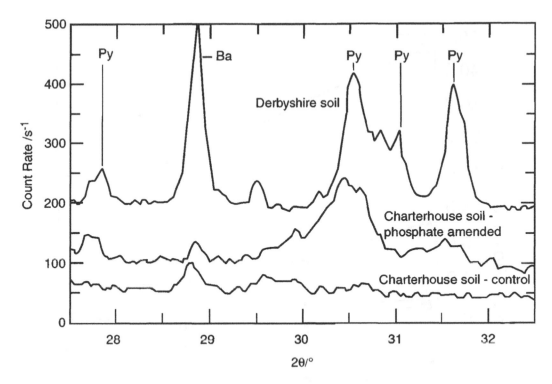

FIGURE 3.3
Portion of XRD trace showing the prominent peaks of CA-rich pyromorphite (Py) identified in the high-density fraction of a Derbyshire soil, together with XRD traces of the high-density fraction from the phosphate amended and control Charterhouse soil. Traces are normalized on barite (Ba), which is present in all three samples. The presence of peaks at similar 2Θ values in the phosphate amended soil, absent in the control soil, suggests the presence of Ca-rich pyromorphite in this sample. (Source: Cotter-Howells, J.D., P.E. Champness, J.M. Charnock, and R.A.D. Pattrick, Identification of pyromorphite in mine-waste contaminated soils by ATEM and EXAFS, *Eur. J. Soil Sci.*, 45, 393, 1994; Cotter-Howells, J.D. and S. Caporn, Remediation of contaminated land by formation of heavy metal phosphates, *Appl. Geochem.*, 11, 335, 1996.)

apatites are greater than that of hydroxypyromorphite and can maintain a sufficient solution phase concentration of phosphorus to inhibit dissolution of the hydroxypyromorphite and reduce lead uptake by the sudax grass.

In environmental conditions, chloropyromorphite forms instead of hydroxypyromorphite (Cotter-Howells and Caporn, 1996; Ruby et al., 1994), and chloropyromorphite has a much lower solubility than hydroxypyromorphite, log Ksp = –4.14 and –25.05, respectively. It is likely that plant uptake from chloropyromorphite would be even less than that observed from hydroxypyromorphite by Laperche et al. (1997). Nevertheless, it seems prudent to maintain an excess quantity of unreacted phosphate in Pb-contaminated soils, even after pyromorphite formation has occurred in order to decrease the uptake of pyromorphite-P by plants.

The association of growing plants and using apatite to remediate a contaminated site has different objectives depending on the use of the plant and the type of polluted sites one is dealing with. First, having a vegetation cover provides pollution control and stability to the soil. Wind erosion is prevented and percolation can be highly reduced. Second, *in situ* immobilization for heavily contaminated bare sites (such as some industrial sites or mining sites) may allow revegetation by plant species with low tolerances for lead. Third, plants can be used not only for phytostabilization of contaminated sites but for agricultural needs. Plant-root exudates induce the formation of lead phosphate in soil, but also some plants can accumulate lead in roots. Apatite applications considerably reduce the lead content in

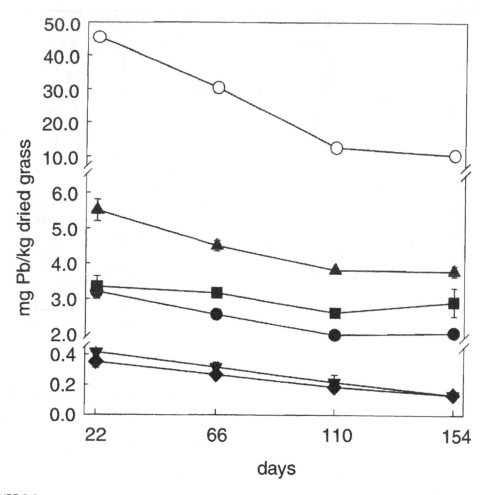

FIGURE 3.4
Sudax lead shoot tissue concentrations in a sand experiment. (○) 12 g of hydroxypyromorphite, (●) 6 g of hydroxypyromorphite + 2.3 g of hydroxylapatite, (▼) 6 g of hydroxypyromorphite + 4.6 g of hydroxylapatite, (■) 6 g of hydroxypyromorphite + 3 g of fine ground phosphate rock, (◆) 6 g of hydroxypyromorphite + 6 g of fine ground phosphate rock, (▲) 6 g of hydroxypyromorphite + 3.5 g of ungrounded phosphate rock.

shoot material, perhaps allowing for use of the aboveground biomass for applications that would be unsuitable if greater tissue concentrations of lead were present. The reuse of contaminated soils treated by apatite in forage or crop production may be viable.

The fact that phosphate added to contaminated soil reduced lead uptake by plants will also reduce food chain contamination, but lead can be involuntarily ingested by livestock from soil. Livestock can be directly exposed to lead, as children can be exposed to lead dust or lead soil. A number of studies have examined lead bioavailability from soils. Researchers have developed estimations of the impact to blood lead levels based upon elevations in soil lead values (Bornschein et al., 1986; Clark et al., 1991). Davis et al. (1992) showed that the geochemical factors that control lead availability (mineralogy, encapsulation, and particle size) in soil or mining waste controlled the bioavailability of lead in the gastrointestinal tract. Consequently, the blood level response noted in the different environments reflects the difference in lead availability based on lead geochemistry. An *in vitro* test to evaluate the *in vivo* bioaccessibility of lead in soil and mine waste lead has been developed by Ruby et al. (1992 and 1993). Macklis (1996) used the test developed by Ruby et al. (1993) to show that the addition of hydroxylapatite decreased lead bioavailability by 8% in a soil contaminated by mining activities and by 14% in a soil contaminated by paint, after 4 months of incubation.

Another experiment on rats and pigs fed with a diet that was supplemented with lead-contaminated soil and contaminated soil treated with hydroxylapatite or phosphate rock showed that lead in blood and bones was reduced in the experiment with the hydroxylapatite or phosphate rock treated soil (Logan et al., 1995). Logan et al. (1995) could not attribute the reduction of lead availability to the formation of insoluble lead phosphate in the soil or in the gastrointestinal track of the animals.

3.3.4 Full-Scale Studies

Field studies at Joplin, Missouri are being conducted by the U.S. EPA, Missouri Department of Natural Resources (MDNR), University of Missouri, and the Department of Agriculture-Agricultural Research Service (USDA-ARS).

Studies involve treatment of contaminated soil with various amendments that may reduce the bioavailability of lead in soil. The contamination of the site is due to smelting activities in the early 1900s. Lead concentrations range from 1000 to 5000 mg/kg. One of the goals of this study is to determine the relative efficacy of various amendments (such as phosphoric acid, hydroxylapatite, rock phosphate) and application techniques in reducing the levels of bioavailable lead in soil. These experiments are not complete and the results have not been published.

More of these full-scale experiments need to be conducted before the use of phosphate rocks for remediation purposes on contaminated sites can be recommended.

3.4 Conclusions

Methods for remediation of lead-contaminated soils have recently become an important issue because of the health risks associated with this pollutant. High concentrations of lead can create serious problems both in agricultural soils and industrial sites and pose a hazard for men, animals, and plants.

The final choice for the remediation of a contaminated site is a function of the nature and the degree of the contamination, the redevelopment of the site, and the technical and financial needs available. Remediation can be achieved by preventing the lead from spreading to the surroundings and to groundwater. One way to prevent such spreading is to immobilize it in soils, through the addition of amendments such as lime, phosphates, apatites, aluminosilicates (zeolites, beringite, clays), iron, and manganese materials.

Phosphate, and in particular apatite, has been shown to have great potential for use in *in situ* immobilization of lead. Laboratory, greenhouse, *in vitro*, and *in vivo* experiments have demonstrated the effectiveness of apatite in remediating contaminated soils, industrial sites, and wastes.

In situ immobilization of lead by phosphate rock with revegetation of the contaminated area can be an efficient, economical, realistic, and cost-effective alternative remediation method for agricultural soils, kitchen gardens, industrial sites, and so on.

References

Adriano, D.C., Lead, in *Trace Elements in the Terrestrial Environment*, Springer-Verlag, New York, 7, 219, 1986.

Berti, W.R., and S.D. Cunningham, In place inactivation of Pb in Pb-contaminated soils, *Environ. Sci. & Technol.*, 31, 1359, 1997.

Bornschein, R.L., P.A. Succop, K.M. Krafft, C.S. Clark, B. Peace, and P.B. Hammond, Exterior surface dust lead, interior house dust lead and childhood exposure in an urban environment. *Conf. Trace Substances Environ. Health*, 11, 322, 1986.

CDC, Preventing Lead Poisoning in Young Children, U.S. Department of Health and Human Services, Atlanta, GA, 1991.

Chaney, R.L., H.W. Mielke, and S.B. Sterrett, Speciation, mobility, and bioavailability of soil lead, *Environ. Geochem. Health*, Suppl. 11, 109, 1989.

Chaney, R.L. and J.A. Ryan, *Risk Based Standards for Arsenic, Lead and Cadmium in Urban Soils*, Kreysa, G. and J. Wieser, Eds., DECHEMA, Frankfurt, 1994.

Charney, E., B. Kessler, M. Farfel, and D. Jackson, Childhood lead poisoning. *N. Engl. J. Med.*, 309(18), 1089, 1983.

Chlopecka, A. and D.C. Adriano, Inactivation of metals in polluted soils using natural zeolite and apatite, *Abstract 4th Int. Conf. Biogeochemistry of Trace Elements*, Berkeley, CA, 1997a, 415.

Chlopecka, A. and D.C. Adriano, Mimicked in-situ stabilization of lead-polluted soils, *Abstract 4th Int. Conf. Biogeochemistry of Trace Elements*, Berkeley, CA, 1997b, 423.

Clark, S., R. Bornschein, P. Succop, S. Roda, and B. Peace, Urban lead exposures of children in Cincinnati, OH, *Chem. Speciation Bioavailability*, 3(3/4), 163, 1991.

Colbourn, P. and I. Thornton, Lead pollution in agricultural soils, *J. Soil Sci.*, 29, 513, 1978.

Cotter-Howells, J.D. and I. Thornton, Sources and pathways of environmental lead to children in a Derbyshire mining village, *Environ. Geochem. Health*, 12, 127, 1991.

Cotter-Howells, J.D., P.E. Champness, J.M. Charnock, and R.A.D. Pattrick, Identification of pyromorphite in mine-waste contaminated soils by ATEM and EXAFS, *Eur. J. Soil Sci.*, 45, 393, 1994.

Cotter-Howells, J.D. and S. Caporn, Remediation of contaminated land by formation of heavy metal phosphates, *Appl. Geochem.*, 11, 335, 1996.

Davies, B.E., Lead, in *Heavy Metals in Soils*, 2nd ed., Alloway, B.J., Ed., Blackie Academic and Professional, London, 1995, 568.

Davis, A., M.V. Ruby, and P.D. Bergstrom, Bioavailability of arsenic and lead in soils from the Butte, Montana, mining district, *Environ. Sci. & Technol.*, 26, 461, 1992.

Epstein, E., R.L. Chaney, C. Henry, and T.J. Logan, Trace elements in municipal solid waste compost, *Biomass Bioenergy*, 3, 227, 1992.

Fisenne, I.M., G.A. Welford, P. Perry, R. Bavid, and H.W. Keller, *Environ. Int.*, 1, 245, 1978.

Gurr, T.M., Phosphate rock, *Mining Eng.*, 47, 550, 1995.

Holmgren, G.G.S., M.W. Meyer, R.L. Chaney, and R.B. Daniels, Cadmium, lead, zinc, and nickel in agricultural soils of the United States of America, *J. Environ. Quality*, 22, 335, 1993.

Koeppe, D.E., The uptake, distribution, and effect of cadmium and lead in plants, *Sci. Total Environ.*, 7, 197, 1977.

Kotuby-Amacher, J.R.P. Gambell, and M.C. Amacher, The distribution and environmental chemistry of lead at an abandoned battery reclamation site, in *Engineering Aspects of Metal-Waste Management*, I.K. Iskandar and H.M. Selim, Eds., Lewis Publishers, London, 1992, 231.

Laperche, V., S.J. Traina, P. Gaddham, and T.J. Logan, Chemical and mineralogical characterizations of Pb in a contaminated soil: reactions with synthetic apatite, *Environ. Sci. & Technol.*, 30, 3321, 1996.

Laperche, V., T.J. Logan, P. Gaddham, and S.J. Traina, Effect of apatite amendments on plant uptake of lead from contaminated soil. *Environ. Sci. & Technol.*, 31, 2745, 1997.

Lindsay, W.L., *Chemical Equilibria in Soils*, John Wiley & Sons, New York, 1979, 449.

Logan, T.J., S.J. Traina, J. Henneghan, and J.A. Ryan, Effects of phosphate addition on bioavailability of lead in contaminated soil fed to rats and pigs. Contaminated soils. Third Int. Conf. on the Biogeochemistry of Trace Elements, Paris, France, May 15–19, 1995.

Ma, Q.Y., S.J. Traina, T.J. Logan, and J.A. Ryan, In-situ lead immobilization by apatite, *Environ. Sci. & Technol.* 27, 1803, 1993.

Ma, Q.Y., S.J. Traina, T.J. Logan, and J.A. Ryan, Effects of aqueous Al, Cd, Cu, Fe(II), Ni, and Zn on Pb immobilization hydroxyapatite, *Environ. Sci. & Technol.*, 28, 1219, 1994a.

Ma, Q.Y., T.J. Logan, S.J. Traina, and J.A. Ryan, Effects of NO_3^-, Cl^-, F^-, SO_4^{2-} and CO_3^{2-} on Pb immobilization hydroxyapatite, *Environ. Sci. & Technol.*, 28, 408, 1994b.

Ma, Q.Y., T.J. Logan, and S.J. Traina, Lead immobilization from aqueous solutions and contaminated soils using phosphate rocks, *Environ. Sci. & Technol.*, 29, 1118, 1995.

Ma, Q.Y. and G.N. Rao, Effects of phosphate rocks on sequential chemical extraction of lead in contaminated soils, *J. Environ. Qual.*, 26, 788, 1997.

Macklis, S.M., Gastrointestinal Absorption of Lead from Soils and Compost and Associated Health Effects, Thesis, Ohio State University, Columbus, 1996, 102.

Mench, M.J., V.L. Didier, M. Löffler, A. Gomez, and P. Masson, A mimicked in-situ remediation study of metal-contaminated soils with emphasis on cadmium and lead, *J. Environ. Qual.*, 23, 58, 1994.

Needleman, H.L., C.E. Gunnoe, A. Leviton, R. Reed, H. Peresie, C. Maller, and P. Barrett, Deficits in psychologic and classroom performance of children with elevated lead levels, *N. Engl. J. Med.*, 300, 689, 1979.

Nriagu, J., Lead orthophosphates. I. Solubility and hydrolysis of secondary lead orthophosphate, *Inorg. Chem.*, 11, 2499, 1972.

Nriagu, J., Lead orthophosphates. II. Stability of at 25°C, *Geochim. Cosmochim. Acta*, 37, 367, 1973.

Page, A.L., Fate and effects of trace elements in sewage sludge when applied to agricultural lands, EPA-670/2-74-005. U.S. EPA, Cincinnati, OH, 1974, 96.

Peterson, P.J., Lead and vegetation, in *The Biogeochemistry of Lead in Environment*, Nriagu, J.O., Ed., Elsevier, New York, 19, 355, 1978.

Preer, J.R., J.O. Akintoe, and J.L. Martin, Metals in downtown Washington, D.C. gardens, in *Biol. Trace Element Res.*, 6, 79, 1984.

Rooney, C.P., R.G. McLaren, and R.J. Cresswell, Forms and phytoavailability of lead in a soil contaminated with lead shot, *Abstract* 4th Int. Conf. Biogeochemistry of Trace Elements, Berkeley, CA, 1997, 289.

Ruby, M.V., A. Davis, J.H. Kempton, J.W. Drexler, and P.D. Bergstrom, Lead bioavailability: dissolution kinetics under simulated gastric conditions, *Environ. Sci. & Technol.*, 26, 1242, 1992.

Ruby, M.V., A. Davis, T.E. Link, R. Schoof, R.L. Chaney, G.B. Freeman, and P.D. Bergstrom, Development of an in vitro screening test to evaluate the in vivo bioaccessibility of ingested minewaste lead, *Environ. Sci. & Technol.*, 27, 2870, 1993.

Ruby, M.V., A. Davis, and A. Nicholson. In-situ formation of lead phosphates in soils as a method to immobilize lead, *Environ. Sci. & Technol.*, 28, 646, 1994.

Santillon-Medrano, J. and J.J. Jurinak, The chemistry of lead and cadmium in soil: solid phase formation, *Soil Sci. Soc. Am. Proc.*, 39, 851, 1975.

Settle, D.M. and C.C. Paterson, Lead in albacore: guide to lead pollution in Americans, *Science*, 207, 1167, 1980.

Sterrett, S.B., R.L. Chaney, C.H. Gifford, and H.W. Mielke, Influence of fertilizer and sewage sludge compost on yield and heavy metal accumulation by lettuce grown in urban soils, *Environ. Geochem. Health*, 18, 135, 1996.

Swaine, D.J. and R.L. Mitchell, Trace-element distribution in soil profiles, *J. Soil Sci.*, 11, 347, 1960.

Tessier, A., P.G.C. Campbell, and M. Bisson, Sequential extraction procedure for the speciation of particulate trace metals, *Anal. Chem.*, 51(7), 844, 1979.

U.S. DHEW, Preventing lead poisoning in young children, A statement by Centers for Disease Control, 1991, 105.

U.S. EPA, Handbook on in-situ treatment of hazardous-waste-contaminated soils. Risk Reduction Engineering Laboratory, Office of Research and Development, Cincinnati, OH. EPA/540/2-90/002, 1990, 155.

U.S. EPA, Revised interim soil lead guidance for CERCLA sites and RCRA corrective action facilities. OSWER directive No. 9355.4-12. Office of Emergency and Remedial Response, Washington, D.C. EPA/540/F-94/043, PB94-963282, 1994, 17.

Xu, Y. and F.W. Schwartz, Sorption of Zn^{2+} and Cd^{2+} on hydroxyapatite surfaces, *Environ. Sci. & Technol.*, 28, 1472, 1994.

Zhang, P., J.A. Ryan, and L.T. Bryndzia, Pyromorphite formation from goethite adsorbed lead, *Environ. Sci. & Technol.*, 31, 2673, 1997.

4

Determinants of Metal Retention to and Release from Soils

Yujun Yin, Suen-Zone Lee, Sun-Jae You, and Herbert E. Allen

CONTENTS

Introduction

Fate and bioavailability of metals in soils are governed by many environmental processes. Without consideration of transport and rhizosphere reactions, the processes influencing a trace metal uptake by a plant are depicted in Figure 4.1. All plants take up metals in soils via pore water; therefore, the partitioning of metals between soil solution and solids is the primary factor determining metal bioavailability. Strong retention affinities for soil solid surfaces would reduce the risk of a metal species to the ecosystem, whereas poor retention by soil particles would result in more metal present in soil solution for uptake by soil organisms. Not all the metal in the solution phase is bioavailable. Both aqueous inorganic ligands and organic ligands, including natural dissolved organic matter and anthropogenic ligands, compete with binding sites on the organism for the metal. The soluble major cations, such as Ca^{2+} and protons, can also compete with the metal of interest for the available binding sites on the organism. In addition, the biotic processes also affect uptake. The combined effects of all of these processes determine biological uptake of metals. Clearly, to determine if a metal contained in the soil at a specific location has potential hazard to the environment, both equilibrium and kinetic aspects of these processes have to be understood. This chapter focuses on equilibrium or pseudo-equilibrium (as soils may not be at equilibrium) aspects of metal retention and release processes in soils.

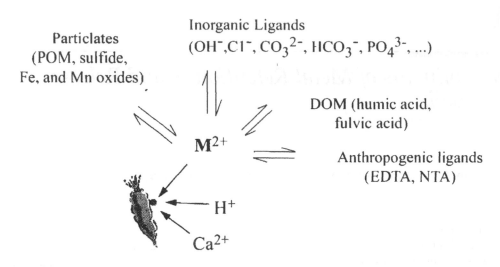

FIGURE 4.1
Metal speciation and bioavailability in soils. M^{2+} is the metal of interest.

Metals can be retained by soils either by adsorption or surface precipitation (Tewari and Lee, 1975; Barrow et al., 1981; Fuller et al., 1993). Based on macroscopic data, it is impossible to differentiate surface adsorption from precipitation (Sparks, 1995). In this chapter, the retention of metals by soils is described in terms of sorption which includes both adsorption and precipitation. Extensive studies have indicated that pH has profound effects on metal sorption to soil colloids (James and Healy, 1972; Sposito, 1984; Stumm, 1992). Both colloid surface charge and aqueous speciation of metals are a function of pH. As discussed later in this chapter, pH also affects the soil-water partitioning of natural organic matter in soils, which in turn affects metal sorption and desorption. At a fixed pH, sorption and desorption of metals in soils depend on both particulate and aqueous composition and characteristics. For example, a larger surface area could result in a greater sorption. The high valent, or even low valent major cations which are present at high concentrations in soil solution, could compete with heavy metals for available surface sites and consequently decrease sorption (Zachara et al., 1993). The presence of inorganic and organic ligands in soil solution, on the other hand, could reduce metal sorption by forming soluble complexes with metals (Huang and Lin, 1981; Yin et al., 1996). The combined effects of all of these factors determine metal sorption and desorption in soils.

This report summarizes some of the results on metal sorption and desorption on soils obtained in our lab over the last several years. The focus of this summary is on the combined effects of major environmental variables on metal sorption and desorption. Sorption of eight metal species on soils are compared. The effects of soil pH and surface properties as well as of soluble organic matter, a major cation, and an anion on metal sorption are elucidated. Likewise, desorption of metals from soils as a function of pH and dissolved organic matter is discussed.

4.2 Experimental Approaches

We have employed 15 representative noncontaminated soils collected from the state of New Jersey to conduct our experiments. The soils were air dried and sieved through a

2 mm screen. Soil aggregates were broken by hand and a wooden mallet before sieving. All experiments were conducted using soil fractions less than 2 mm which were thoroughly mixed before use. The soils were characterized in terms of surface properties and chemical composition. Detailed soil characteristics have been reported elsewhere (Yin et al., 1996). The texture of these soils varies from sand to silt-clay loam with sand (630 µm to 2 mm) content ranging from 20 to 92%, silt (2 to 630 µm) from 3 to 49%, and clay (< 2 µm) from 6 to 37%. Soil pH ranges from 4.2 to 6.4, CEC from 800 to 9500 mmol/kg, organic matter from 0.2 to 8.6%, and surface area from 1100 to 11,590 m^2/kg.

Sorption experiments were conducted by adding metal nitrates to soil-water suspensions which were mixed at a ratio of 1 g/100 mL in 125 ml Erlenmeyer flasks or polyethylene plastic bottles. Sodium nitrate or $Ca(NO_3)_2$ (or a mixture of both) was used to maintain an ionic strength of 0.01 M. The suspension pH was adjusted using either 0.1 N HNO_3 or NaOH. The samples were equilibrated by shaking for 24 h on a rotatory shaker (Orbit, no. 3590, Lab-line Instruments) at 100 rpm at a room temperature of 25 ± 2°C. After equilibration, the samples were filtered through a 0.45 µm pore size membrane filter (Costar). The syringe and filter holder as well as the membrane filter were rinsed twice with the sample solution before the filtrates were collected for analysis. The metal concentrations in the filtrates were analyzed by cold vapor atomic absorption spectrometry (Perkin Elmer MHS 10) for Hg and by an ICP (Spectro EOP) or an atomic absorption spectrometer (Perkin Elmer 5000, flame or graphite furnace) for other metals. The metal sorbed for each sample was calculated based on the analyzed soluble metal concentration, the total added metal concentration, and the metal naturally present in the soil. The concentrations of the metals initially present in the soils were determined by acid microwave digestion. The digestion was carried out in a mixture of 6% HNO_3 and 4% HCl following the method described by Shi (1996).

Desorption of metals from soils was investigated by mixing each soil sample with 0.01 M $NaNO_3$ at a ratio of 25 g/20 ml in 50 ml polyethylene centrifuge tubes. Five parallel mixtures were prepared for each soil, and the pH values of the mixtures were adjusted with 1 N HNO_3 or NaOH to cover a range of 4 to 8 so that the effect of pH on metal desorption could be evaluated. The soil mixtures were equilibrated by shaking for 24 h under the same conditions described previously. After equilibration, the samples were centrifuged and the supernatants were filtered through a 0.45 µm pore size membrane filter. The metal concentrations in the filtrates were then determined by an ICP.

Clean techniques were employed throughout the experimental process. All glassware and plastic containers were cleaned by acid-soaking overnight followed by thorough rinse with distilled and then deionized water. Metal-free ultrapure water generated by a NANO pure system (Burnstead) was used for all experiments. Trace metal grade or ACS certified chemicals were used. All persons conducting experiments wore non-talc, class-100 plastic gloves. Duplicate metal-free deionized water blanks were run for each batch of experiment to ensure that no contamination occurred during the experiment.

4.3 Results and Discussion

4.3.1 Sorption of Metals to Soils

As expected, sorption of metals on soils was highly pH-dependent. The response of metal sorption to the pH change depended on both soil properties and the nature of the metal. As an example, Figure 4.2 shows sorption of metals on the Freehold subsurface sandy loam as

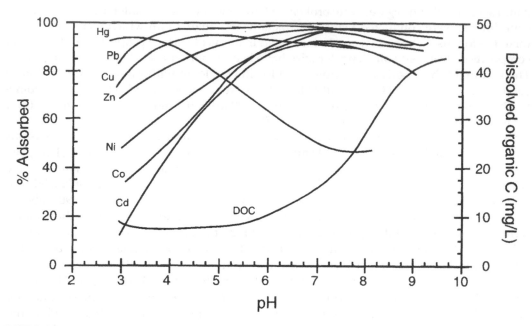

FIGURE 4.2

Sorption of metals on a Freehold sandy soil (subsurface). DOC refers to dissolved organic C. Soil solution = 1 g/100 mL; I = 0.01 M NaNO₃; T = 25 ± 2°C. The concentration of metals added was 1×10^{-6} mol/L.

a function of pH. At pH less than 3.5, sorption followed a sequence of Hg(II) > Pb(II) > Cu(II) > Zn(II) > Ni(II) > Co(II) > Cd(II), which is similar to the sequence of the first hydrolysis constants (log KOH) of these metals, i.e., Hg (10.6) > Pb (6) ~ Cu (6) > Zn (5) > Ni (4.1) ~ Co (4.3) > Cd (3.9). The sorption difference mainly resulted from the variable nature of metals. The greater the log K_{OH} value, the stronger the Lewis acid metal is. The metals of stronger Lewis acids not only have stronger binding affinity for OH⁻, but also for the surface sites (-O⁻) that can be considered as bases. Thus, at low pH values, while the sorption of metals of weaker acids was small due to the competition of protons, the sorption of metals of stronger acids was large because of their stronger binding affinity for the surfaces.

Increases in pH increased sorption principally because the surface potential decreased. Further increases in pH beyond about pH 7 decreased sorption slightly, except for Hg which decreased substantially. The pH value at which the sorption of a metal started to decrease mainly depended on the acidity of a metal. Generally speaking, for metals that are stronger Lewis acids, sorption reached maximum at lower pH values and then decreased. As discussed later in detail, decreases in metal sorption at higher pH values mainly resulted from the complexation of metals by soluble ligands. The metals that are stronger Lewis acids tended to be affected to a greater extent by soluble ligands; thus, the sorption of these metals decreased to a greater extent as pH increased. Cadmium(II) and Hg(II) represented the two ends in terms of metal sorption behavior; the rest of the metals of this study fell in-between. Cadmium sorbed the least to the soil at low pH and was also affected the least by soluble ligands, while Hg(II) sorbed the most to the soil at low pH and was also affected the most by soluble ligands at high pH. In this report, we use Cd(II) and Hg(II) as examples to illustrate the effects of metal nature and soil properties on sorption of metals.

4.3.1.1 Sorption of Cadmium

Sorption of Cd(II) on 15 soils as a function of pH is shown in Figure 4.3a. The partition coefficient (log K_d) increased linearly as pH increased. For a given pH, sorption of Cd(II) on different soils varied by a factor of up to 100. Clearly, soil composition had a dramatic effect on metal sorption. Efforts have been made to correlate metal sorption with soil properties, including soil components and pH (Anderson and Christensen, 1988). Soil components, however, are not independent from pH with regard to sorption/desorption; rather, the pH affects the surface acidity of soil components and the hydrolysis of the metal and consequently affects the sorption ability of these components for metals. Therefore, pH should be incorporated into the speciation of the surface binding sites and the metal. Alternatively, regressions should be done at fixed pH with only independent parameters as variables. In this study, we obtained K_d values for all soils at fixed pH 4, 5, and 6 based on the sorption curves plotted as K_d vs. pH. The obtained K_d value at fixed pH was then correlated to soil components, including the organic matter content and Fe, Mn, and Al oxides. Both mono- and multi-linear regressions were performed. A very good linear relationship between the K_d value and the soil organic matter content was obtained at each fixed pH value. As an example, Figure 4.3b shows the linear relationship for pH 6.

The close correlation of Cd(II) sorption to the organic matter content suggests that soil organic matter is the most important component determining Cd(II) sorption. It is expected that a normalization of the K_d values with the soil organic matter content would reduce the variance of the K_ds at fixed pH shown in Figure 4.3a. As shown in Figure 4.3c, the normalization of K_d values indeed shrank the data to a line. The regression coefficient R^2 increased from 0.799 to 0.927 when K_{om}, instead of K_d, was used for regression. Similar results have been obtained for hydrophobic organic compounds (Karickhoff et al., 1979; Oepen et al., 1991). We applied the regression results shown in Figure 4.3c to the field data collected from the Netherlands by Janssen et al. (1995). The K_{om} values predicted based on the regression equation obtained in this study agreed within one order of magnitude with the measured ones. The mean deviation of $\log K_{om}$ is 0.241.

4.3.1.2 Sorption of Mercury

As shown in Figure 4.2, sorption of Hg(II) reached maximum at pH around 4 and then decreased. The sorption isotherms of Hg(II) on these soils at the natural pH followed an S-shape rather than the L-shape usually observed for sorption of metals on mineral surfaces. The S-shaped isotherm has been ascribed to the complexation of soluble organic matter for metals (Sposito, 1989; Yin et al., 1997a). As shown in Figure 4.4, at low Hg(II) loading, a large amount of Hg(II) was complexed by soluble organic matter; thus, sorption was small. Increases in Hg(II) loading level increased the amount of Hg(II) available for sorption, and therefore increased sorption. When the soluble organic matter was saturated, further increases in Hg(II) loading resulted in rapid increases in Hg(II) sorption, and eventually saturated the surface binding sites.

Based on Hg(II) sorption isotherms, we speculated that decreases in Hg(II) sorption at higher pH might also result from complexation of Hg(II) by soluble organic matter. This was supported by the fact that the dissolved organic matter increased as pH increased (Figure 4.2). We further studied the effect of dissolved organic matter on Hg(II) sorption by adding varying amount of organic matter extracted from soil to soil suspensions at fixed pH 6.5. When the dissolved organic matter (measured as organic C) increased from 1.4 to 61.1 mg/L, sorption of Hg(II) decreased from near 60 to 28% at pH 6.5, implying strong complexation of Hg(II) by the dissolved organic matter.

When $Ca(NO_3)_2$ was used as the background electrolyte and the ionic strength was still maintained at 0.01 M, sorption of Hg(II) significantly increased comparing with that in $NaNO_3$ (Figure 4.5a). This is not because the competition of Ca with Hg(II) for the available surface sites; otherwise, sorption would have decreased. We measured the dissolved organic matter concentrations in both electrolytes and found that addition of Ca^{2+} signifi-

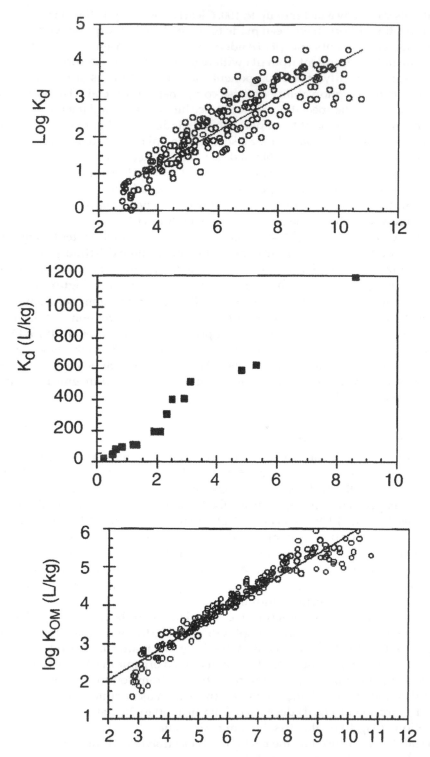

FIGURE 4.3
Soil-water partitioning of Cd(II) as a function of pH and soil organic matter content. K_d is the partitioning coefficient and K_{OM} is the soil organic matter normalized partitioning coefficient. The regression equation for the middle graph is $\log K_{om} = 1.084 + 0.477\ pH$. (Top and bottom graphs are from Lee et al., *Environ. Sci. Technol.*, 3418–3424, 1996.)

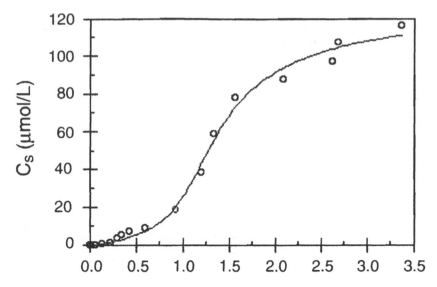

FIGURE 4.4
Sorption isotherm of Hg(II) on the Boonton Bergen loam. Soil solution = 1 g/100 mL; I = 0.01 M NaNO$_3$; T = 25 ± 2°C.

cantly decreased the solubility of organic matter (Figure 4.5b). Calcium has been suggested to be able to decrease the solubility of organic matter either by coagulation or complexation in which Ca serves as a bridge between solid surfaces and organic matter (Schnitzer, 1986). Because the dissolved organic matter decreased in Ca electrolyte, more Hg(II) became available for surface sorption; therefore, Hg(II) sorption increased.

In the presence of high concentrations of Cl$^-$ in the solution, Hg(II) sorption could be reduced at the lower pH range in which the calculated Hg-Cl complexes are the dominant aqueous species (Yin et al., 1996). The efficacy of Cl$^-$ in reducing Hg(II) sorption, however, also depends on soil composition. It is the competition between Cl$^-$ and surface binding sites for Hg(II) that determines the efficacy of Cl$^-$ effect on Hg(II) sorption. For a low organic matter soil, the predominant surface binding sites are inorganic. The binding affinity of these inorganic sites for Hg(II) is weaker than that of Cl$^-$ for Hg(II) (MacNaughton and James, 1974; Barrow and Cox, 1992). Hence, as the Cl$^-$ concentration increased, Hg(II) sorption significantly decreased (Figure 4.6). In contrast to inorganic binding sites, the binding sites on organic matter tend to have stronger affinity for Hg(II) (Yin et al., 1997b). Consquently, an increase in Cl$^-$ concentration had almost no or only slight effect on Hg(II) sorption on the soils with high organic matter contents (Figure 4.6). At the high pH range in which the calculated Hg-OH complexes become predominant over Hg-Cl complexes, addition of Cl$^-$ had almost no effect on Hg(II) sorption (Yin et al., 1996).

Because of the significant effects of soluble ligands on Hg(II) sorption, both surface and aqueous reactions have to be considered in modeling Hg(II) sorption on soils. In this study, we developed a model to describe these reactions. In the solution phase of this study without addition of extra Cl$^-$, the concentration of Cl$^-$ ranged from nearly 1×10^{-6} to 1×10^{-5} mol/L. The effect of Cl$^-$ on Hg(II) aqueous speciation is small compared with that of soluble organic matter (Yin et al., 1996). To make the model simple, only dissolved organic matter was considered in the aqueous speciation calculation. Both free and hydroxo Hg species were assumed to react with the surface sites and the dissolved organic matter, and the binding constants for all Hg(II) species were assumed to be equal. For both surface and aqueous reactions, proton competition with Hg(II) for available binding sites was considered, and 1:1 complexation reactions were assumed.

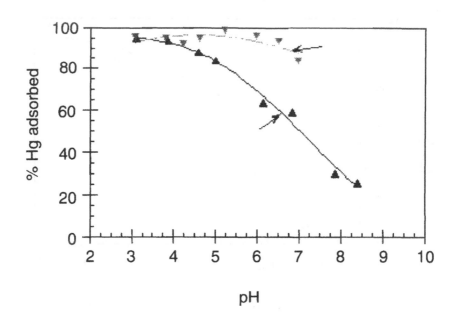

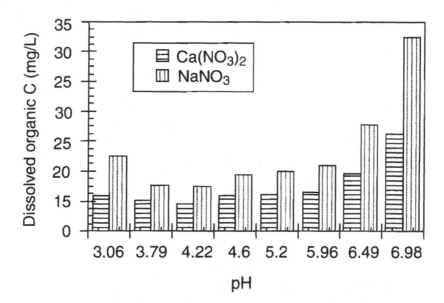

FIGURE 4.5
Effect of Ca^{2+} on Hg(II) sorption onto the Booton Union loam as a function of pH. (Top) Fractional Hg sorption; (bottom) solubility of organic matter. Soil solution = 1 g/100 mL; I = 0.01 M; T = 25 ± 2°C.

The reactions for protons and Hg(II) binding to the dissolved organic matter are expressed by Equations 1 to 4 in which charges are omitted for simplicity:

$$L + Hg = HgL \tag{1}$$

$$K_{HgL} = \frac{[HgL]}{[Hg][L]} = \frac{C_w - [Hg]}{[Hg][L]} \tag{2}$$

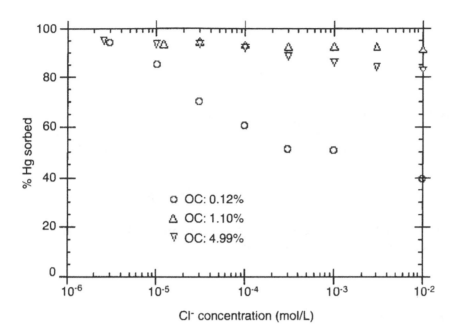

FIGURE 4.6
Effect of Cl⁻ on Hg(II) sorption. OC: soil organic C content; soil solution = 1 g/100 mL; I = 0.01 *M* NaNO₃; T = 25 ± 2°C.

$$L + H^+ = HL \tag{3}$$

$$K_{HL} = \frac{[HL]}{[H^+][L]} \tag{4}$$

where L is the organic ligand, Hg is the inorganic Hg which includes both free and hydroxo Hg species, HgL is the organically bound Hg, K_{HgL} and K_{HL} are the stability constants for Hg and proton binding with ligand (L/µmol), [ ] denotes concentration (µmol/L), and C_w is the total aqueous Hg(II) concentration (µmol/L).

The total concentration of organic ligand, which is expressed by the total dissolved organic C (DOC), is

$$DOC = [L] + [HL] + [HgL] \tag{5}$$

Since the total Hg(II) concentration employed in this study (1×10^{-7} mol/L) was 1000 or more times lower than the DOC concentration ($> 1 \times 10^{-4}$ mol/L), the third term in Equation 5 is negligible, which yields:

$$DOC \cong [L] + [HL] \tag{6}$$

The reactions for proton and Hg binding with surface sites are expressed by Equations 7 to 10:

$$\equiv SO^- + Hg = \ \equiv SOHg \tag{7}$$

$$K_{\equiv SOHg} = \frac{[\equiv SOHg]}{[\equiv SO^-][Hg]} = \frac{T_{Hg} - C_w}{[\equiv SO^-][Hg]} \tag{8}$$

$$\equiv SO^- + H^+ = \equiv SOH \tag{9}$$

$$K_{\equiv \text{SOH}} = \frac{[\equiv SOH]}{[\equiv SO^-][H^+]} \tag{10}$$

where $\equiv SO^-$ is the surface binding site, $\equiv SOHg$ is the Hg sorbed by the surface, and $K_{\equiv \text{SOHg}}$ is the binding constant of Hg with solid surface (L/μmol), $\equiv SOH$ is the protonated surface site, and $K_{\equiv \text{SOH}}$ is the proton binding constant for the surface (L/μmol). The total surface site density Γ_t (μmol/L) is

$$\Gamma_t = [\equiv SO^-] + [\equiv SOH] + [\equiv SOHg] \tag{11}$$

The third term in Equation 11 can be neglected compared to the total surface binding sites due to low mercury loading level. This gives:

$$\Gamma_t = [\equiv SO^-] + [\equiv SOH] \tag{12}$$

Solving Equations 2, 4, 6, 8, 10, and 12 gives:

$$C_w = \frac{1 + K_{\text{HL}}[H^+] + K_{\text{Hgl}}.DOC}{1 + K_{\text{HL}}[H^+] + K_{\text{Hgl}}.DOC + K_{\equiv \text{SOHg}}\Gamma_t \dfrac{1 + K_{\text{HL}}[H^+]}{1 + K_{\equiv \text{SOH}}[H^+]}} T_{\text{Hg}} \tag{13}$$

Both $K_{\equiv \text{SOHg}}$ and Γ_t are constant for a given soil under the experimental conditions. The product of these two parameters determines Hg binding effectiveness to soils. Defining $K_{\equiv \text{SOHg}}\Gamma_t = A$, Equation 13 becomes

$$C_w = \frac{1 + K_{\text{HL}}[H^+] + K_{\text{Hgl}}.DOC}{1 + K_{\text{HL}}[H^+] + K_{\text{Hgl}}.DOC + A\dfrac{1 + K_{\text{HL}}[H^+]}{1 + K_{\equiv \text{SOH}}[H^+]}} . \tag{14}$$

We fit the experimental data for Hg(II) sorption on each soil with Equation 14 using a multivariant nonlinear regression method (Wilkinson, 1993). The predicted soluble Hg concentration based on the model agreed well with the measured value for all soils with a regression coefficient R^2 ranging from 0.911 to 0.981. We correlated the model fitting parameter A with soil properties and found a good linear relationship between *A* and soil organic C content (Figure 4.7):

$$A = 0.81 \ SOC \tag{15}$$

Equation 15 again suggests that soil organic matter is the major binding component. The model fitting also indicated that the competition of protons with Hg for the surface sites was unimportant under the experimental conditions where the pH ranged from 3 to 10. Therefore, the term $K_{\equiv \text{SOHg}}[H]$ in Equation 14 is negligible. Based on the model fitting result for each soil, the average values for proton and Hg binding with dissolved organic matter were calculated. Substitution of these average values and Equation 15 into Equation 14 gives a universal equation:

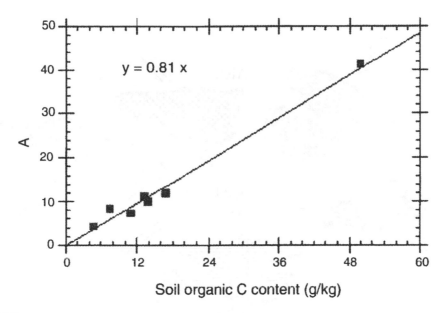

FIGURE 4.7
Correlation between model fitting parameter A for Hg(II) sorption on soils and the soil organic C content.

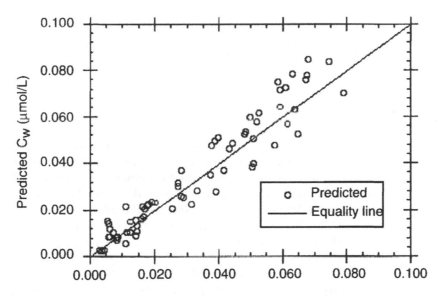

FIGURE 4.8
Predicted vs. measured soluble Hg(II) concentrations. The predicted values were calculated based on Equation 16.

$$C_w = \frac{1 + 0.32[H^+] + 1.096 \times 10^{-2} DOC}{1 + 0.32[H^+] + 1.096 \times 10^{-2} DOC + 0.81 SOC(1 + 0.32[H^+])} T_{Hg} \qquad (16)$$

We calculated the soluble Hg concentration using Equation 16 based on the total Hg and the experimentally determined DOC and pH values. The calculated and measured values agree well with a correlation coefficient R^2 of 0.915 and a residual mean square of 0.0019 (Figure 4.8).

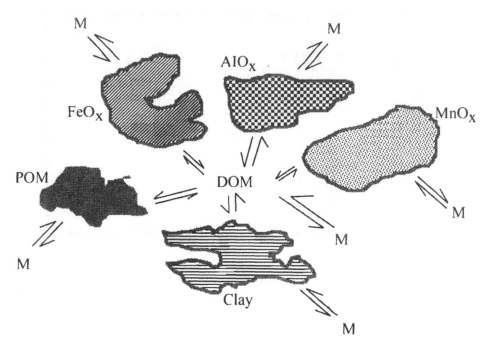

FIGURE 4.9
Hypothetical coating of mineral surfaces by organic matter in soils. POM denotes particulate organic matter; DOM denotes dissolved organic matter; M denotes soluble inorganic metal. The rim around clay mineral and oxides is organic matter coating.

4.3.1.3   Soil Sorption Phases

Sorption of metals by soil components has been extensively documented. Almost all soil components, especially natural organic matter, oxides, and 2:1 clay minerals, were effective sorbents for metals. In natural systems, not all minerals are necessarily important in metal sorption. Some of the mineral surfaces may be coated by other components and become unavailable. For example, clay minerals are generally coated by metal oxides and organic matter (Hart, 1982; Davis, 1984; Jenne, 1988), while the metal oxides can also be coated by organic matter (Stumm, 1992). Our studies have indicated that sorption of both Cd(II) and Hg(II) correlated very well with the soil organic matter content. This suggests that a large quantity of the inorganic minerals was probably coated with organic matter in the soils of this study as hypothesized in Figure 4.9. As a result, the soil-water partitioning of organic matter as well as the affinity of a metal for both particulate and dissolved organic matter determined the partitioning of the metal in soils.

Based on this study, it seems that the organic functional groups that dissolved were different from those remaining in the particulate phase. Both particulate and dissolved organic matter dominated metal sorption. The maximum amount of organic matter dissolved accounted for less than 30% of the total soil organic matter content; however, the dissolved organic matter could reduce Hg(II) sorption by 60 to 70%, indicating strong affinity of Hg(II) to the dissolved organic matter. In case of Cd(II), the effect of the dissolved organic matter on sorption was very small. It needs to be pointed out that all of the 15 soils used in this study were acidic agricultural soils with pH ranging from 4.2 to 6.4. The conclusion regarding the soils of this study may not apply to mineral soils in which the organic matter content is small.

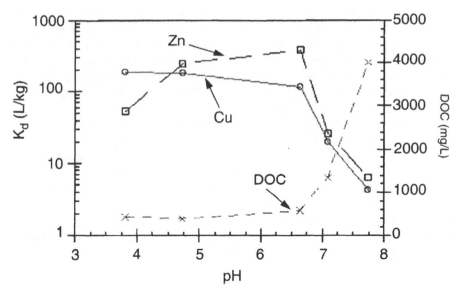

FIGURE 4.10
Desorption of Cu and Zn from the Freehold subsurface sandy loam. DOC is dissolved organic C. Soil solution
= 25 g/20 mL; I = 0.01 M NaNO$_3$; T = 25 ± 2°C.

4.3.2 Metal Desorption from Soils

Desorption of metals from soils was investigated as a function of pH following the procedure described previously. Under the experimental conditions, only Cu and Zn among the desorbed metals were within the instrumental detection limit; thus, desorption of Cu and Zn was examined. As shown in Figure 4.10 for the Freehold subsurface sandy loam, desorption of both metals followed a similar trend as that obtained by sorption experiments. At pH around 4, the partitioning coefficient (ratio of metal sorbed/soluble metal concentration) of Cu was greater than that of Zn because of the stronger Lewis acid nature of Cu than Zn as explained previously. As pH increased, the K_d value of Cu slightly increased until pH 6.7 and then abruptly decreased with increasing pH. For Zn, the K_d value initially increased until pH 6.7 and then abruptly decreased with further increases in pH.

We determined the dissolved organic C concentration and found that the change of DOC concentration as a function of pH matched the desorption change of both metals. Because Cu has stronger affinity for the dissolved organic matter, desorption of Cu slightly increased as the DOC concentration increased. Zinc, however, has relatively weaker affinity for the dissolved organic matter (Tipping and Hurley, 1992); thus slight increases in DOC concentration did not cause noticeable effect on its desorption. Consequently, as the proton competition decreased and the surface potential increased with increasing pH, Zn desorption decreased. At pH above 6.7, the significant increases in DOC concentration caused desorption of both metals to abruptly increase due to the complexation of metals by the dissolved organic matter.

4.4 Conclusions

This review indicates that metal retention to and release from soils was affected by many interdependent environmental factors. Among these factors, soil pH and organic matter are the most important. The pH not only affected the surface potential and the competition of protons for metal binding in both particulate and solution phases, but also affected the partitioning of organic matter between soil solid and solution. The latter further affected soil-water partitioning of metals due to the strong complexation of organic matter for metals. Based on our results, it seems that the nature of the dissolved and the remaining particulate organic matter at a given pH was different. Thus, the binding affinity of the organic matter for metals in two phases is different.

The presence of high concentrations of divalent major cations has previously been shown to decrease trace metal sorption by competing binding (Zachara et al., 1993) in clay mineral soils. In this study, we showed that divalent major cations present in solution could also increase trace metal sorption by decreasing the solubility of organic matter in acidic agricultural soils. Chloride and other inorganic ligands, which have strong affinity for metals, could increase metal release from soils at the pH values where the metal–ligand complexes are dominant over metal–OH complexes in the solution phase. The efficacy of the effects of these inorganic ligands on metal release also depends on the content of soil organic matter, which contains sites of greater affinity for metals than does Cl^-.

References

Anderson, P.R. and T.H. Christensen, Distribution coefficients of Cd, Co, Ni, and Zn in soils, *J. Soil Sci.*, 39, 15, 1988.

Barrow, N.J. and V.C. Cox, The effects of pH and chloride concentration on mercury sorption. I. By goethite, *J. Soil Sci.*, 43, 295, 1992.

Barrow, N.J., J.W. Bowden, A.M. Posner, and J. P. Quirk, Describing the adsorption of copper, zinc and lead on a variable charge mineral surface, *Aust. J. Soil Res.*, 19, 309, 1981.

Davis, J.A., Complexation of trace metals by adsorbed natural organic matter, *Geochim. Cosmochim. Acta*, 48, 679, 1984.

Fuller, C.C., J.A. Davis, and G.A. Waychunas, Surface chemistry of ferrihydrite. II. Kinetics of arsenate adsorption and coprecipitation, *Geochim. Cosmochim. Acta*, 57, 2271, 1993.

Hart, B.T., Uptake of trace metals by sediments and suspended particles: a review, in *Sediment/Freshwater Interaction*, Sly, P.G., Ed., Dr. W. Junk Publishing, Boston, 1982, 293.

Huang, C.P. and Y.T. Lin, Specific adsorption of Co(II) and [Co(III)EDTA]⁻ complexes on hydrous oxides surfaces, in *Adsorption from Aqueous Solutions*, Tewari, P.H., Ed., Plenum, New York, 1981, 61.

James, R.O. and T.W. Healy, Adsorption of hydrolyzable metal ions at the oxide-water interface. III. A thermodynamic model of adsorption, *J. Colloid Interface Sci.*, 40, 65, 1972.

Janssen, R.P.T., P.J. Pretorius, W.J.G.M. Peijnenburg, and M.A.G.T. van den Hoop, *Proc. Int. Conf. Heavy Metals in the Environment*, Hamburg, Germany, CEP Consultants, Edinburgh, 1995, 153.

Jenne, E.A., Controls on Mn, Fe, Co, Ni, Cu, and Zn concentrations in soils and water: the significant role of hydrous Mn and Fe oxides, in *Trace Inorganics in Water*, Baker, R.A., Ed., Adv. Chem. Ser. 73, American Chemical Society, Washington, D.C., 1988, 337.

Karickhoff, S.W., D.S. Brown, and T.A. Scott, Sorption of hydrophobic pollutants on natural sediments, *Water Res.*, 13, 241, 1979.

Lee, S.Z., H.E. Allen, C.P. Huang, D.L. Sparks, P.F. Sanders, and W.J.G.M. Peijnenberg, Predicting soil-water partition coefficients for cadmium, *Environ. Sci. Technol.*, 30, 3418, 1996.

MacNaughton, M.G. and R.O. James, Adsorption of aqueous mercury (II) complexes at the oxide/water interface, *J. Colloid Interface Sci.*, 47, 431, 1974.

Oepen, B.von, W. Kordel, and W.J. Klein, Sorption of nonpolar and polar compounds to soils: processes, measurements and experience with the applicability of the modified OECD-Guideline 106, *Chemosphere*, 22, 285, 1991.

Schnitzer, M., Binding humic substances by soil mineral colloids, in *Interaction of Soil Minerals with Natural Organics and Microbes*, Huang, P.M. and M. Schnitzer, Eds., SSSA Spec. Publ. No. 17, Soil Sci. Am., Madison, WI, 1986, 77.

Shi, B., Fate and Effects of Metals in Aquatic Environment: Evaluation for POTW Effluent Discharges and Metal Partitioning in Surface Water, Ph.D. thesis, University of Delaware, Newark, 1996.

Sparks, D.L., *Environmental Soil Chemistry*, Academic Press, New York, 1995.

Sposito, G., *The Surface Chemistry of Soils*, Oxford University Press, New York, 1984.

Sposito, G., *The Chemistry of Soils*, Oxford University Press, New York, 1989.

Stumm, W., *Chemistry of the Solid-Water Interface: Processes at the Mineral-Water and Particle-Water Interface in Natural Systems*, John Wiley, New York, 1992.

Tewari, P.H. and W. Lee, Adsorption of Co(II) at the oxide — water interface, *J. Colloid Interfacial Sci.*, 52, 77, 1975.

Tipping, E. and M.A. Hurley, A unifying model of cation binding by humic substances, *Geochem. Cosmochim. Acta*, 56, 3627, 1992.

Wilkinson, L., *SYSTAT: The System for Statistics*, SYSTAT, Evanson, IL, 1988.

Yin, Y., H.E. Allen, C.P. Huang, Y. Li, and P.F. Sanders, Adsorption of Hg(II) by soil: effects of pH, chloride concentration, and organic matter, *J. Environ. Qual.*, 25, 837, 1996.

Yin, Y., H.E. Allen, C.P. Huang, and P.F. Sanders, Adsorption/desorption isotherms of Hg(II) by soil, *Soil Sci.*, 162, 35, 1997a.

Yin, Y., H.E. Allen, C.P. Huang, D.L. Sparks, and P.F. Sanders, Kinetics of Hg(II) adsorption and desorption on soil, *Environ. Sci. Technol.*, 31, 496, 1997b.

Zachara, J.M., S.C. Smith, J.P. McKinley, and C.T. Resch, Cadmium sorption on specimen and soil smectites in sodium and calcium electrolytes, *Soil Sci. Soc. Am. J.*, 57, 1491, 1993.

5

Chemical Remediation Techniques for the Soils Contaminated with Cadmium and Lead in Taiwan

Zueng-Sang Chen, Geng-Jauh Lee, and Jen-Chyi Liu

CONTENTS

5.1   Introduction

Soils can be regarded as a major sink of heavy metals that were discharged from different kinds of anthropogenic pollution sources (Nriagu, 1991; Mench et al., 1994). Once heavy metals were released and adsorbed by the soil, most would be persistent because of their fairly immobile nature. Cadmium is known to be more mobile and bioavailable than most other heavy metals, but lead is demonstrated to be fairly immobile and unavailable for plant uptake in soil systems. Lead deposited on the surface of plant tissue can be of concern (Adriano, 1986).

Many researchers have indicated that agricultural soils contaminated by heavy metals may result in foliar damage, reducing growth yield of crops. Heavy metals in soils may also be taken up by crops and adversely affect the human health through the food chain (Alloway, 1990).

Several methods have been used to immobilize metals present in contaminated soils. One general chemical technique is to apply dolomite, phosphates, or organic matter residues into the polluted soils to reduce the soluble metal concentration by precipitation, sorption, or complexation (Impents, 1991; Mench et al., 1994; Chen and Lee, 1997). Different soil remediation techniques including engineering, chemical, and biological treatments were proposed and tested to remediate these two contaminated cadmium and lead sites in northern Taiwan (Wang et al., 1989; Chen et al., 1992a; 1992b; Chen, 1994; Chen et al., 1994; Lee and Chen, 1994; Lo and Che, 1994; Wang et al., 1994; Cheng, 1996; Chen and Lee, 1997). Based on economic assessment, engineering and chemical remediation techniques are most efficient and save time for changing the land uses of contaminated sites (Chen et al., 1994). The most effective remediation techniques of the engineering method are removing the polluted surface soils and replacing them with uncontaminated soil. Then, the removed contaminated soils can be washed with some chemical extractants or chelating reagents (Chen et al., 1994). The remediation techniques of chemical treatments include adding some chemical material into the polluted soils to reduce the concentration of Cd and Pb in the soil solution, such as lime material, manure or composts, phosphate materials, and hydrous iron and manganese oxides (Mench et al., 1995; Chen and Lee, 1997; Chen et al. 1997). The application of lime materials can significantly reduce the solubility of heavy metals in contaminated sites (McBride and Blasiak, 1979; McBride, 1980; Sommers and Lindsay, 1979; Kuo et al., 1985; Liu et al., 1998). Some reports also indicated that the application of hydrous iron or manganese oxides in contaminated soils could reduce the concentration of Cd or Pb in the soil solution (McKenzie, 1980; Kuo and McNeal, 1984; Tiller et al., 1984; Khattak and Page, 1992; Mench et al., 1994). High quantity applications of phosphate in polluted soils also can reduce the solubility of zinc in the soil solution by precipitation (Bolland et al., 1977; Saeed and Fox, 1979; Barrow, 1987). Some vegetation species, flowers, and trees planted in the polluted soils are also effective in removing the heavy metals from the sites (Lee and Liao, 1993; Lee and Chen, 1994).

The objectives of this chapter are (1) to evaluate the effects of different chemical remediation treatments on the reduction of Cd and Pb soluble in the soils, and (2) to evaluate their bioavailability for wheat grown in the contaminated soils.

5.2   Materials and Methods

5.2.1   The Contaminated Sites

Two rural soils, Chunghsing clayey soil (including sites A and B) and Chaouta sandy soil (including sites C and D), were selected from contaminated sites irrigated with discharged water from chemical plants in northern Taiwan (Chen, 1991). The total area of contaminated sites in these two regions is about 100 ha.

The mean total cadmium concentrations of brown rice and soils in these two sites are 1.49 to 2.99 mg/kg and 4.7 to 378 mg/kg and mean total lead concentrations of brown rice and soils are 1.13 to 8.37 mg/kg and 25.8 to 3145 mg/kg, respectively (Lu et al., 1984; Chen, 1991). These two sites were designated as contaminated sites by the Taiwan government in 1984 because the concentration of Cd in brown rice was higher than the critical health concentration of Cd of 0.5 mg/kg issued by Department of Health of Taiwan.

TABLE 5.1

The Physical and Chemical Properties of Four Contaminated Soils

Site	Soils	pH H₂O	pH KCl	O.C. (g/kg)	Sand (g/kg)	Silt (g/kg)	Clay (g/kg)	CEC (cmol(+)/kg soil)	Exch. Base (cmol(+)/kg soil)	Base Saturation (%)
Chungfu	A	5.0	4.3	23.5	113	481	406	12.4	3.82	31
	B	5.5	4.5	15.2	102	524	374	9.9	3.77	38
Tatan	C	5.5	4.7	12.1	726	57	217	4.5	3.23	72
	D	5.4	4.8	12.9	742	123	135	4.5	3.71	82

From Lee, G. J., The Assessment of Remediation Techniques by Chemical Treatments for Soils Contaminated with Cadmium and Lead, Master's thesis, Graduate Institute of Agricultural Chemistry, National Taiwan University, Taipei, 1996. With permission.

5.2.2 Analysis of Basic Soil Properties

The particle size distribution of polluted soils was determined by the pipette method (Gee and Bauder, 1986). Soil pH value was determined by a glass electrode in a soil/water ratio of 1:1 and a soil/1 M KCl ratio of 1:2.5 (McLean, 1982). Total organic carbon content was determined by the Walkley-Black wet combustion method (Nelson and Sommer, 1982). Exchangeable cations (K, Na, Ca, and Mg) and cation exchangeable capacity (CEC) were exchanged by ammonium acetate (pH 7) (Thomas, 1982). The percentage of base saturation (BS%) was calculated by the sum of exchangeable cations divided by CEC. The basic soil properties of these four sites are shown in Table 5.1.

Based on the databases of the heavy metals in rural soils of Taiwan, sites A and B are only moderately contaminated with Cd, but sites C and D are seriously contaminated with Cd and Pb. The bioavailability concentrations of Cd and Pb in polluted soils were determined by different extraction solutions, such as distilled water by shaking 2 h, 0.1 M HCl by shaking 1 h (EPA/ROC, 1991), 0.005 M DTPA (pH 5.3) by shaking 1 h (Norvell, 1984), and 0.05 M EDTA (pH 7.0) by shaking 1 h (Mench et al., 1994). Then the extraction solution was filtered with Whatman no. 42 filter paper and 0.45 µm Millipore filter paper. The concentrations of Cd and Pb in the extraction solution were determined by flame atomic absorption spectroscopy (Hitachi 180-30 type). Total concentration analysis of Cd and Pb in the polluted soil was digested with concentrated HCl and HNO₃ (3:1, v/v) and filtered with Whatman no. 42 filter paper and 0.45 µm Millipore filter paper, then the concentrations of Cd and Pb also determined by flame atomic absorption spectroscopy (Hitachi 180-30 type) (EPA/ROC, 1991).

5.2.3 Treatments in Pot Experiments

Seven chemical treatments were used to compare and evaluate the remediation techniques for soils from the two contaminated sites. Five hundred grams of soils were treated as follows and placed in polyethylene pots. The treatments included (1) liming with calcium carbonate to increase soil pH to 7.0, (2) applying high phosphate of 10 g P/500 g soils, (3) applying 2% composts, (4) applying 1% iron oxide, (5) applying 1% manganese oxide, (6) applying 1% zeolite, and (7) maintaining a control treatment. Each treatment was replicated three times. The treated and controlled soils were incubated for 2 months at room temperature and field capacity. Wheat (*Triticum aestivum*) was planted in the treated soils to evaluate the effectiveness of the chemical treatments on uptake of Cd and Pb. Plants were harvested after 1 month.

5.2.4   Bioavailability to Wheat

Several extractants were used to extract the concentration of heavy metals for prediction of bioavailability of metal in the polluted soils (Lakanen and Ervio, 1971; Norvell, 1984). The changes of the bioavailable concentration of Cd and Pb are extracted and evaluated with different extraction reagents including distilled water, 0.1 M Ca(NO$_3$)$_2$ (Mench et al., 1994), 0.05 M EDTA (pH 7.0) (Norvell, 1984), 0.43 M HOAC (Mench et al., 1994), and 0.1 M HCl (EPA/ROC, 1991). Changes in the forms of heavy metals in the polluted soils before and after chemical treatments were used to compare the differences among chemical treatments. The extraction solution was filtered by Whatman no. 42 filter. The concentration of Cd and Pb was determined by the atomic adsorption spectrophotometer (Hitachi 180–30 type).

The harvested samples of wheat were dried at 60°C for 2 days and digested with concentrated sulfuric acid mixed with perchloric acid. The concentration of Cd and Pb in the digestion solution was determined by atomic adsorption spectrophotometer.

5.2.5   Sequential Fractionation of Heavy Metals in Soils

To evaluate chemical forms of heavy metals in treated contaminated soils, a sequential extraction technique based on the method of Mench et al. (1994) was used. Briefly, the sequential fraction procedure was performed in four steps with the assumption that the four chemical forms of metals existing in the contaminated soils were (1) water soluble, exchangeable, weakly bounded to organic matter, carbonate fractions, extracted with 0.11 M acetic acid (HOAC) and shaken for 16 h; (2) occluded Fe or Mn oxide fraction, extracted with 0.1 M hydroxyl ammonium chloride (HONH$_3$Cl) and shaken for 16 h; (3) organically bound and sulfide fraction, extracted with 1 M ammonium acetate and shaken for 16 h; and (4) structurally bound in residual fraction, digested in 3:1 v/v 12 M HCl and 14 M HNO$_3$ (aqua regia). All extracts were stored in polyethylene tubes and retained at 4°C for analysis. The concentration of Cd and Pb in these fractions of soils was determined by atomic absorption spectrophotometer.

5.2.6   Statistical Analyses

The analysis of variance and significant differences of concentration of Cd and Pb in the different chemical treatments for four polluted soils was performed by SAS (SAS, 1982). The statistical significance was defined at $p < 0.10$.

5.3   Results and Discussion

5.3.1   Cd and Pb Concentration Extracted by Different Reagents in Untreated Soils

The concentrations of Cd and Pb in these four contaminated soils extracted with different extraction reagents are shown in Table 5.2. Results indicated that the most serious contaminated site was site D with sandy soils showing the highest concentration of Cd (18.6 mg/kg) and Pb (611 mg/kg) (Table 5.2). For the clayey soils, site A is a more seriously polluted site (5.47 mg/kg total Cd and 39.2 mg/kg total Pb) than site B.

TABLE 5.2

Cd and Pb Concentration in Soils Extracted by Different Single Solutions for Four Contaminated
Soils before Chemical Treatments

Soils	Water	Ca(NO$_3$)$_2$	EDTA	HOAC	HCl	Aqua Regia
			(mg/kg)			
Cd						
A	ND	3.70	3.72	3.52	4.56	5.47
B	ND	0.73	0.82	0.58	1.36	2.06
C	ND	3.79	4.24	4.25	5.02	5.31
D	ND	12.4	14.21	7.8	17.6	18.6
Pb						
A	ND	8.82	11.9	4.63	13.2	39.2
B	ND	4.97	6.50	4.61	8.80	29.6
C	ND	15.3	19.8	10.4	25.1	50.4
D	1.31	202	269	322	398	611

Note: A and B soils (clayey soils) are located at Chungfu contaminated site; C and D soils (sandy soils) are
located at Tatan contaminated site. ND: not detectable.

From Lee, G. J., The Assessment of Remediation Techniques by Chemical Treatments for Soils Contaminated
with Cadmium and Lead, Master's thesis, Graduate Institute of Agricultural Chemistry, National Taiwan University, Taipei, 1996. With permission.

5.3.2 Changes on the Bioavailability of Cd and Pb after Chemical Treatments

An index of the bioavailability of heavy metals in these polluted soils can be evaluated by
different extractants, such as distilled water, 0.1 M Ca(NO$_3$)$_2$, 0.05 M EDTA (pH 7.0), 0.43 M
HOAC, and 0.1 M HCl. Results of pot experiments in these soils indicate that the decreasing sequences for different chemical materials to reduce the extraction concentration of Cd
in these polluted soils are, at first, calcium carbonate, manganese oxide, or zeolite, and then
composts based on the concentration extracted by EDTA (p <0.10); there are no effects on
the other chemical treatments (Table 5.3). These results support some other papers that
indicated that the application of lime materials can significantly reduce the solubility of Cd
in soil solutions (McBride and Blasiak, 1979; Sommers and Lindsay, 1979; McBride, 1980;
Christensen, 1984; Kuo et al., 1985; Mench et al., 1994; Liu et al., 1998). These results also
support results that the application of hydrous iron or manganese oxide materials can significantly reduce the solubility of Cd in the soil solutions (McKenzie, 1980; Kuo and
McNeal, 1984; Tiller et al., 1984; Fu et al., 1991; Khattak and Page, 1992; Mench et al., 1994;
Lee, 1996). In this study of Taiwan polluted sites, control of soil pH is a key factor to control
the extractability (or bioavailability) of Cd in clayey or sandy polluted soils.

The decreasing sequence of different methods to reduce the extraction concentration of Pb
in these polluted soils consists of manganese oxide, calcium carbonate, or zeolite based on
the concentration extracted by Ca(NO$_3$)$_2$ or HOAC (p <0.10), and there are no effects for the
other chemical treatments (Table 5.4). These results support the results that the application
of lime materials can significantly reduce the solubility of Pb in the soil solutions of contaminated sites (McBride and Blasiak, 1979; McBride, 1980; Sommers and Lindsay, 1979; Kuo et
al., 1985; Mench et al., 1994; Lee, 1996). These results also support results that the application
of hydrous iron or manganese oxide materials can significantly reduce the solubility of Pb
in the soil solutions (McKenzie, 1980; Kuo and McNeal, 1984; Tiller et al., 1984; Khattak and
Page, 1992; Mench et al., 1994; Lee, 1996). These results also support results that the application of zeolite can significantly reduce the solubility of Pb in the soil solutions
(Gworek, 1992). In this study of Taiwan polluted soils, application of hydrous manganese

TABLE 5.3

Cadmium Concentrations Extracted by Single Extractant for Four Soils Treated with Different Chemical Materials

	Cd Concentrations in Single Extraction Solution				
Treatment	Water	Ca(NO₃)₂	EDTA	HOAC	HCl
			(mg/kg)		
Site A after 2 Months					
Control	ND	4.31 a	4.09 a	3.75 a	4.56 a
FO	ND	4.11 b	4.09 a	3.13 bc	4.45 ab
MO	ND	4.11 b	3.86 ab	2.92 c	4.23 bc
Lime	ND	3.95 b	3.42 c	3.13 bc	4.17 c
Phosphate	ND	3.34 b	3.98 ab	3.34 b	4.34 abc
Compost	ND	3.13 b	3.87 ab	3.13 bc	4.56 a
Zeolite	ND	4.05 b	3.76 b	2.92 c	4.17 c
Site B after 2 Months					
Control	ND	1.11 bc	0.86 a	0.65 a	1.23 a
FO	ND	1.13 b	0.85 a	0.44 a	1.06 b
MO	ND	1.08 c	0.85 a	0.44 a	1.06 b
Lime	ND	1.08 c	0.86 a	0.44 a	1.01 b
Phosphate	ND	1.11 bc	0.75 a	0.44 a	1.01 b
Compost	ND	1.21 a	0.86 a	0.44 a	1.01 b
Zeolite	ND	1.11 bc	0.86 a	0.44 a	1.06 b
Site C after 2 Months					
Control	ND	4.36 a	4.48 a	4.46 a	4.77 a
FO	ND	4.31 a	4.21 a	4.06 ab	4.66 a
MO	ND	4.41 a	4.42 a	3.66 b	4.42 b
Lime	ND	4.26 a	3.34 c	4.26 a	4.45 b
Phosphate	ND	4.26 a	4.21 a	3.66 b	4.66 a
Compost	ND	4.16 a	3.77 b	3.66 b	4.45.b
Zeolite	ND	4.21 a	3.55 bc	3.66 b	4.23 c
Site D after 2 Months					
Control	0.12 a	13.7 a	15.1 a	17.0 a	15.3 a
FO	0.12 a	13.7 a	15.1 a	16.0 bc	15.3 a
MO	0.12 a	13.7 a	14.0 ab	15.3 c	15.8 a
Lime	ND b	13.7 a	11.3 c	16.2 c	15.8 a
Phosphate	ND b	13.2 a	14.0 ab	16.2 bc	15.8 a
Compost	ND b	13.2 a	12.4 bc	14.9 c	15.3 a
Zeolite	ND b	13.7 a	12.9 bc	17.0 ab	15.8 a

Note: Data are expressed as mean value and with the same letter within a column ($p < 0.10$) are not significantly different. FO: iron oxides; MO: manganese oxides; ND: not detectable.

From Lee, G. J., The Assessment of Remediation Techniques by Chemical Treatments for Soils Contaminated with Cadmium and Lead, Master's thesis, Graduate Institute of Agricultural Chemistry, National Taiwan University, Taipei, 1996. With permission.

oxides is also another key factor to control the extractability (or bioavailability) of Pb in clayey or sandy polluted soils.

5.3.3 Transformation of Chemical Forms of Cd and Pb in the Amended Soils

The changes of sequential fractions of Cd and Pb in soils B and D after chemical treatments in 2 months are shown in Figure 5.1 and 5.2 (the results of site A and site C are not shown here).

TABLE 5.4

Lead Concentration Extracted by Different Single Solution for Four Soils Treated with Different Chemical Materials

Treatment	Pb Concentrations in Single Extraction Solution				
	Water	Ca(NO$_3$)$_2$	EDTA	HOAC	HCl
	(mg/kg)				
Site A after 2 Months					
Control	ND	14.6 a	12.2 a	1.27 a	11.3 a
FO	ND	12.9 b	10.5 ab	1.27 a	11.3 a
MO	ND	8.59 c	6.13 c	1.27 a	6.14 b
Lime	ND	12.9 b	11.3 ab	1.27 a	10.5 a
Phosphate	ND	13.3 ab	12.2 a	1.27 a	11.3 a
Compost	ND	14.2 ab	10.5 ab	1.27 a	11.3 a
Zeolite	ND	12.9 b	8.72 b	1.27 a	10.5 a
Site B after 2 Months					
Control	ND	8.13 a	7.,84 a	1.26 a	7.84 a
FO	ND	6.42 ab	6.11 b	1.26 a	6.12 b
MO	ND	4.29 c	3.53 c	1.26 a	4.39 c
Lime	ND	6.42 ab	7.81 a	1.26 a	6.12 b
Phosphate	ND	6.42 ab	7.84 a	1.26 a	6.12 b
Compost	ND	7.28 ab	6.97 ab	1.26 a	6.12 b
Zeolite	ND	5.57 bc	4.39 c	1.26 a	5.25 bc
Site C after 2 Months					
Control	ND	26.7 a	21.2 a	7.90 a	23.7 a
FO	ND	25.1 a	18.6 bc	4.57 a	21.2 b
MO	ND	23.4 a	13.6 d	4.57 a	14.4 d
Lime	ND	25.1 a	17.8 c	4.57 a	21.2 b
Phosphate	ND	26.7 a	19.5 a	7.90 a	22.8 a
Compost	ND	26.7 a	17.8 c	7.90 a	21.2 b
Zeolite	ND	24.2 a	11.9 e	4.57 a	19.5 c
Site D after 2 Months					
Control	1.31 a	347 a	415 a	282 a	398 a
FO	ND b	343 a	381 b	282 a	398 a
MO	ND b	330 b	381 b	235 c	398 a
Lime	ND b	343 a	389 b	282 a	398 a
Phosphate	ND b	343 a	381 b	275 b	398 a
Compost	ND b	339 ab	398 ab	242 c	398 a
Zeolite	1.31 a	347 a	372 b	275 b	372 b

Note: Data are mean of two duplicates and with the same letter within a column are not significantly different (p <0.10). Treatments: FO: iron oxides; MO: manganese oxide; Liming: applying calcium carbonate to increase soil pH to 7.0; ND: non-detectable.

From Lee, G. J., The Assessment of Remediation Techniques by Chemical Treatments for Soils Contaminated with Cadmium and Lead, Master's thesis, Graduate Institute of Agricultural Chemistry, National Taiwan University, Taipei, 1996. With permission.

Results from sequential fractionations of metals indicate that Cd in polluted soil B can be transformed to Fe-, Mn-bound, or residual forms from soluble forms after these chemical treatments (Figure 5.1), but Pb in polluted soil D can be transformed to Fe- and Mn-bound forms, or organic compound bound forms from soluble or residual forms after these chemical treatments (Figure 5.2). These results also indicate that Cd and Pb in the available (or soluble) form in the contaminated soils can be significantly transformed into fixed (or unavailable) forms after these chemical treatments, especially when treated with manganese oxide, lime material, or zeolite applied with 1% of these materials ($p < 0.10$).

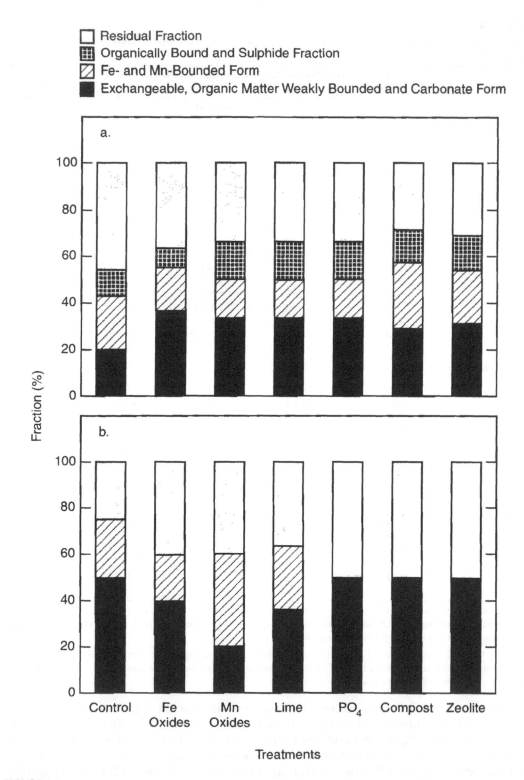

FIGURE 5.1
The distribution of Cd in different soil fractions in soil B treated in 1 month (a) and in 2 months (b).

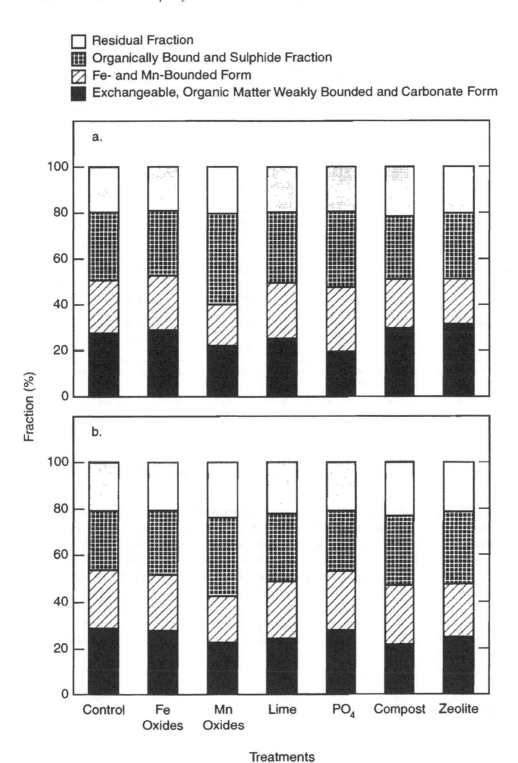

FIGURE 5.2
The distribution of Pb in different soil fractions in soil D treated in 1 month (a) and in 2 months (b).

TABLE 5.5

Cd and Pb Concentrations in Wheat Growing in Four Soils with Different Chemical Treatments

Treatments	Soil A	Soil B	Soil C	Soil D
	(mg/kg dry weight)			
Cd				
Control	45.0 a	32.7 a	95.5 a	115 a
FO	35.5 ab	35.5 a	80.5 a	115 a
MO	21.8 bc	13.6 b	61.4 b	101 ab
Lime	15.5 c	5.45 b	20.5 c	99.6 ab
Phosphate	34.1 ab	13.6 b	84.6 a	108 ab
Compost	34.1 ab	8.20 b	56.0 b	115 a
Zeolite	28.7 b	4.68 b	24.6 c	69.6 b
Pb				
Control	108 a	65.3 a	43.5 ab	305 a
FO	43.5 b	43.5 ab	43.5 ab	218 ab
MO	43.5 b	43.5 ab	21.8 b	196 b
Lime	43.5 b	21.8 b	21.8 b	152 b
Phosphate	21.8 b	21.8 b	43.5 ab	196 b
Compost	21.8 b	21.8 b	21.8 b	174 b
Zeolite	21.8 b	21.8 b	65.3 a	152 b

Note: Data are expressed as mean value and with the same letter within a column (p <0.10) are not significantly different. FO: iron oxides; MO: manganese oxides.

From Lee, G. J., The Assessment of Remediation Techniques by Chemical Treatments for Soils Contaminated with Cadmium and Lead, Master's thesis, Graduate Institute of Agricultural Chemistry, National Taiwan University, Taipei, 1996. With permission.

5.3.4 Effect of Chemical Treatments on the Concentration of Cd and Pb Uptake by Wheat

The concentration and total uptake of Cd and Pb in wheat species (*Triticum aestivum*) growing in the chemical treatments are shown in Tables 5.5 and 5.6. Results indicated that concentrations of Cd in the leaves of wheat for different chemical treatments in these four soils, except treatment of iron oxide, were all significantly lower than that of control treatment (p <0.10). But there is not any effect on the concentration of Cd in the leaf of wheat when the concentration ranged from 70 to 115 mg/kg in very serious Cd polluted sandy soils. Tables 5.5 and 5.6 also showed that only manganese oxide treatment can significantly reduce the concentration and total uptake of Pb in the leaf of wheat growing in sandy or clayey polluted soils, even in the soils A and B in which the concentrations of Pb are close to background level.

5.4   Conclusions

Results indicate that the chemical forms of Cd and Pb in the polluted soils can be transformed to unavailable forms from available forms after these chemical treatments. Applications of manganese oxides, calcium carbonate, and zeolite slightly reduced the concentration of Cd and Pb extracted from contaminated soils and also reduced the concentration and uptake of Cd and Pb in the tested wheat species.

TABLE 5.6

Total Uptake of Cd and Pb in Wheat Grown in Four Soils with Different Chemical Treatments

Treatments	Soil A	Soil B	Soil C	Soil D
	(mg/20 plants/pot)			
Cd				
Control	0.05 a	0.03 a	0.09 a	0.13 a
FO	0.02 bcd	0.03 a	0.10 a	0.13 ab
MO	0.02 bcd	0.01 b	0.06 b	0.10 b
Lime	0.01 d	ND b	0.03 c	0.10 b
Phosphate	0.04 ab	0.02 b	0.09 a	0.12 ab
Compost	0.03 abc	0.01 b	0.06 b	0.12 ab
Zeolite	0.02 cd	0.01 b	0.02 c	0.06 c
Pb				
Control	0.14 a	0.10 a	0.08 a	0.31 a
FO	0.02 b	0.05 ab	0.05 ab	0.25 ab
MO	0.05 b	0.07 ab	0.02 b	0.21 b
Lime	0.05 b	0.03 b	0.03 ab	0.15 c
Phosphate	0.02 b	0.03 b	0.07 ab	0.23 b
Compost	0.02 b	0.03 b	0.03 ab	0.18 bc
Zeolite	0.02 b	0.03 b	0.05 ab	0.14 c

Note: Data are expressed as mean value and the mean value with the same letter within a column ($p < 0.10$) are not significantly different. FO: iron oxides; MO: manganese oxides.

From Lee, G. J., The Assessment of Remediation Techniques by Chemical Treatments for Soils Contaminated with Cadmium and Lead, Master's thesis, Graduate Institute of Agricultural Chemistry, National Taiwan University, Taipei, 1996. With permission.

Acknowledgments

This research was supported by funds provided by the National Science Council of Republic of China (Grants no. NSC85-2621-P-002-004 and NSC86-2621-P-002-006).

References

Adriano, D.C., *Trace Elements in the Terrestrial Environment*, Springer-Verlag, New York, 1986.

Alloway, B.J., *Heavy Metals in Soils*, Blackie Publishers, London, 1990.

Barrow, N.J., Effect of phosphorus on zinc sorption by a soil, *J. Soil Sci.*, 38, 453, 1987.

Bolland, M.D.A., A.M. Posner, and J.P. Quirk, Zinc adsorption by geothite in the absence and presence of phosphate, *Aust. J. Soil Res.*, 15, 179, 1977.

Chen, Z.S., Cadmium and lead contamination of soils near plastic stabilizing materials producing plants in northern Taiwan, *Water, Air, & Soil Pollut.*, 57, 745, 1991.

Chen, Z.S., Soil properties of soil contaminated with Cd and Pb and their remediation tests in Taiwan, in *Proc. Remediation of Polluted Soils and Sustainable Land Use*, Taipei, Taiwan, 1994, 3-1.

Chen, Z.S. and D.Y. Lee, Evaluation of remediation techniques on two cadmium polluted soils in Taiwan, in *Remediation of Soils Contaminated with Metals*. Special Volume of 2nd Int. Conf. on the Biogeochemistry of Trace Elements, Taipei, Taiwan, September 5–10, 1993, I.K. Iskandar and D.C. Adriano, Eds., *Science Reviews*, London, 1997, 209.

Chen, Z.S., D.Y. Lee, D.N. Huang, and Y.P. Wang, The effect of chemical methods treated in polluted soils on the Cd uptake by the vegetables, in *Proc. 3rd Workshops of Soil Pollution and Prevention*, Taipei, Taiwan (in Chinese, English abstract and tables), 1992a, 277.

Chen, Z.S., D.Y. Lee, T.K. Lin, and Y.P. Wang, Selecting the suitable vegetation growing in the soils contaminated with Cd in northern Taiwan, in *Proc. 3rd Workshops of Soil Pollution and Prevention*, Taipei, Taiwan (in Chinese, English abstract and tables), 1992b, 247.

Chen, Z.S., G.J. Lee, and J.C Liu, Chemical remediation treatments for soils contaminated with cadmium and lead, in *Proc. Fourth Int. Conf. Biogeochemistry of Trace Elements*, June 22–26, 1997, Berkeley, CA, 1997, 421.

Chen, Z.S., S.L. Lao, and H.C. Wu, Summary Analysis and Assessment of Rural Soils Contaminated with Cd in Taoyuan. Project Report of Scientific Technology Advisor Group, Executive Yuan. Taipei, Taiwan (in Chinese, English abstract and tables), 1994, 153.

Cheng, H.F., Remediation of Soils Contaminated with Heavy Metals by Soil Conditioners: Cd, Zn, and Cu. Project Report of Council of Agriculture (in Chinese, English abstract and tables), Taipei, Taiwan, 1996.

Christensen, T.H., Cadmium soil sorption at low concentrations. I. Effect of time, cadmium load, pH, and calcium, *Water, Air & Soil Pollut.*, 21, 106, 1984.

EPA/ROC, Water content and determination of As, Cr, Cu, Cd, Hg, Ni, Pb, and Zn in soils (1991.8.12.(80), NIEA announcement No. 27038), Taipei, Taiwan, 1991.

Fu, G., H.E. Allen, and C.E. Cowan, Adsorption of cadmium and copper by manganese oxide, *Soil Sci.*, 152, 72, 1991.

Gee, G.W. and J.W. Bauder, Particle-size analysis, in *Methods of Soil Analysis, Part 1. Physical and Mineralogical Methods*, 2nd ed., Klute, A. et al., Eds., Agronomy Monograph 9, 383, 1986.

Gworek, B., Lead inactivation in soils by zeolites, *Plant Soil*, 143, 71, 1992.

Impents, R., J. Fagot, and C. Avril, Gestion des sols contaminé par les métaux lourds. Association Francaise Interprofessionnelle du Cadmium, Paris, France, 1991.

Jackson, A.P. and B.J. Alloway, The transfer of cadmium from agricultural soils to the human food chain, in *Biogeochemistry of Trace Metals*, D.C. Adriano, Ed., Lewis Publishers, Boca Raton, FL, 1992, 109.

Khattak, R.A. and A.L. Page, Mechanism of manganese adsorption on soil constituents, in *Biogeochemistry of Trace Metals*, D.C. Adriano, Ed., Lewis Publishers, Boca Raton, FL, 1992, 383.

Kuo, S. and B.L. McNeal, Effects of pH and phosphate on cadmium sorption by a hydrous ferric oxide, *Soil Sci. Soc. Am. J.*, 48, 1040, 1984.

Kuo, S., E.J. Jellum, and A.S. Baker, Effects of soil type, liming, and sludge application on Zn and Cd availability to swiss chard, *Soil Sci.*, 139, 122, 1985.

Lakanen, E. and R. Ervio, A comparison of eight extractants for the determination of plant available micronutrients in soils, *Acta Agral. Fenn.*, 123, 223, 1971.

Lee, D.Y. and Z.S. Chen, Plants for cadmium polluted soils in northern Taiwan, in *Biogeochemistry of Trace Elements*, Special issue of *Environmental Geochemistry and Health*, D.C. Adriano, Z.S. Chen, and S.S. Yang, Eds., Sci. Technol. Lett., 16, 161, 1994.

Lee, F.Y. and C.T. Liao, Removal of Cd and Cu from the polluted soils by vegetations, in *Proc. 4th Workshop on Soil Protection and Remediation*, Taipei, Taiwan (in Chinese, English abstract and tables), 1993, 471.

Lee, G.J., The Assessment of Remediation Techniques by Chemical Treatments for Soils Contaminated with Cadmium and Lead, Master's thesis, Graduate Institute of Agricultural Chemistry, National Taiwan University, Taipei (in Chinese, English abstract and tables), 1996.

Liu, J.C., K.S. Looi, Z.S. Chen, and D.Y. Lee, The effects of composts and calcium carbonate on the uptake of cadmium and lead by vegetables grown in polluted soils, *J. Chinese Inst. Environ. Eng.*, 8, in press.

Lu, S.T., H.L. Chang, C.C. Houng, and K.C. Yee, Cadmium and Lead in the Discharged Water, Sediments, and Brown Rice of Polluted Region, Project report of Environmental Protection Bureau of Taiwan Province (in Chinese, English abstract and tables), 1984.

Lo, S.L. and P.T. Che, The assessment of suitability of soil remediation techniques for soils contaminated with heavy metals, in *Proc. Remediation of Polluted Soils and Sustainable Land Use*, Taipei, Taiwan (in Chinese, English abstract and tables), 1994, 3-1.

Jackson, A.P. and B.J. Alloway, The transfer of cadmium from agricultural soils to the human food chain, in *Biogeochemistry of Trace Metals*, D.C. Adriano, Ed., Lewis Publishers, Boca Raton, FL, 1992, 109.

McBride, M.B. and J.J. Blasiak, Zinc and copper solubility as a function of pH in an acid soil, *Soil Sci. Soc. Am. J.*, 43, 866, 1979.

McBride, M.B., Chemisorption of Cd(II) on calcite surfaces, *Soil Sci. Soc. Am. J.*, 44, 26, 1980.

McKenzie, R.M., The adsorption of lead and other heavy metals on oxides of manganese and iron, *Aust. J. Soil Res.*, 18, 61, 1980.

McLean, E.O., Soil pH and lime requirement, in *Methods of Soil Analysis, Part 2, Chemical and Microbiological Properties*, 2nd ed., A.L. Page et al., Eds., Agronomy Monograph 9, 199, 1982.

Mench, M.J., V.L. Didier, M. Loffler, A. Gomez, and P. Masson, A mimicked in situ remediation study of metal-contaminated soils with emphasis on cadmium and lead, *J. Environ. Qual.*, 23, 58, 1994.

Naidu, R., N.S. Bolan, R.S. Kookana, and K.G. Tiller, Ionic-strength and pH effects on the sorption of cadmium and the surface charge of soils, *Eur. J. Soil Sci.*, 45, 419, 1994.

Nelson, D.W. and L.E. Sommer, Total carbon, organic carbon, and organic matter, in *Methods of Soil Analysis, Part 2, Chemical and Microbiological Properties*, 2nd ed., A.L. Page et al., Eds., Agronomy Monograph 9, 539, 1982.

Norvell, W.A., Comparison of chelating agents as extractants for metals in diverse soil materials, *Soil Sci. Soc. Am. J.*, 48, 1285, 1984.

Nriagu, J.D., Ed., *Changing Metal Cycles in Human Health*, Springer, Berlin, 1984, 445.

Rhoades, J.D., Cation exchange capacity, in *Methods of Soil Analysis, Part 2, Chemical and Microbiological Properties*, 2nd ed., A.L. Page et al., Eds., Agronomy Monograph 9, 149, 1982.

Saeed, M. and R.L. Fox, Influence of phosphate fertilization on zinc adsorption by tropical soils, *Soil Sci. Soc. Am. J.*, 43, 683, 1979.

SAS Institute, SAS User's Guide, Statistics, SAS Inst., Cary, NC, 1982.

Sommers, L.E. and W.L. Lindsay, Effect of pH and redox on predicted heavy metal-chelate equilibria in soils, *Soil Sci. Soc. Am. J.*, 43, 39, 1979.

Thomas, G.W., Exchangeable cations, in *Methods of Soil Analysis, Part 2, Chemical and Microbiological Properties*, 2nd ed., A.L. Page et al., Eds., Agronomy Monograph 9, 149, 1982.

Tiller, K.G., J. Gerth, and G.W. Bruemmer, The relative affinities of Cd, Ni, and Zn for different clay fraction and goethite, *Geoderma*, 34, 17, 1984.

Wang, Y.P., K.C. Li, Z.S. Chen, Y.M. Huang, and C.L. Liu, Assessment of Soil Pollution and Soil Remediation Techniques Tested in the Polluted Soils, Project paper of EPA/ROC (EPA-83-H105-09-04), Taipei, Taiwan (in Chinese, English abstract and tables), 1994.

Wang, Y.P., M.C. Wang, and Z.S. Chen, The Soil Remediation Studies in Soils Contaminated with Cd and Pb, Project paper of EPA/ROC (EPA-78-004-09-109), Taipei, Taiwan (in Chinese, English abstract and tables), 1989.

6

Soil pH Effect on the Distribution of Heavy Metals Among Soil Fractions

C.D. Tsadilas

CONTENTS

6.1 Introduction

Some decades ago the interest about the metal nutrients that are essential for plant growth focused basically on the investigation of factors causing deficiency to the plants (Thorne, 1959; Brown, 1961). The main goal was the mobilization of trace metals to plant roots. Nowadays, the interest has shifted into the opposite direction, i.e., to removing excess metals or transferring them to immobile phases. That is because trace metals became important environmental contaminants seriously affecting the whole ecosystem. Practices associated with mining and smelting of ores, secondary smelting of scrap metals, and industrial and municipal wastes caused a high accumulation of potentially toxic metals to soils that can enter the food chain and affect human health. Metal sources related especially to the food chain are municipal sewage sludge, composts, swine and poultry manure, industrial wastes, coal fly ash, and P fertilizers (Adriano et al., 1997). For mitigation of the consequences caused by the use of metal containing materials, the responsible authorities imposed restrictions in their use. The U.S. Environmental Protection Agency (U.S. EPA, 1993) as well as the European Union (CEC, 1986) imposed upper limits in the amount of heavy metals permitted to be applied in the soils with sewage sludge. These restrictions in general reflect soil factors that affect metal retention such as soil pH and cation exchange capacity.

Remediation of soils contaminated with heavy metals is an important problem for many countries throughout the world and concentrates the efforts of many authorities and scientists. Treatment of soil contaminated with heavy metals is classified in three main categories, i.e., physical, chemical, and biological treatments (Iskandar and Adriano, 1997). Physical processes include physical separation, carbon adsorption, frozen ground processes, and thermal processes such as vitrification, incineration, cyclone furnace, and roasting. Chemical processes aim at removing the metals or decreasing their availability to the plants through which they enter the food chain. They include neutralization, precipitation, solidification/stabilization, encapsulation, ion exchange, and washing. Finally, biological processes utilize the ability of some plants to accumulate high amounts of heavy metals into their tissues ("hyperacummulators") for removing them from contaminated soils.

Neutralization of acidic or alkali soils is one of the most simple and inexpensive methods used for immobilization of heavy metals. Solubility of all metals is strongly dependent on the redox potential and pH (Sims and Patrick, 1978). With an exception of As, Se, and Mo, the solubility of most metals decreases as pH increases reducing their availability. Zinc behavior is sometimes different than the other metals. For example, McBride and Blasiak (1979) reported that at pH values >7.5 soil solution Zn increases due to the formation of soluble-Zn organic matter complexes.

Total heavy metal concentration is not a good indicator of metal availability to the plants. Heavy metals in soils occur in various chemical forms with a different degree of availability to the plants. Separation of various forms of heavy metals in soils is carried out using sequential extraction techniques. Several such sequential extraction techniques are used for studying the availability of metals to plants and their mobility and reactivity in soils and sediments. These procedures utilize a number of selective extractants to solubilize metals associated with various soil component fractions. By these techniques metals are usually partitioned into exchangeable, carbonate, organic, iron and manganese oxides, and residual fractions (Shuman, 1979; Iyengar et al., 1981; Sposito et al., 1982; Emmerich et al., 1982; Hickey and Kittrick, 1984; LeClaire et al., 1984; Tsadilas et al., 1995). From all the fractions being determined, the one available to plants was found to be mainly the exchangeable one, extracted usually with KNO_3 (Pierzynski and Schwab, 1993; LeClaire et al., 1984; Sims, 1986), the organically bound fraction (Sims, 1986; Samaras and Tsadilas, 1997) or the carbonate fraction (LeClaire et al., 1984; Samaras and Tsadilas, 1997).

The above-mentioned heavy metal forms do not remain constant in the soils. They dramatically change because of the change of many soil factors affecting their distribution such as organic matter addition, metal addition, time, pH or Eh. These factors strongly influence heavy metal forms, causing redistribution of them among the various soil components. Therefore, in soil remediation practices, it is extremely important to know the influence of each one factor on the distribution of heavy metals in order to transfer the maximum amount of them into forms unavailable to plants.

Soil pH is one of the most important factors affecting heavy metal distribution among the soil constituents, as was suggested by several workers. Sims (1986) found, for four soils varied widely in organic matter content and cation exchange capacity, that below pH 5.2 the dominant species of Mn, Cu, and Zn was the exchangeable one, while at pH values greater than 5.2 the organically complexed and Fe-oxide bound forms dominated. Shuman (1986) reported that liming decreased exchangeable Zn and increased organic fraction of Zn and Mn. Neilsen et al. (1986) found in 20 orchard soils from British Columbia that soil acidification caused a redistribution of soil Zn from the residual fraction into the exchangeable and organic fractions. As soil pH manipulation through liming is a relatively simple and low cost practice, it can be considered for remediation of soils contaminated with heavy metals. The purpose of this chapter is to discuss soil pH influence on the distribution of some heavy metals, including lead (Pb), nickel (Ni), zinc (Zn), copper (Cu), and manganese (Mn), in some strongly acid Greek soils to which sewage sludge was applied for their amelioration.

6.2 Materials and Methods

6.2.1 Soils and Measurements

Four surface (0 to 15 cm) soils, classified as Ultic Haploxeralfs, were selected from the Elassona area located in the western part of central Greece. In this area, because of the high rainfall and the continuous application of acidifying nitrogen fertilizers for a long time, the soils became strongly acid and their productivity was dramatically reduced. Soil liming is a common practice in this area for the improvement of soil productivity. In recent years, farmers started to use sewage sludge as a soil amendment. The samples were air dried, crushed to pass a 2-mm sieve, and analyzed for the following: pH in a 1:1 water/soil suspension (McLean, 1982), organic matter content with the wet oxidation procedure (Nelson and Sommers, 1982), exchangeable aluminum with the aluminum method (Hsu, 1963) after extraction with 1 M KCl (Barnishel and Bertsch, 1982), and cation exchange capacity with the NaOAc (pH 8.2) method (Rhoades, 1982). Total content of Pb, Ni, Cu, Zn, and Mn was determined in extracts obtained from digestion of 2-g soil samples by 12.5 mL 4 M HNO_3 at 80°C overnight (Sposito et al., 1982).

6.2.2 Soil pH Adjustment

Soil pH was adjusted to the desirable level using calcium oxide, and 200-g subsamples of each soil were put in plastic pots and thoroughly mixed with various amounts of calcium oxide to obtain the pH values of 4.0 to 8.4 (Table 6.2). Each treatment was replicated three times. The samples were wetted up to the field capacity and incubated for 30 days at room temperature in a complete randomized block design. During the incubation period soil moisture was kept constant by adding deionized water after weighting. At the end of the incubation period, soil pH was determined in a soil/water suspension of 1:1.

6.2.3 Trace Metal Fractionation

Trace metal fractionation was carried out using the procedure proposed by Emmerich et al. (1982), but slightly modified. In brief, the procedure included the following: triplicate 2-g samples were sequentially extracted with 0.5 M KNO_3 for 16 h (exchangeable fraction), 0.5 M NaOH for 16 h (organic fraction), 0.05 M Na_2-EDTA for 6 h (carbonate fraction), and 4 M HNO_3 at 70 to 80°C for 16 h (residual fraction). Preliminary investigation showed that extraction with water after the 0.5 M KNO_3 extraction, as proposed by Emmerich et al. (1982), did not extract a detectable amount of the metals studied, so this step was not included in the fractionation procedure. Separate subsamples were also extracted with 0.005 *M* DTPA solution adjusted to a pH 7.3 (Lindsay and Norvell, 1978). Heavy metals in the extraction solutions were determined by atomic absorption spectrometry (AAS, Perkin-Elmer model 5000).

6.2.4 Statistical Analysis

For the evaluation of the influence of pH on the various metal forms the data were analyzed using conventional analysis of variance considering them separately for each soil. In studying the relationship between soil pH and heavy metal fractions simple regression analysis or stepwise variable selection, forward elimination techniques were used.

6.3 Results and Discussion

6.3.1 Soil Characteristics

The basic physicochemical characteristics are shown in Table 6.1. The soils were sandy loamy to loamy, strongly acidic, with a high concentration of exchangeable aluminum, low organic matter content, and low cation exchange capacity. Soil pH after liming was raised from 4.0 up to 8.4 (Table 6.2).

6.3.2 DTPA-Extractable Metals

The DTPA extraction procedure was proposed by Lindsay and Norvell (1978) for simultaneous extraction of the available iron, manganese, copper, and zinc mainly for near-neutral and calcareous soils. Several workers reported a very good correlation between the concentration of heavy metals extracted by this method and the respective concentrations in the plants (Samaras and Tsadilas, 1997; Tsadilas et al., 1995). The correlation, however, was more or less specific to the soil. The procedure was also successfully used for the determination of an index of the availability of Ni and cadmium (Cd) (Baker and Amacher, 1982). In the present study, this procedure, in addition to the above-mentioned metals, was also

TABLE 6.1

Selected Physicochemical Properties of the Soils Studied

Property	Soils			
	S1	S2	S3	S4
Texture	Sandy loam	Sandy loam	Loam	Sandy loam
pH (H$_2$0 1:1)	4.0	4.4	4.3	4.2
Exchangeable Al, mg kg^{-1}	95	87	92	94
Organic matter, g kg^{-1}	8.5	12.3	9.2	10.2
Cation exchange capacity, cmol(+) kg^{-1}	8.72	7.53	9.62	8.75
Total Pb, mg kg^{-1}	41.8	44.9	43.8	39.5
Total Ni, mg kg^{-1}	31.2	42.9	35.6	34.5
Total Zn, mg kg^{-1}	31.4	40.6	37.8	37.6
Total Cu, mg kg^{-1}	16.2	29.2	36.4	21.3
Total Mn, mg kg^{-1}	355	640	495	372

TABLE 6.2

Soil pH and Concentration of Pb, Ni, Zn, Mn, and Cu (mg kg^{-1}) Extracted by DTPA[a]

Soil	pH	Pb	Ni	Zn	Mn	Cu
S1	4.00d[b]	0.82a	0.55a	0.82a	40.51a	0.75a
	6.60c	0.53b	0.17b	0.55b	15.50b	0.75a
	7.50b	0.53b	0.17b	0.43b	9.41c	0.65a
	8.20a	0.62b	0.23b	0.48b	7.19c	0.68a
S2	4.40d	0.84a	1.01a	0.92a	42.37a	1.88a
	6.60c	0.65b	0.50b	0.53b	20.89b	2.02a
	7.50b	0.70ab	0.42b	0.59b	17.54b	1.96a
	7.90a	0.74ab	0.51b	0.51ab	12.83c	2.05a
S3	4.30d	1.16a	1.03a	1.14a	59.54a	2.59a
	6.20b	0.98a	0.77ab	0.72b	33.01b	3.28a
	7.30c	1.04a	0.51b	0.63b	26.66c	3.20a
	7.70a	0.98a	0.54b	0.50b	22.28c	3.12a
S4	4.20d	0.92a	0.62a	0.99a	49.27a	1.99a
	6.40c	0.60b	0.33b	0.44b	16.20b	1.47a
	7.60b	0.57b	0.32a	0.49b	12.03c	1.18a
	8.40a	0.66ab	0.38a	0.40b	10.94c	1.52a

[a] Mean values of the three replicates.
[b] Soil numbers in the same column followed by different letters differ significantly at the probability level *P* <0.05 according to the LSD test.

used for Pb extraction. Concentrations of heavy metals extracted by DTPA are presented in Table 6.2. Soil pH increases, because of the soil liming, significantly affected the concentration of all the metals extracted except for Cu (Table 6.2). A very good negative correlation was recorded between pH and their concentration. The higher correlation coefficient was found for Mn (Figure 6.1). Similar observations were reported by several investigators (Shuman, 1991). However, Schwab et al. (1990) found a positive relationship between soil pH and Zn extracted by DTPA. They attributed this positive relationship to the chemistry of the extracting solution rather than the chemistry of soil Zn.

6.3.3 Metal Fractions

There was a very good agreement between total concentration of heavy metals measured directly by the HNO$_3$ acid digestion procedure and total heavy metal concentration calculated by the summation of the separate fractions (Tables 6.1 and 6.3). In all cases total heavy metals measured by HNO$_3$ digestion tended to be a little lower than the summation of separate fractions. The same observations were reported by Sposito et al. (1982), who attributed these small differences in the higher effectiveness of the HNO$_3$ in the sequential extraction procedure to previous extraction with NaOH and EDTA in the sequential extraction procedure.

6.3.4 Lead Fractions

Distribution of lead fractions is presented in Table 6.3. Exchangeable fraction in soils in their original form ranged between 15 and 24%, organic matter fraction between 4 and 6%, carbonate fraction between 23 and 25%, and the residual fraction between 41 and 43%. Soil liming had a very small influence on the distribution of Pb fractions. Exchangeable and organic fractions tended to decrease, the carbonate fraction remained almost constant, and the residual fraction tended to increase. It seems therefore that a small amount of Pb tended to shift from the exchangeable and organic fractions into the residual fraction. A positive relationship was observed between residual Pb and soil pH (*r* = 0.60**).

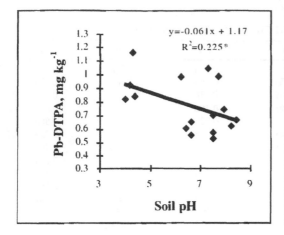
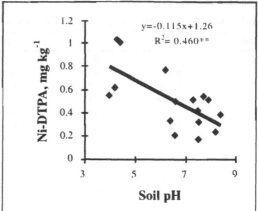
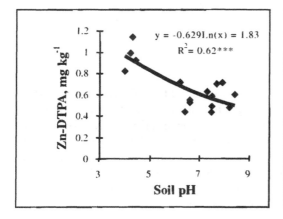
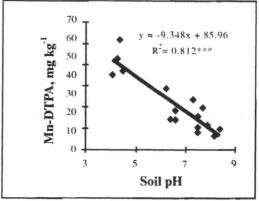

FIGURE 6.1
Relationship between DTPA extractable Pb, Ni, Zn, and Mn with soil pH.

6.3.5   Nickel Fractions

Total Ni content ranged from 31.2 to 42.4 mg kg^{-1} soil (Table 6.1). The distribution of Ni fractions is shown in Table 6.3. The highest percentage of the total Ni (67 to 68%) was found in the residual fraction. The exchangeable fraction ranged from 14 to 16%, the organic fraction from 9 to 12%, and the carbonate fraction from 6 to 9%. Soil pH increase mainly affected the exchangeable fraction, which was substantially reduced in all the soils. For example, in soil S2, the exchangeable fraction at the original pH value of 4.4 decreased from 6.60 to 3.26 mg kg^{-1} in the highest pH value. A significant negative relationship was found between soil pH and exchangeable Ni (Figure 6.2). All the rest of the fractions tended to increase slightly (Table 6.3).

6.3.6   Zinc Fractions

Total Zn ranged from 31.4 to 40.6 (Table 6.1) in the soils S1 and S2, respectively. The relative distribution of Zn fractions is shown in Table 6.3. A percentage of 7 to 12% was found in the exchangeable fraction, only 4% in the organic fraction, 11 to 14% in the carbonate fraction, and the most in the residual fraction. The soil pH increase strongly affected exchangeable

Zn in all the soils. It was reduced mainly in the pH increase up to about 7.5. A strong negative relationship was found between soil pH and exchangeable Zn (Figure 6.2). Similar results were reported by others (Sims, 1986; Shuman,1986). Organic fraction of Zn remained unaffected, while carbonate fraction tended to increase and the residual fraction was significantly increased. It seems, therefore, that soil pH increase causes a shifting of the exchangeable Zn fraction into the residual fraction. These results are similar to those reported by Neilsen et al. (1986), who found that a soil pH decrease with acidification caused a shifting from the residual Zn fraction to the exchangeable fraction.

6.3.7 Copper Fractions

The main amount of Cu was found in organic and residual fractions. The organic fraction of Cu ranged from 32 to 42% and the residual fraction from 31 to 46%. The exchangeable fraction covered about 8 to 12% and the carbonated fraction 10 to 16% (Table 6.3). Several workers found soil Cu to be associated with organic matter (Shuman, 1991). The increase in soil pH significantly affected exchangeable and organic Cu fractions. The exchangeable fraction was decreased significantly in all soils. The drop in this fraction was mainly from pH 4.4 to 6.6. (Table 6.3, Figure 6.3). These results are in agreement with those reported by Sanders et al. (1986) and Elsokkary and Lag (1978). Organic fraction was also decreased in all soils except soil S3. A strong relationship was found between the concentration of Cu in the exchangeable fraction and soil pH (Figure 6.2). The carbonate fraction was not significantly influenced by pH change. It tended to increase but the increase was not statistically significant except in soil S3. The same trend observed in carbonate fraction was recorded for residual Cu fraction. The increase was significant only in the case of soil S4 and it was observed in the soil at pH 6.4 (Table 6.3, Figure 6.3). So it seems that in soil a pH increase decreases exchangeable and organic Cu fractions, causing a shifting of part into the carbonate and residual fractions. However, because the amount of exchangeable and organic fractions redistributed was not high, they didn't significantly increase the other fractions.

6.3.8 Manganese Fractions

A significant percentage of Mn was found in the exchangeable fraction, which ranged from 16 to 32% (Table 6.3). The organic fraction covered only a very small percentage of Mn, ranging from 0.1 to 2%. The carbonate fraction was found in a percentage of 22 to 47%, while the most abundant fraction was the residual covering from 36 to 50% of the soil Mn. Soil pH increase strongly affected Mn fraction distribution, causing a sharp decrease in exchangeable fraction in the pH values from 4.0 to about 6.5. Above pH 6.5 very little exchangeable Mn was detected. These findings are in close agreement with those reported by Sims (1986). However, Sims (1986) as well as Shuman (1986) found that exchangeable Mn was transformed into organic forms, while in the present study organic form was not significantly affected. Exchangeable Mn was mainly transformed into carbonate and residual forms, which were significantly increased (Table 6.3, Figure 6.3). In agreement with results reported by Sims (1986), a very strong relationship was recorded between soil pH and exchangeable Mn (Figure 6.2).

6.3.9 Relationship between DTPA-Extractable and Metal Fractions

Several workers found that the DTPA test (Lindsay and Norvell, 1978) is effective in prediction of plant available metals (Sims and Jonson, 1991). The knowledge of the pools, from

TABLE 6.3

Distribution of Soil Fractions of Pb, Ni, Zn, Cu, and Mn in Relation to Soil pH

Soil	pH	Exchangeable	Organic	Carbonate	Residual	Sum
				(mg kg^{-1})		
Pb						
S1	4.00d[a]	6.14ab	6.55a	9.83a	18.42b	40.94a
	6.60c	6.41a	5.21a	9.22a	19.23b	40.07a
	7.50b	5.79a	4.97b	9.52a	21.11b	41.38a
	8.20a	5.36a	5.36ab	9.08a	21.45b	41.25a
S2	4.40d	8.69a	6.08a	9.99a	18.68b	43.44a
	6.60c	7.80a	4.33ab	10.83a	20.36ab	43.32a
	7.70c	7.58a	4.01b	11.14a	21.84b	44.57a
	7.90a	7.53a	3.99b	10.68a	22.16b	44.32a
S3	4.30d	9.28a	4.22a	10.55a	18.15a	42.20a
	6.20b	8.56a	3.40b	10.28a	20.55a	42.82a
	7.30c	8.00a	3.37b	10.11a	20.64a	42.13a
	7.70a	7.75a	3.01b	11.20a	21.10a	43.07a
S4	4.20d	9.39a	4.70a	9.00a	16.04b	39.13a
	6.40c	8.35a	3.42b	7.97a	18.22a	37.95a
	7.60b	8.84a	3.46b	8.07a	18.07a	38.45a
	8.40a	8.70a	3.40b	7.94a	17.78a	37.82a
Ni						
S1	4.00d[a]	4.67a	2.92a	1.75a	19.85a	29.19a
	6.60c	3.39a	3.39a	2.26a	19.22a	28.26a
	7.50b	3.54a	3.54a	2.07a	20.35a	29.5a
	8.20a	3.60a	3.60a	2.40a	20.40a	30.00a
S2	4.40d	6.60a	3.71a	3.30b	27.64a	41.25a
	6.60c	4.05ab	4.05a	4.45ab	27.91a	40.45a
	7.50b	3.77b	4.61a	5.03b	28.51a	41.93a
	7.90a	3.26b	4.07a	5.29b	28.08a	40.7a
S3	4.30d	4.73a	3.71a	3.04a	22.28a	33.75a
	6.20b	3.07ab	4.44a	3.41a	23.21a	34.13a
	7.30c	2.03b	4.40a	3.72a	23.67a	33.82a
	7.70a	2.13b	4.61a	4.25a	24.45a	35.44a
S4	4.20d	4.28a	3.67a	2.45a	20.18a	30.57a
	6.40c	2.64b	3.96a	3.30a	23.11a	33.01a
	7.60b	2.89b	3.86a	3.21a	22.17a	32.13a
	8.40a	2.59b	3.88a	3.23a	22.63a	32.33a
Zn						
S1	4.00d[a]	2.09a	1.20a	3.29a	23.31a	29.88a
	6.60c	1.13b	1.13a	4.24a	21.75a	28.25a
	7.50b	0.86b	1.15a	3.74a	23.00a	28.75a
	8.20a	0.88b	1.46a	4.09a	22.77a	29.19a
S2	4.40d	3.19a	1.59a	5.18a	29.90a	39.87a
	6.60c	1.13b	1.13a	5.65a	29.74a	37.64a
	7.50b	0.76c	1.15a	6.50a	29.84a	38.26a
	7.90a	0.75c	1.50a	6.74a	28.45a	37.44a
S3	4.30d	4.41a	1.47a	5.15a	25.73a	36.75a
	6.20b	1.79b	1.44a	5.74a	26.91a	35.88a
	7.30c	1.09c	1.45a	6.16a	27.56a	36.26a
	7.70a	0.78c	1.56a	7.40a	29.21a	38.95a
S4	4.20d	3.25a	1.45a	5.06a	26.37a	36.13a
	6.40c	0.75b	1.49a	5.59a	29.43a	37.25a
	7.60b	0.71b	1.43a	5.35a	28.16a	35.64a
	8.40a	0.71b	1.42a	5.68a	27.69a	35.5a

TABLE 6.3 CONTINUED

Distribution of Soil Fractions of Pb, Ni, Zn, Cu, and Mn in Relation to Soil pH

Soil	pH	Exchangeable	Organic	Carbonate	Residual	Sum
				(mg kg^{-1})		
Cu						
S1	4.00d[a]	2.05a	5.46a	1.71a	7.85a	17.07a
	6.60c	1.36b	5.00a	1.36a	7.42a	15.14a
	7.50b	1.24b	4.65ab	1.55a	8.06a	15.50a
	8.20a	1.23b	3.85b	1.99a	8.00a	15.38a
S2	4.40d	2.28a	11.38a	3.98a	10.81a	28.44a
	6.60c	1.34b	10.15ab	4.01a	11.21a	26.70a
	7.50b	1.39b	11.15a	4.18a	11.15a	27.88a
	7.90a	1.37b	9.58b	4.93a	11.22a	27.37a
S3	4.30d	4.24a	14.83a	3.89b	12.36a	35.32a
	6.20b	1.41b	15.11a	4.94b	13.70a	35.27a
	7.30c	1.42b	15.31a	5.34ab	12.46a	35.60a
	7.70a	1.49b	15.29a	6.71b	13.43a	37.30a
S4	4.20d	2.27a	8.66a	3.30a	6.40b	20.63a
	6.40c	1.44b	7.59ab	2.26a	9.23a	20.50a
	7.60b	1.38b	6.92b	2.37a	9.09a	19.77a
	8.40a	1.42b	6.71b	3.05a	8.95a	20.33a
Mn						
S1	4.00d[a]	94.20a	3.49a	76.78b	174.50b	349a
	6.60c	10.50b	3.50a	164.50a	161.50b	350a
	7.50b	7.04b	3.52a	144.32a	197.12b	352a
	8.20a	2.82c	0.70b	137.28a	211.20a	352a
S2	4.40d	102.10a	6.38a	299.86b	229.68c	638a
	6.60c	19.10b	6.37a	382.20a	229.32c	637a
	7.50b	0.63c	0.63b	374.06a	259.94b	634a
	7.90a	0.63c	0.63b	347.60a	284.40a	632a
S3	4.30d	155.52a	9.72a	140.94b	184.68c	486a
	6.20b	67.34b	4.81b	202.02a	206.83b	481a
	7.30c	28.80c	4.80b	244.80a	201.60b	480a
	7.70a	14.70d	4.90b	249.90a	220.50a	490a
S4	4.20d	110.10a	7.34a	99.09b	154.14b	367a
	6.40c	7.40b	3.70b	177.60a	181.30a	370a
	7.60b	3.65b	3.65b	178.85a	178.85a	365a
	8.40a	3.68b	3.68b	169.28a	191.36a	368a

[a] Soil numbers in the same column followed by different letters differ significantly at the probability level $P < 0.05$ according to the LSD test.

[b] Mean values of the three replicates.

which DTPA solution extracts metals, is useful in managing soils with regard to the metal nutrients. Therefore, prediction equations of the DTPA extractable metal concentrations as a function of the metal fractions were developed, using a stepwise variable selection, forward elimination, technique. In these equations, DTPA extractable metal concentration was the independent variable and metal fractions concentration and soil pH were the dependent variables. Only the variables that increased R^2 significantly entered the equation. The equations derived are shown in Table 6.4. For Pb extracted by DTPA, only the exchangeable fraction entered the equation. This fraction explained about 36% of its variance. It seems therefore that DTPA extracts Pb mainly from forms residing in exchangeable sites. In the case of Ni, 82.3% extracted by DTPA was explained by soil pH, and exchangeable and organic fraction. From the relevant equation of Table 6.4 it is obvious that exchangeable and

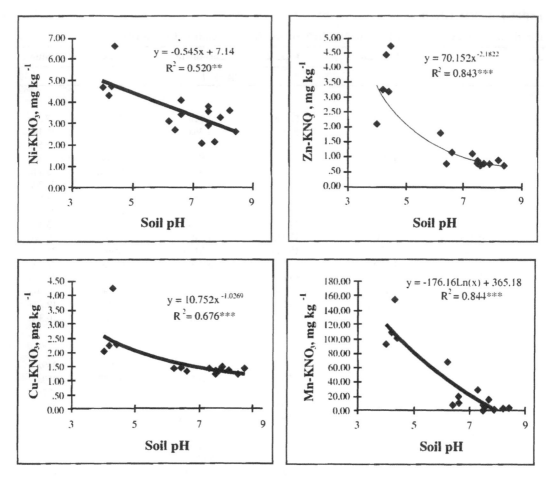

FIGURE 6.2
Relationship between exchangeable Ni, Zn, Cu, and Mn with soil pH.

organic fractions are the main available Ni forms. Nearly the whole variance of the DTPA-extractable Zn (a percentage 98.8%) was explained by soil pH, carbonate, and residual fractions. The relevant equation suggests that exchangeable and carbonate fractions are sources of available Zn, while residual Zn contains unavailable Zn forms. For Cu, the main source of the DTPA extractable is organic fraction, suggesting that this fraction is an available form. Finally 90% of the variation of Mn extractable by DTPA was explained by pH and the exchangeable fraction.

6.4   Conclusions

Soil pH increase by liming substantially decreases the exchangeable fractions of the metals studied, especially of Mn, shifting it mainly into carbonate, organic, and residual fractions. DTPA-extractable metals originate mainly from the exchangeable and organic fraction and in the case of Zn, from the carbonate fraction. Soil liming decreases DTPA extractable metal fractions.

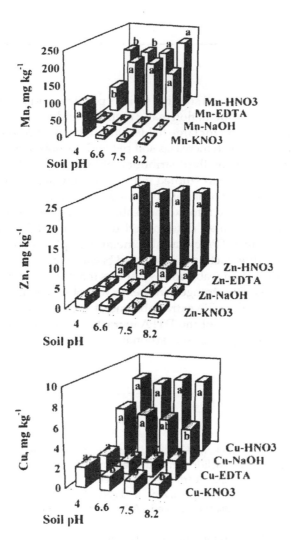

FIGURE 6.3
Soil pH influence on the distribution of Mn, Zn, and Cu fractions (soil 1).

TABLE 6.4

DTPA Extractable Metal Concentration as a Function of Metal Fraction Concentration[a]

DTPA Extractable	Linear Stepwise Multiple Regression Equations[a]	R^2
Pb	$Pb_{DTPA} = 0.08\ Pb_{KNO_3}$[b]	0.360[a]
Ni	$Ni_{DTPA} = 0.12pH + 0.08\ Ni_{KNO_3} + Ni_{NaOH}$	0.823[c]
Zn	$Zn_{DTPA} = 0.04\ pH + 0.22\ Zn_{KNO_3} + 0.12\ Zn_{EDTA} - 0.02\ Zn_{Res}$	0.988[c]
Cu	$Cu_{DTPA} = -0.36 + 0.23\ Cu_{NaOH}$	0.946[c]
Mn	$Mn_{DTPA} = -47.31 + 7.92\ pH + 0.53\ Mn_{KNO_3}$	0.900[c]

[a] Only independent variables significant at the <0.05 probability level are included in the equations.

[b] KNO₃, NaOH, EDTA, Res denote the exchangeable, organic, carbonate, and residual fraction, respectively.

[a,c] Significant at probability level 0.05 and 0.001, respectively.

6.5 Summary

Remediation of metal-contaminated soils is a serious problem and has attracted the interest of many researchers. For soil remediation many methods and technologies are in use, most of which are successful but costly for large fields. One of the most commonly used inexpensive methods is soil liming, which increases soil pH and immobilizes heavy metals. In this chapter the influence of soil pH on the distribution of lead (Pb), nickel (Ni), zinc (Zn), copper (Cu), and manganese (Mn) was studied in four strongly acid Ultic Haploxeralfs from Greece which were limed to a pH of about 4.0 to 8.4. After liming the above-mentioned metals in these soils were fractionated into exchangeable, organic, carbonate, and residual fractions. The same metals were also determined using the DTPA method. Soil liming significantly decreased all the DTPA extractable metals except Cu. From the metal fractions determined, the exchangeable fraction was reduced because of the pH increase in all metals except Pb. The amount of exchangeable forms lost was mainly transformed into the carbonate or the residual fractions. The strongest influence was observed in Mn fractions. Soil pH was negatively correlated with DTPA extractable metal and exchangeable fractions. Significant percentages of the variance of the DTPA extractable metal concentration were predicted by the exchangeable, organic, and carbonate fractions.

References

Adriano, D.C., J. Albright, F.W. Whicker, I.K. Iskandar, and C. Sherony, Remediation of soil contaminated with metals and radionuclide-contaminated soils, in *Remediation of Soils Contaminated with Metals*, I.K. Iskandar and D.C.Adriano, Eds., Science Reviews, Northwood, 1997, 27.

Baker, D.L. and M.C. Amacher, Nickel, copper, zinc, and cadmium, in *Methods of Soil Analysis, Part 2, Chemical and Microbiological Properties*, 2nd ed., A.L. Page et al., Eds., ASA SSSA, Madison, WI, 1982, 323.

Barnishel, R.B. and P.M. Bertsch, Aluminum, in *Methods of Soil Analysis, Part 2, Chemical and Microbiological Properties*, 2nd ed., A.L. Page et al., Eds., ASA SSSA, Madison, WI, 1982, 275.

Brown, J., Iron chlorosis in plants, in *Advances in Agronomy*, 13, 329, 1961.

(CEC) Council of the European Communities, Council Directive of 12 June 1986 on the Protection of the environment, and in particular of the soil, when sewage sludge is used in agriculture (86/278/EEC), *Official J. Eur. Communities*, No. L.181, 6, 1986.

Elsokkary, I.H. and J. Lag, Distribution of different fractions of Cd, Pb, Zn, and Cu in industrially polluted and non-polluted soils of Odda Region, Norway, *Acta Agric. Scand.*, 28, 262, 1978.

Emmerich, W.E., L.J. Lund, A.L. Page, and A.C. Chang, Solid phase forms of heavy metals in sewage sludge-treated soils, *J. Environ. Qual.*, 11, 178, 1982.

Hickey, M.G. and J.A. Kittrick, Chemical partitioning of cadmium, copper, nickel and zinc fractions in soils and sediments containing high levels of heavy metals, *J. Environ. Qual.*, 13, 372, 1984.

Hsu, P.H., Effect of initial ph, phosphate, and silicate on the determination of aluminum with aluminon, *Soil Sci.*, 96, 230, 1963.

Iskandar, I.K. and D.C. Adriano, Remediation of soils contaminated with metals — a review of current practices in the U.S.A., in *Remediation of Soils Contaminated with Metals*, I.K. Iskandar and D.C. Adriano, Eds., Science Reviews, Northwood, 1997, 1.

Iyengar, S.S., D.C. Martens, and W.P. Miller, Distribution and plant availability of soil zinc fractions, *Soil Sci. Soc. Am. J.*, 45, 735, 1981.

LeClaire, J.P., A.C. Chang, C.S. Levesque, and G. Sposito, Trace metal chemistry in arid-zone field soils amended with sewage sludge. IV. Correlations between zinc uptake and extracted soil zinc fractions, *Soil Sci. Soc. Am. J.*, 48, 509, 1984.

Lindsay, W.L. and W.A. Norvell, Development of a DTPA test for zinc, iron, manganese, and copper, *Soil Sci. Soc. Am. J.*, 42, 421, 1978.

McBride, M.B. and J.J. Blasiak, Zinc and copper solubility as a function of pH in an acid soil, *Soil Sci. Soc. Am. J.*, 43, 1137, 1979.

McLean, E.O., Soil pH and lime requirement, in *Methods of Soil Analysis, Part 2, Chemical and Microbiological Properties*, 2nd ed., A.L. Page et al., Eds., ASA SSSA, Madison, WI, 1982, 199.

Neilsen, D., P.B. Hoyt, and A.F. MacKenzie, Distribution of soil zinc fractions in British Columbia interior orchard soils, *Can. J. Soil Sci.*, 66, 445, 1986.

Nelson, D.W. and L.E. Sommers, Total carbon, organic carbon, and organic matter, in *Methods of Soil Analysis, Part 2, Chemical and Microbiological Properties*, 2nd ed., A.L. Page et al., Eds., ASA SSSA, Madison, WI, 1982, 539.

Pierzynski, G.M. and A.P. Schwab, Bioavailability of zinc, cadmium, and lead in a metal-contaminated alluvial soil, *J. Environ. Qual.*, 22, 247, 1993.

Rhoades, J.D., Cation exchange capacity, in *Methods of Soil Analysis, Part 2, Chemical and Microbiological Properties*, 2nd ed., A.L. Page et al., Eds., ASA SSSA, Madison, WI, 1982, 149.

Samaras, V. and C.D. Tsadilas, Distribution and availability of six heavy metals in a soil treated with sewage sludge, in Proc. Int. Conf. Biogeochemistry of Trace Elements, Berkeley, CA, 1997, 145.

Sanders, J.R., T.M. Adams, and B.D. Christensen, Extractability and bioavailability of zinc, nickel, cadmium, and copper in three Danish soils sampled 5 years after application of sewage sludge, *J. Soil Food Agric.*, 37, 1155, 1986.

Schwab, A.P., C.E. Owensby, and S. Kulyingyong, Changes in soil chemical properties due to 40 years of fertilization, *Soil Sci.*, 149(1), 35, 1990.

Shuman, L.M., Zinc, manganese, and copper in soil fractions, *Soil Sci.*, 127(1), 10, 1979.

Shuman, L.M., Effect of liming on the distribution of manganese, copper, iron, and zinc among soil fractions, *Soil Sci. Soc. Am. J.*, 50, 1236, 1986.

Shuman, L.M., Chemical forms of micronutrients in soils, in *Micronutrients in Agriculture*, J.J. Motvedt, F.R. Cox, L.M. Shuman, and R.M. Welsh, Eds., ASA SSSA, Madison, WI, 1991, 113.

Sims, J.T., Soil pH effects on the distribution and plant availability of manganese, copper, and zinc, *Soil Sci. Soc. Am. J.*, 50, 367, 1986.

Sims, J.L. and J.L. Patrick, Jr., The distribution of micronutrient cations in soil under conditions of varying redox potential and pH, *Soil Sci. Soc. Am. J.*, 42, 258, 1978.

Sims, J.T. and G.V. Jonson, Micronutrient soil test, in *Micronutrients in Agriculture*, J.J. Motvedt, F.R. Cox, L.M. Shuman, and R.M. Welsh, Eds., ASA SSSA, Madison, WI, 1991, 427.

Sposito, G., L.J. Lund, and A.C. Chang, Trace metal chemistry in arid-zone field soils amended with sewage sludge. I. Fractionation of Ni, Cu, Zn, Cd, and Pb in solid phases, *Soil Sci. Soc. Am. J.*, 46, 260, 1982.

Thorne, W., Zinc deficiency and its control, in *Adv. Agron.*, 9, 31, 1959.

Tsadilas, C.D., Th. Matsi, N. Barbayiannis, and D. Dimoyiannis, Influence of sewage sludge application on soil properties and on the availability of heavy metals fractions, *Commun. Soil Sci. Plant Anal.*, 26(15-16), 2603, 1995.

(U.S. EPA) Environmental Protection Agency, Process Design Manual for Land Application of Municipal Sludge, Center for Environmental Research, 1993.

7

Physical Separation of Metal-Contaminated Soils

Clint W. Williford, Jr. and R. Mark Bricka

CONTENTS

1-56670-457-X/00/$0.00+$.50
© 2000 by CRC Press LLC

7.1   Introduction

7.1.1   The Problem of Metals Contamination

Numerous industrial, construction, and military practices have contaminated soil and water with heavy metals and organics. Examples include use of lead-based paints, firing ranges, electroplating, and nuclear materials manufacture (Bricka et al., 1993). Heavy metals frequently disrupt metabolic processes and produce toxic effects in the lungs, kidneys, and central nervous system. Organometallic forms such as dimethyl mercury are highly toxic. Heavy metals contamination threatens both industrial sites and heavily populated areas. Furthermore, the "indestructible" nature of metals has limited options for remediation to solidification/stabilization, "dig and haul," and to a lesser extent soil flushing. The 1993 EPA Status Report on Innovative Treatment Technologies (U.S. EPA, 1993a) states that of 301 innovative treatment applications (as of June 1993), only 20 involved metals. Remediation costs on the order of $500 per cubic meter, and more for radioactive materials, motivate research to minimize volumes requiring costly treatment and to improve the efficiency of those treatments.

The physical separation approach reviewed here uses minerals processing technologies to deplete soil fractions of contaminants. The depleted soil should require less aggressive follow-up treatment, and cost effectiveness should be improved for solidification or soil flushing. Research is needed to assess, select, and integrate separations technologies for partitioning contaminants among soil fractions.

7.1.2   The Purpose and Scope of This Chapter

Here we review and provide guidance for the adaptation of minerals processing technologies for the separative remediation of heavy metal contaminated soils. An enriched fraction is obtained for intense treatment, as well as a depleted fraction, for disposal of onsite or simpler treatment. Remediation can be simplified and dollar resources used more effectively.

This review acquaints the reader with (1) the extent and nature of metal contamination in soil; (2) soil characterization needs; (3) principles, unit operations, and experimental results for remediation technologies based on physical separation; and finally (4) descriptions and applications of integrated process trains.

Though not exhaustive, the discussion of recent research and applications covers significant and representative methods. Most are adaptations of placer mining techniques in which moving water (or air) is used to selectively carry smaller-sized, less dense components of the soil away from larger-sized, denser components that settle more quickly. Separation methods reviewed are based on size, density, and surface hydrophobicity. Specific technologies include screening, mineral tabling, hydroclassification, and flotation. Integrated systems are discussed incorporating, for example, a barrel trommel, screens, an attrition scrubber, and hydrocyclones. Discussion is organized in the following sections:

- Extent and Nature of Contamination
- Soil Characteristics and Heavy Metal Contaminants
- Soil Property Data Required for Investigation and Remediation
- Physical Separation
- Integrated Process Trains
- Summary

7.2 Extent and Nature of Contamination

Here, we briefly describe contamination at military installations as a representative example of the magnitude and nature of the problem. It is estimated that the Department of Defense (DOD) has about 1900 installations worldwide, containing about 11,000 individual sites, that will require some form of active remedial action (Table 7.1). As of 1994, 93 of these were listed on the EPA's Superfund National Priorities List (U.S. Department of Defense, 1993).

The end of the Cold War accelerated downsizing and closure of a number of military facilities. The pressures to convert these properties to civilian purposes has grown more imperative. Some facilities, e.g., Fort Ord, CA, occupy properties with high economic value. Of the 165 federal facility sites on the NPL, 35 are also Base Realignment and Closure Sites (U.S. EPA, 1998a).

Metals-contaminated sites include artillery and small arms impact areas, battery disposal areas, burn pits, chemical disposal areas, contaminated marine sediments, disposal wells and leach fields, electroplating/metal finishing shops, firefighting training areas, landfills and burial pits, leaking collection and system sanitary lines, leaking storage tanks, radioactive and mixed waste disposal areas, oxidation ponds/lagoons, paint stripping and spray booth areas, blasting areas, surface impoundments, and vehicle maintenance areas (Bricka et al., 1993; Marino et al., 1997).

Typically, heavy metals contamination occurs in sludges, contaminated soil and debris, surface water, and groundwater. Sandblasting, lead-based paints, and firing range operations have produced soils with discrete metal-rich particles. In contrast, electroplating and cooling tower discharges have produced ionic forms of heavy metal contaminants that

TABLE 7.1

Examples of Types of Physical and Chemical Partitioning

Physical Factors	Chemical Interactions	Chemical Phase Groups
Grain size	Adsorption	Interstitial water
Surface area	Precipitation or coprecipitation	Carbonates clay minerals
Specific gravity	Organmetallic bonding	Hydrous Fe and Mn oxides
Surface charge	Cation exchange	Sulfides
Water content	Incorporation in minerals lattices	Silicates

From Horowitz, A.J., *A Primer on Sediment-Trace Element Chemistry*, 2nd ed., Lewis Publishers, Chelsea, MI, 1991.

associate with soil particles. A survey conducted by Bricka et al. (1994) indicates the most frequently cited metal contaminants at military installations are lead, cadmium, and chromium. Mercury and arsenic occur to a lesser extent, but are of concern due to their extreme toxicity. Of particular concern are abandoned firing ranges. Very high levels of lead (1000s of ppm) are generally found in the berms and soils surrounding such areas, requiring remediation activities.

7.3   Soil Characteristics and Heavy Metal Contaminants

7.3.1   Soil Characteristics

7.3.1.1   Definition/Properties of Soils

In remediation, we focus on the geochemical/geotechnical properties of soil vs. the agricultural. Soil occurs naturally at or near the surface. It combines mineral matter from the breakdown of rocks and organic matter from the decomposition of plants and animals. A liquid phase consists primarily of water with dissolved solids, and a gaseous phase consists primarily of air with carbon dioxide from plant and animal respiration (Briggs, 1977). Soils result from three types of processes: autogenic processes (weathering and biological) may form a soil at a given location; detrital (suspension in air and water) may move soil from one location to another; and anthropogenic (human) activities may move the soil or modify it, for example by compaction, tilling, or addition of fertilizer, lime, or aggregate. The following sections on soil characteristics summarize the major parameters describing the soil and terms for classifying.

Four classes of properties describe soils:

1. *Physical properties* include soil texture, structure, aggregate stability (consistency), density, and porosity.
2. *Hydrological properties* include the classification of soil water, capacity, chemical content, and interaction with oxidation/reduction reactions and soil structure (clay moisture regime).
3. *Chemical properties* include pH, buffering capacity, cation exchange capacity, organic content, and surface substrates.
4. *Biological properties* include the nature of the flora (e.g., bacterial, fungal, and actionomycetes) and fauna (e.g., earthworms, protozoa) communities and how they interact with organic matter decomposition and nutrient cycling (U.S. Department of Agriculture, Soil Conservation Service, 1988; Briggs, 1977).

7.3.2   Properties and Behavior of Metals/Inorganics

Selection of remediation technologies may be immediately narrowed, based on the presence and form of one or more contaminants, e.g., discrete metal fragments or adsorbed species (U.S. EPA, 1998b). Likewise, relative amounts of each may tend to favor certain technologies. Metals may be found sometimes in the elemental form, but more often they are found as salts mixed in the soil. Metals, unlike organic contaminants, cannot be destroyed (or mineralized) through treatment technologies such as bioremediation or incineration. Once a metal has contaminated a soil, it will remain a threat to the environment until it is removed or rendered immobile. The fate of the metal depends on its physical and chemical properties, the associated waste matrix, and the soil. Significant transport of metals from the soil surface occurs when the metal retention capacity of the soil is exceeded or when metals are solubilized (e.g., by low pH). As the concentration of metals exceeds the ability of the soil to retain them, the metals may travel downward with leaching waters. Surface transport through dust and erosion of soils is also a common transport mechanism. The extent of vertical contaminant distribution intimately relates to the soil solution and surface chemistry. Currently, treatment options for radioactive materials are generally limited to volume reduction/concentration and immobilization. Properties and behavior of specific inorganics (e.g., chromium, lead, mercury, etc.) and inorganic contaminant groups are readily available online and are summarized in the Remediation Technologies Screening Matrix and Reference Guide Version 2.0 (U.S. EPA, 1998b).

7.3.3   Toxicity

The toxicities of metals are presented at length elsewhere in this text. Major toxic effects of a number of compounds referred to as heavy metals are also described in Amdur et al. (1991) and Manahan (1990).

7.3.4   Heavy Metal Interactions with Soil Particles

7.3.4.1   Parameters Affecting Association with Soil

The primary parameters affecting the association of a heavy metal with soil and sediment include grain size and surface area, the nature of the geochemical substrate, metal species, and affinity of the metal for the soil. Most organic and inorganic contaminants tend to bind chemically or physically to clay and silt particles. These are attached to sand and gravel by physical processes, primarily compaction and adhesion. Table 7.1 presents factors and characteristics of physical and chemical partitioning of metals between soil and surrounding media (Horowitz, 1991).

Physical factors subdivide sediments or soils according to their physical properties: grain-size distribution, surface area, surface charge, density, or specific gravity. Chemical phase groups describe the different geochemical substrates that form the basis of the soil, such as carbonates, clay minerals, organic matter, iron and manganese oxides, and hydroxides, sulfides, or silicates. Chemical interactions characterize the different types of association between metals and the geochemical substrates. The most important interactions are adsorption, precipitation, organometallic bonding, and incorporation into crystal lattices (Horowitz, 1991).

7.3.4.2   Surface Area Effects

Heavy metals, in ionic form, predominantly associate with smaller, higher surface area particles. Clay-sized sediments (<2 to 4 μm) have surface areas of tens of square meters per gram. Sand-sized particles have surface areas of tens of square centimeters per gram (Grim, 1968; Jones and Bowser, 1978). A very strong correlation exists between decreasing grain size and the amount of heavy metal held by the soil fraction.

Horowitz (1991) reported the concentration of copper in a marine sediment having its highest value for the smallest clay particles. The <2-μm fraction had a concentration of 750 mg/kg, about seven times higher than for any other fraction. While it comprised 20 wt% it held about 75% of the copper. Such selective concentration of metals supports the application of physical separations. These observations also support a need to determine metal distribution of particle size as well as physical and chemical state. For example, the lead in firing range soil would consist of particles and smears, while a sample from a battery reworking operation would have adsorbed and ion-exchanged lead species. The form of the contaminant and its association with the soil would be very different. These differences would strongly impact the choice of a treatment process.

7.3.4.3   Mechanisms for Accumulation

Adsorption can take place by physical adsorption, chemical adsorption, and ion exchange (Lieser, 1975). Physical adsorption on a particle surface results from van der Waals forces or relatively weak ion-dipole or dipole-dipole interactions and is reversible. These occur with iron oxides, aluminum oxides, clay minerals, and molecular sieves, such as zeolites (Calmano and Forstner, 1983).

The solid phase also has a certain exchange capacity (CEC) for holding and exchanging cations. In soil components this effect is primarily due to the adsorptive properties of negatively charged anionic sites such as $Si(OH)_2$ and $Al(OH)$ (clay minerals), $FeOH$ (iron hydroxides), and $COOH$ and OH (organic matter) (Forstner and Wittman, 1981; Horowitz, 1991). The type of adsorption is affected by the composition of the geochemical substrate, and thus its composition.

7.3.4.4   Geochemical Substrates

The geochemical substrates that are most important in collecting and retaining heavy metals occur in abundance and have large surface areas, ion exchange capacities, and surface charges. They also tend to predominately occur in the smaller size fraction material. These substrates include iron and manganese oxides, organic matter, and clay minerals.

Iron and manganese oxides are well-known scavengers of heavy metals (Goldberg, 1954; Krauskopf, 1956). Surface areas are on the order of 200 to 300 m/g (Fripiat and Gastuche, 1952; Buser and Graf, 1955).

Organic matter in soils and suspended and bottom sediments have a large capacity to concentrate heavy metals (Goldberg, 1954; Krauskopf, 1956; Horowitz and Elrick, 1987; and Hirner et al., 1990). Organic surface coatings tend to concentrate in the smaller size fractions, while discrete particles tend to concentrate in the ore coarse size fraction (Horowitz and Elrick, 1987).

The main role of clays in metals collection may not, however, stem directly from its surface properties, but from its high surface area, supporting other substrates (Horowitz, 1991).

7.4   Soil Property Data Required for Investigation and Remediation

The vertical and horizontal contaminant profiles clearly define the overall range and diversity of contamination across the site. Obtaining this information generally requires taking sampling and analysis of physical and chemical characteristics. This conveys the specific data needs (for remediation) that can be met during the initial stages of the investigation.

7.4.1   Physical Properties

Physical properties of soil significantly affecting the application of physical separation include the following (U.S. EPA, 1993b; 1994):

- Soil particle size distribution
- Clay content
- Heterogeneity
- Geochemical makeup (organic content, humic content, other organics)

Site soil conditions frequently limit the selection of a treatment process. Process-limiting characteristics such as pH or moisture content may sometimes be adjusted. In other cases, a treatment technology may be eliminated based upon the soil classification (e.g., particle-size distribution) or other soil characteristics.

Usually, properties vary much more vertically than horizontally. This results from the variability in the processes that originally formed the soils. Soil variability results in variability in the distribution of water and contaminants and their transport within, and removal from, the soil at a given site.

Soil particle-size distribution may be the key factor in many soil treatment technologies. In general, coarse, unconsolidated materials, such as sands and fine gravels, are easiest to treat. Soil washing may be ineffective where high percentages of silt and clay inhibit separation of the adsorbed contaminants from fine particles and wash fluids.

The bulk density of soil is the weight of the soil per unit volume, including water and voids. It is used in converting weight to volume in materials-handling calculations and estimating whether proper mixing and heat transfer will occur.

Particle density is the specific gravity of a soil particle. Differences in particle density are important in heavy mineral/metal separation processes (heavy media separation). Particle density is also important in soil washing and in determining the settling velocity of suspended soil particles in flocculation and sedimentation processes.

Other important parameters include clay content, organics (humic materials), and iron. Clay content affects soil processing in several respects. High clay content will lead to low permeabilities, inhibiting any *in situ* procedure. Clay increases the plasticity of the soil leading to clumping and mechanical handling problems. The large surface area of the particles contributes to contaminant adsorption. Finally, fine clay particles will remain suspended in process water, thus requiring dewatering techniques. These can represent a significant portion of the hardware requirement.

Humic content (organic fraction) is the decomposing part of the naturally occurring organic content of the soil. High humic content will act to bind metals to the soil, decreasing their mobility and the threat to groundwater; however, high humic content can inhibit soil vapor extraction (SVE), steam extraction, soil washing, and soil flushing as a result of strong adsorption of the contaminant by the organic material. Mercury is strongly sorbed to humic materials. Inorganic mercury sorbed to soils is not readily desorbed; therefore, freshwater and marine sediments are important repositories for inorganic mercury.

Clay carbonates, or hydrous oxides, readily adsorb zinc (Zn). The greatest percentage of total zinc in polluted soil and sediment is associated with iron (Fe) and manganese (Mn) oxides. Rainfall removes zinc from soil because the zinc compounds are highly soluble.

Table 7.2 summarizes physical and chemical soil characteristics required for planning treatability studies (U.S. EPA, 1990).

7.4.2 Site and Soil Characterization

The successful implementation of a physical separation remediation requires a thorough characterization protocol of the site, soil, and contaminant. Hansen (1991) compared the steps in planning mineral extraction to those for remediation and provided the following outline (Figure 7.1).

TABLE 7.2

Waste Soil Characterization Paramenters

	Parameter	Purpose and Comment
Key physical	Particle size distribution	Oversize pretreatment requirements
	>2 mm	Effective soil washing
	0.25–2 mm	Limited soil washing
	0.063–0.25 mm	Clay and silt fraction — difficult soil washing
	<0.063 mm	
Other physical	Type, physical form, handling properties	Affects pretreatment and transfer requirements
	Moisture content	Affects pretreatments and transfer requirements
Key chemical	Organics	Determine contaminants and assess separation
	Concentration	and washing efficiency, hydrophobic
	Volatility	interaction, washing fluid compatibility,
	Partition coefficient	changes in washing fluid with changes in contaminants; may require preblending for consistent feed; use the jar protocol to determine contaminant partitioning
	Metals	Concentration and species of constituents (specific jar test) will determine washing fluid compatibility, mobility of metals, posttreatment
	Humic acid	Organic content will affect adsorption characteristics of contaminants on soil important in marine/wetlands sites
Other chemical	pH, buffering capacity	May affect pretreatment requirements, compatibility with equipment materials of construction, wash fluid compatibility

From U.S. EPA, Soil Washing Treatment, Eng. Bull., Office of Emergency and Remedial Response, EPA/540/2-90/017, 1990.

The U.S. EPA (1991) developed a two-tier protocol, focusing on soil (waste) characterization for radioactively contaminated soils. Tier 1 analysis includes finding the concentration of the contaminant; size classification to determine the mass and contaminant distributions according to size; and petrographic analysis to identify the mineral species and determine shape, hardness, weathering, coatings, and aggregation. A density separation is made on sand and silt size fractions. Tier II tests focus on coatings or materials requiring more precise instrumentation. Tests are performed to assess particle separation, particle liberation (physical debonding), and chemical extraction. This provides the basis to assess applicability of specific treatment technologies. Specific treatability procedures appear in "Superfund Treatability Study Protocol: Bench Scale Level of Soils Washing For Contaminated Soil" (U.S. EPA, 1989a).

7.4.3   Implications for Treatment Methods

The strong tendency of metals to associate with distinct soil/sediment fractions offers opportunities to selectively separate heavy metal from contaminated soil. For example, more than 90 mass percent of lead in a firing range soil may occur in the >2.0-mm fractions.

Chemical, physical, and biological methods can immobilize the metals, separate them from the particle, or separate and concentrate the most contaminated particles. The enrichment of adsorbed contaminants, generally in the finer size fractions, means this fraction will probably require follow-up treatment. In addition to separating solid particles, contaminants may be mobilized into solution, requiring water treatment with precipitation or ion exchange.

Physical separation can be used standing alone or with other treatment processes. It may achieve acceptable levels alone, but in other cases is most effective combined with other

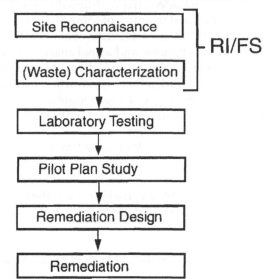

Environment site Remediation

Site Reconnaisance

↓

(Waste) Characterization

} RI/FS

↓

Laboratory Testing

↓

Pilot Plan Study

↓

Remediation Design

↓

Remediation

FIGURE 7.1
Flowchart for environmental site remediation
process development.

treatment processes. It may reduce volume or may convert soil to a more homogeneous condition improving further processing. It is most effective with sandy soil; performance declines with increasing clay and silt content, especially as a stand-alone technology. Soils with high percentages of silt and clay tend to strongly adsorb contaminants.

Soil washing alone is not advised. Hydrophobic contaminants generally require surfactants or organic solvents for their removal. Complex contaminant mixes including metals and nonvolatile organics and semivolatile organics and frequent changes in composition make selection/formulation of washing fluids difficult. Surfactants and chelators may improve contaminant removal efficiencies, but may also interfere with downstream water treatment (U.S. EPA, 1989a; 1989b).

Finally, the use of soil slurries generates significant volumes of water with suspended solids. Removal and concentration of the suspended soils can require a third (on a size basis) of the unit operations brought to a site. Use of "dry" pneumatic systems eliminates this problem. These systems generally separate more slowly and less efficiently, relative to water slurry systems (Silva, 1986). However, they have been successfully employed, for example to remediate firing range soil at a police firing range in New York City (MARCOR Remediation, Inc., 1997).

7.5 Physical Separation

7.5.1 Background

In 1993, the U.S. Army Corps of Engineers (USACE) Waterways Experiment Station (WES) Environmental Laboratory (EL) reviewed technologies for treatment of metals-contaminated soil that warranted further development and implementation (Bricka et al., 1993). The project report concluded that few advanced technologies were widely practiced for heavy metals-contaminated soil. Questions existed for many technologies (major concerns were production of residual streams and long-term stability of treated metals left in the soil). It was also concluded that additional research was needed to resolve concerns and better understand fundamentals of some processes.

A second WES report (Bricka et al., 1994) integrated the first report, a survey of contamination at installations, and a final analysis by WES-EL, Restoration Branch staff. This report prioritized technologies and identified research needs to field one or more technologies in 5 years. It concluded that (1) extraction methods coupled with physical separation offered the most promising and appropriate area for continued research; and (2) a limited number of precipitation and thermal processes (roasting and enhanced volatilization) warranted further research support.

These reports and Web-based material by U.S. EPA (1998d) provide a wide-ranging review of technologies (presumptive and innovative), including descriptions, modes of action, applications, and limitations. Based on reviews such as these, and a growing awareness in the late 1980s to early 1990s of the need for metals-remediation alternatives, a number of organizations began to explore and develop systems for physical separations.

7.5.2   Fundamentals of Physical Separation

Heavy metals can exist as discrete particles, adsorbed species, or dissolved species. Lead paint deterioration, sand blasting, and firing range operations produce discrete fragments of metallics smear on soil particles. Electroplating, battery reworking, and cooling tower discharge can produce ionic metals associated with soil particles.

Each form of metal contamination exhibits different physical properties: particle size, density, and surface charge depending upon the metallic particles, soil characteristics, and contaminant. To the extent that these particles differ from those of the soil, the contamination will not occur uniformly in the soil, but will associate disproportionately with particular soil fractions, e.g., fines. The major parameters affecting the association of a heavy metal with soil and sediment include grain size, surface area, geochemical substrate, and metal affinity (Horowitz, 1991).

The general approach in physical separations remediation is to use unit operations commonly applied in the minerals processing industry. Most exploit differences in particle size, density, and surface properties to effect a separation. Other methods exploit magnetic and electrostatic properties. Ideally, the "cleaned" fraction will require no further treatment, and the "concentrated" fraction can be more economically processed. A conceptual process train (U.S. Bureau of Mines, 1991) appears in Figure 7.2. The following material presents the principles of operation of a number of major unit operations, along with experimental results that inform us of their performance and limitations. Examples are given of how these individual unit operations have been integrated into process trains. The flowsheet in Figure 7.2 will be described in further detail at that point. Table 7.3 shows categories of technologies subdivided according to the principle of separation, e.g., size or density. Significant technologies and applications are listed.

7.5.3   Size-Based Separation

7.5.3.1   Screening

Screening uses size exclusion through a physical barrier. Although simple in concept, screening has often been described as more art than science. A wide range of screens exists, both stationary and vibrating, and each screen has a specific purpose and application.

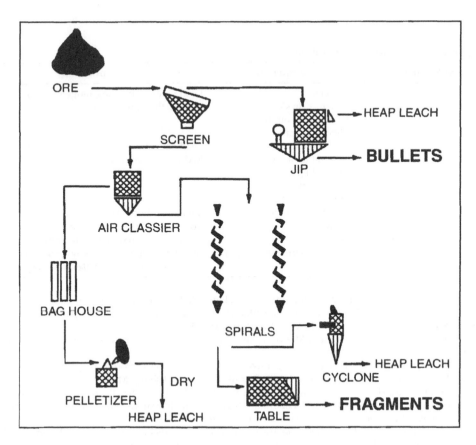

FIGURE 7.2
Conceptual process train for remediation of lead-contaminated firing range soil. (From U.S. Bureau of Mines, Heavy Metal Removed from Small Arms Firing Ranges, R. McDonald, Ed., prepared for U.S. Naval Civil Engineering Research Laboratory, Salt Lake City Research Center, Salt Lake City, UT, 1991.)

Once the possibility of a separation has been established, estimation of screen performance requires estimation of screen efficiency and screen capacity. In environmental remediation, the goal may often be to recover oversized material while smaller, more highly contaminated fractions pass through the screen. In such cases the screen efficiency can be calculated as the ratio of oversized recovery to the oversize feed.

Screen capacity may be estimated on a unit area basis as the ratio of flow-through capacity (tons) to overall unit capacity (tons/h-ft²), modified by some correction factor. Correction factor derivations vary widely on the basis of the nature of the material to be screened, the application, screen opening size, and technical reference. The empirical nature of screening technology requires laboratory and pilot analysis and field experience in the initial phase of screen selection to estimate performance beyond preliminary design.

Accordingly, bench-scale screening (sieving) assesses the mass and contaminant distribution among the soil size fractions. This is one of the key assessments in treatability studies for evaluating the feasibility of physical separation and the choice of separation unit operations. Figures 7.3 and 7.4 show the equipment and procedures for wet sieving of a metal-contaminated soil. With some modification, these procedures involve a vigorous agitation with a water/soil ratio of 4:5 five (Figure 7.3), followed by wet-sieving (Figure 7.4), with screen openings of 10 mm down to 63 μm, and with several intermediate sizes, 100 μm.

TABLE 7.3

Unit Operations for Physical Separation

Basis of Separation	Technology	Application
Size	Screens	Lead slugs from firing range (2, 4, 5)
	Bench-scale sieves (1,2)	Radioactive sand (1, 3)
Density	Stokes law settling hydroclassification (1, 2)	
	Hindered settling	Lead slugs from firing range (2, 4, 5)
	Mineral table (3)	Radioactive sand (1, 3)
	Mineral jig (3)	
	Hydrocyclone (1)	
	Spiral concentrator	
Hydrophilic/hydrophobic	Flotation (3, 6, 8)	Metal sulfide-contaminated tailings
Magnetic	HGMS	Iron oxide removal from clay
Disaggreation/declumping	Trommel (1)	Preliminary separation to remove debris and
	Attrition scrubber (1, 2, 7)	break up soil clumps; remove smaller (e.g., radioactive) particles from larger ones

Sources:

1. Williford, C.W., Trip report (and photographs) of Volume Reduction and Chemical Extraction System (VORCE), National Air and Radiation Environmental Laboratory, Montgomery, AL, prepared for U.S. Army Corps of Engineers (WES-EE-R: R.M. Bricka), Vicksburg, MS, 1991a.
2. Williford, C.W., Jr., Z. Li, Z. Wang, and R.M. Bricka, Vertical column hydroclassification of metal-contaminated soils, *J. Hazardous Mater.*, 66, 1, 1999.
3. Williford, C.W., Trip report (and photographs) of AWC Lockheed TRUClean process, prepared for Restoration Branch, U.S. Army Corps of Engineers (WES-EE-R: R.M. Bricka), Vicksburg, MS, 1991b.
4. U.S. Bureau of Mines, Heavy Metal Removed from Small Arms Firing Ranges, R. McDonald, Ed., prepared for U.S. Naval Civil Engineering Research Laboratory, Salt Lake City Research Center, Salt Lake City, UT, 1991.
5. U.S. Navy Civil Engineering Laboratory and U.S. Bureau of Mines, Heavy Metal Removal from Small Arms Ranges: A Pilot-Scale Demonstration at Marine Corps Base, Camp Pendleton, CA, 1993.
6. U.S. EPA, Toronto Harbor Commissioners (THC) Soil Recycle Treatment-Technology Demonstration Summary, EPA/540/SR-93/517, 1993c.
7. Marino, M.A., R.M. Bricka, and C.N. Neale, Heavy metal soil remediation: the effects of attrition scrubbing on a wet gravity process, *Environ. Progress*, 16(3), 208, 1997.
8. Mann, M. and J. Besch, Divide and conquer, *Soils*, March, 20, 1992.

During wet sieving, the soil on the screen is washed with and agitated in water to facilitate the separation and passage of smaller particles through the screen openings.

Commercial-scale stationary screen designs range from grizzlies, used to scalp cobbles and debris, to wedge-bar screens and hydro sieves used for size separation down to 10 mesh (Weiss, 1985). Moving screens or vibrating screens are most commonly used and can be arranged in series for progressively finer screening. Vibrating screens are most commonly used to separate material ranging from 1/8 in. to 6 in., with high speed vibrating screens available for separation of material ranging from 4 to 325 mesh (Averett et al., 1990). Beyond size separation, screens are also used for dewatering, scalping, media recovery, and the removal of very fine particles in wet or dry media, known as desliming or dedusting, respectively. The following photographs in Figures 7.5 and 7.6 show a commercial screen system with a close-up of the screen deck. Figure 7.7 shows a grizzly with a hopper on top and a conveyor belt exiting on the lower left. A close-up of the bars across the top of the hopper appears in Figure 7.8. The entrance to a barrel trommel appears in Figure 7.9, showing the water spray and internal baffles that break up and mix the soil. Figure 7.10 shows the overall view of a small pilot-scale system. Undersize material falls through the screen around the drum and then into a chute for further processing. The oversize material then continues to the lip of the drum and then into another chute at a right angle to the drum.

FIGURE 7.3
Soil and water slurry shaken before separation.

These systems, and a mineral jig, have been combined by AWC Lockheed for remediation of radioactivity-contaminated soils (Williford, 1991b).

Screening serves as a key preliminary unit operation, dividing the soil stream into appropriate size fractions for subsequent mineral processing unit operations, e.g., hydrocyclones or mineral jigs. The quantity, size, and density distribution must be known to size and plan the layout of these unit operations. Figure 7.11 lists separations unit operations and their applicable particle size ranges (Mular and Bhappu, 1980). Figure 7.12 shows the applicable size ranges for soil washing, with respect to physical separation (U.S. EPA, 1990). For soils in Regime I, these coarse soils are very amendable to soil washing; most contaminated soils have a size distribution that falls within Regime II (the types of contaminants will govern the composition of the washing fluid and the overall efficiency of the washing process). In Regime III, the soils consist of finer sand, silt and clay fractions, and those with highly humic content. These materials strongly adsorb organics and inorganics and generally do not respond well to systems attempting to dissolve or suspend the contaminants. They may respond to soil washing that separates the contaminant-rich fraction into a smaller volume.

7.5.3.2 Sample Results for Size Separation of Contaminated Soil and Sediment

Results of a representative soil/contaminant characterization appear in Figures 7.13 and 7.14. These show the overall mass and lead distribution produced by sieving and hydro-classification of a firing range soil. On a mass basis (Figure 7.13), this soil appears fine, with over 40% in the silt and clay range (<63 μm). On a lead basis (Figure 7.14), there is a bimodal distribution of the lead to the larger and smaller size extremes. For the whole soil, the lead distribution is even more dramatic, with 90+% of the lead slugs and fragments screened out

FIGURE 7.4
Wet sieving of metals-contaminated
soil through an 8-in. sieve.

before the assessment of the "soil" fraction smaller than 10 mm in size. The differing results
between sieving and hydroclassification are presented in the following section.

Notice that almost 50% of the mass of the soil (for the size range separated) lies in the
intermediate size fractions. However, the lead distribution is only 2 to 9% in these fractions,
revealing substantial depletion of lead in these fractions.

In other work, sediment from the Great Lakes region was separated by the variety of physical
separation methods (Allen, 1993). Wet sieving enriched a sample identified as "Saginaw #2"
with distribution of 48 to 60% of the chromium, cadmium, and lead to the -400 mesh size
fraction, which comprised just 12.4% of the mass of the feed.

7.5.4   Gravity-Based (Density) Separation

7.5.4.1   *Vertical Column Hydroclassification*

Many physical separation process trains use gravity-based unit operations such as spiral
hydroclassifiers, hydrocyclones, and mineral tables. These systems operate on the principle
that the settling behavior is a function of size and density. A simple treatability test, hydroclassification, proves useful to assess the application of such unit operations. Hydroclassification determines free settling characteristics governed by Stokes' law, as opposed to
"hindered settling." In the latter case, a slurry of water and particles forms a dense media
in which separation occurs. In this section, we describe the general principles of operation
for bench-scale hydroclassification. Of course, industrial-scale operations will use larger
appropriate unit operations. Accordingly, we also describe the equipment and principles of
operation for spiral hydroclassifiers, hydrocyclones, and mineral tables.

FIGURE 7.5
Two deck screen with cast grizzly bar top deck and rod deck on the bottom.

FIGURE 7.6
Close-up of screen (rod) deck.

Vertical column hydroclassification uses upward flowing water in a small column to elute a series of contaminated soil fractions, producing depletion and enrichment of metals among the fractions. Results provide a best case separation for gravity-based methods, for example, minerals table, hydrocyclone, mineral jig, spiral concentrator, or hydroclassifier.

FIGURE 7.7
Grizzly with hopper and conveyor belt (lower left).

According to Stokes' law, at low Reynolds number, particles of uniform shape and density settle through water at rate proportional to their density and the square of their diameter. Stokes' law for spherical particles falling slowly through water appears in the following equation (McCabe, et al., 1993):

$$u_t = gD_p^2\rho\frac{(\rho_p - \rho)}{18\mu}$$

Figure 7.15 illustrates the forces acting on a settling particle. These include buoyancy, drag, and gravity. The particle depicted in the center is in balance (equilibrium), with the actual fluid velocity matching the potential terminal velocity. If the net force is nonzero, the particle will accelerate until increasing drag force balances the other forces and the particle reaches its terminal velocity, u_t. Relative to the particle in balance, a more dense particle sinks and a smaller particle rises. This explanation, while essentially correct, does not account for higher terminal velocities and irregular shapes, which generate Reynolds numbers and settling behavior beyond Stokes' law. Correlations have been adapted using drag coefficients to correct for these effects. Likewise, experimental data have been collected to provide empirical relationships. Figure 7.16 shows terminal velocities in water for various minerals (McCarter, 1982).

As the upward flow of water exceeds the terminal velocity of select particles, they will be swept out the top of the column to a collection tank. Smaller or less dense particles will be preferentially removed. Larger, more dense particles will remain in the column until the flow rate is adjusted upward. Contaminant particles having a size or density distribution different from that of the host soil will distribute differently than the mass for the soil. This leads to enrichment and depletion of the contaminant in the resulting particle size fractions.

FIGURE 7.8
Closeup of grizzly bars at top of hoppe.

The greater the enrichment/depletion, the easier the separation of a contaminant from the soil using gravity-based separation technologies. Vertical column separation serves as a simple, economical method for finding a "best case" result for gravity-based separations (U.S. EPA, 1993d).

Hydroclassification can be carried out simply and economically on soil fractions smaller than 600 mm. Figure 7.17 shows a small vertical hydroclassification column used at the University of Mississippi to separate several metals-contaminated soils on behalf of the U.S. Army Corps of Engineers (Williford et al., 1997). Figure 7.18 shows the overhead effluent of a sand-sized fraction.

7.5.4.2 Spiral Classifiers

Spiral classifiers operate on the same principle of settling. A typical spiral classifier consists of a steel trough with an inclined section. The feed slurry enters the trough where small particles remain suspended and are carried out over a weir. Larger particles sink and are removed by a spiral conveyor. Figure 7.19 shows a spiral classifier (Mular and Bhappu, 1980).

7.5.4.3 Sample Results for Vertical Column Hydroclassification

Results are presented here for two vertical column hydroclassification studies. The first study was performed by the U.S. EPA at its National Air and Radiation Environmental Laboratory (NAREL) (Williford, 1991a). Investigators separated a soil contaminated by

FIGURE 7.9
Closeup of entrance of barrel trommel with wash nozzles.

FIGURE 7.10
Barrel trommel (pilot-scale) note screen around drum and product chutes on right.

low-level radioactive sand particles (Hay et al., 1991). The second study was performed by the University of Mississippi, on behalf of the U.S. Army Corps of Engineers, to separate metals-contaminated soils from firing ranges, an electroplating facility, and a popping furnace used for ordnance destruction (Williford et al., 1997). Both studies used adaptions of an approach developed by NAREL with its contractor, Sandy Cohen and Associates.

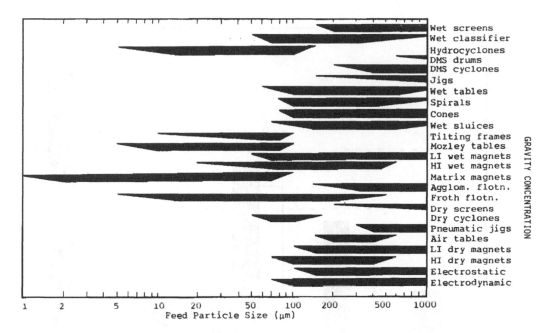

FIGURE 7.11
Particle size ranges for mineral processing unit operations.

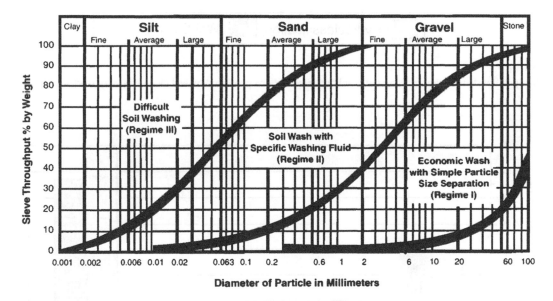

FIGURE 7.12
Applicable particle size ranges for soil washing.

The NAREL study assessed the feasibility of physical separation to reduce the volume of a low-level radioactively contaminated soil from the Wayne Interim Storage site. Petrographic examination was performed to identify the waste forms of the radioactivity and the distribution of the waste forms within the various size fractions. The soil was an unconsolidated glacial till, best described as gravelly, silty sand. The radioactive contaminants were monazite and zircon, high-density, hard, smooth-surfaced, projectile-shaped radioactive minerals.

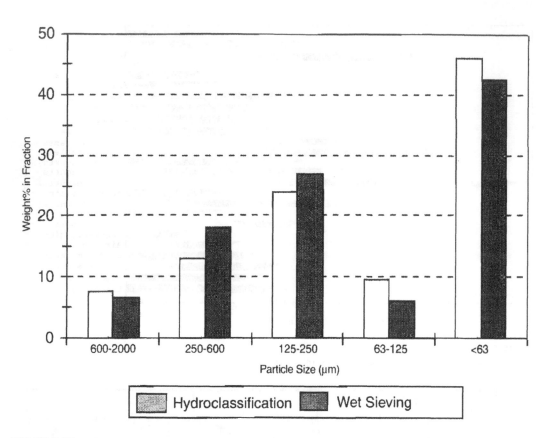

FIGURE 7.13

Mass percent distribution of firing range soil into size factions produced by wet sieving and hydroclassification.

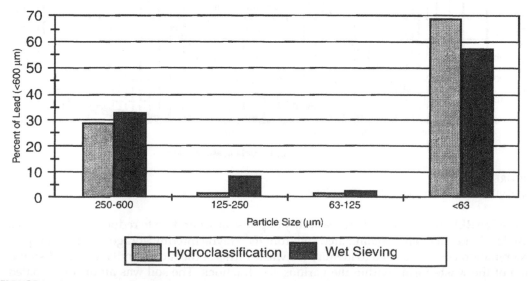

FIGURE 7.14

Lead percent distribution of firing range soil into size fractions produced by wet sieving and hydroclassification.

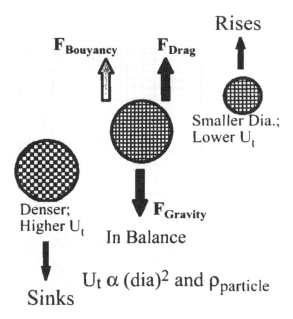

FIGURE 7.15
Graphic explanation of Stokes' law settling. Terminal velocity u_t proportional to particle density and the square of diameter.

The petrographic analysis indicated the following relationships between the radioactive minerals and the host material:

- The monazite and zircon are concentrated in the fine sand to upper silt-size range, -200 ± 2000 mesh, or 10 to 100 μm.
- The specific gravity of monazite is 4.7 to 5.4, 3.9 to 4.8 for zircon, and 2.6 to 2.7 for the host material. Approximately 2% of the host material has a specific gravity above 3.0.
- The monazite and zircon are from placer deposits and are not water soluble.

Figure 7.20 graphically presents the separation of the soil at 200 and 325 mesh by both wet sieving and hydroclassification. (Front row bars are for the bottom row of numerical data.) Separation results agree between the two methods. Up to 52.9% of the soil by weight is reported to the +200 mesh fractions, and up to 66.6% to the +325 mesh fraction. This figure also shows that the radioactivity is enriched by an order of magnitude in the smaller size fraction. Samples of wash water showed very little radioactivity.

In the second study (Williford et al., 1997), the distribution of mass and heavy metals was compared for wet sieving and hydroclassification of four soils.

The popping furnace and electroplating soils were sandy with only 15 to 18 wt% in the <63-μm fraction, while the firing range soils were finer, with 45 to 48 wt% in the <63-μm (silt/clay) fraction. Hydroclassification and wet sieving generally produced mass distributions that tracked each other closely, as shown in Figure 7.13. However, for the furnace soil, hydroclassification shifted more material (relative to wet sieving) from the 250- to 600-μm to the 600- to 2000-μm fraction. This indicated that dense material, potentially rich in lead, which would pass the 600-mm screen, was retained with larger material during hydroclassification.

The firing range soils exhibited two characteristics significant for separations. First, a substantial fraction of the lead mass concentrates in the coarse, >600-μm material, about 75 wt% of the lead for the firing range 1 soil. Further, the <600-μm material exhibits a

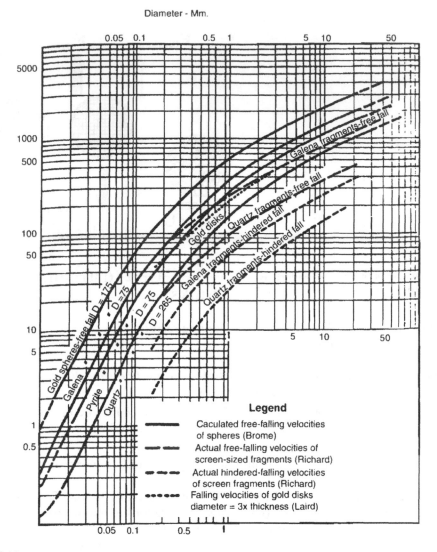

FIGURE 7.16

Chart showing settling velocities in water of gold, galena, pyrite, and quartz grains of various sizes and shapes (McCarter, W.A., Placer Recovery, in Yukon Placer Mining Industry 1978–1982, R.L. Debicki, Ed., Exploraton and Geological Services, Northern Affairs Program, Indian and Northern Affairs Canada, presented at the D.I.A.N.D.-K.P.M.A. Placer Mining Short Course, Whitehorse, Yukon, 1982. Reproduced with permission of the Ministry of Public Works & Government Service, Canada, 1999.)

bimodal pattern, with substantially less lead distribution to the 63- to 250-μm fractions (Figure 7.21). In contrast, the plating soil exhibits a strong distribution of both chromium (up to 93.2 wt%) to the <63-μm fraction, which composes 25.1 wt% of the mass. Again, this enrichment could be exploited to facilitate separation and remediation.

Extraction with 5% acetic acid indicated that lead was mobile in all size fractions of the furnace soil. Extraction of hydroclassified, midrange, 63- to 250-μm fractions of firing range 1 soil yielded extracts an order of magnitude lower in concentration. This is consistent with the sharp depletion of lead for this fraction. Chromium and lead were relatively immobile in the <63-μm plating soil, yielding an extract concentration of 45 mg/L. This represented mobilization of only about 2% of the chromium in the sample (over 40,000 mg/kg).

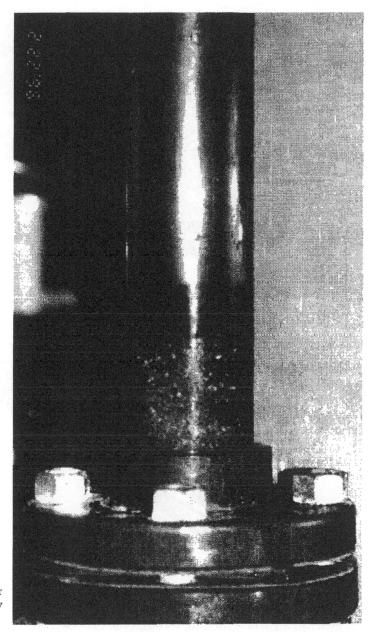

FIGURE 7.17
Vertical column hydroclassifier
with glass beads and settled slurry
(2-in. diameter).

7.5.4.4 *Hydrocyclones*

Hydrocyclones are simple mechanical devices whose theoretical complexity borders on the frontiers of mathematics and physics. "There still exist no systems or series of equations that exactly describe either capacity or separation for any given cyclone or operating circumstance. The reasons are the large number of variables, known and unknown, design, and operation, and the fact that by its very nature the liquid cyclone analyze, in spite of its apparent simplicity" (Weiss, 1985). This lack of understanding is not the result of shortage of effort. There has been a large number of theoretical investigations during the past 20 years. This research has been driven by the enormous number of industrial applications. These range from mineral processing to potato chip production. Designs are extremely varied.

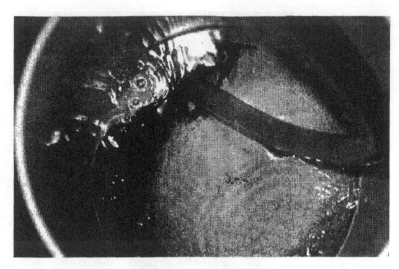

FIGURE 7.18
Effluent from vertical column hydroclassification (sandy-sized product).

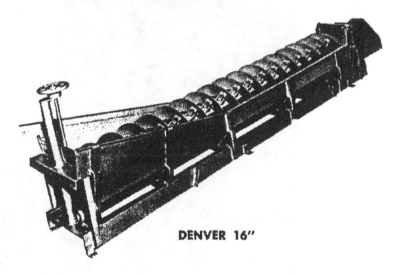

DENVER 16"

FIGURE 7.19
Spiral classifier.

In general, a particle moving through the cyclone is held in field of centrifugal force. Small lightweight particles will be influenced by the mass flow of water leaving the cyclone close to the feed point. These small particles will leave in the overflow. Heavy particles will be held by high centrifugal forces away from the overflow. These particles will report to the underflow as gravity drags them, within their fields of rotation, toward the bottom of the cyclone (Figures 7.22 and 7.23) (Mular and Bhapu, 1980).

Hydrocyclones in parallel permit more efficient, smaller units to handle high flows, while units in series increase overall recoveries. Usually, hydrocyclones have low capital and operational costs and are small, compared with other separation equipment.

7.5.4.5   Sample Results for Hydrocyclones Separation of Contaminated Soil

Hydrocyclones were employed in the Toronto Harbor project separation system (U.S. EPA, 1993) as well as in the volume reduction and chemical extraction system (VORCE) assembled by

WET SIEVING

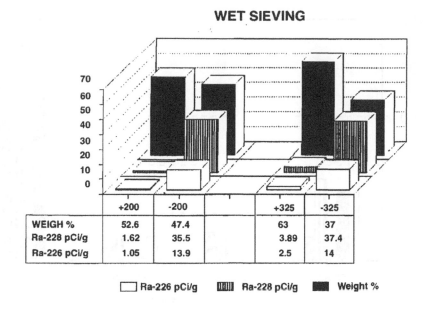

	+200	-200		+325	-325
WEIGH %	52.6	47.4		63	37
Ra-228 pCi/g	1.62	35.5		3.89	37.4
Ra-226 pCi/g	1.05	13.9		2.5	14

☐ Ra-226 pCi/g   ▥ Ra-228 pCi/g   ■ Weight %

VERTICAL-COLUMN HYDROCLASSIFICATION

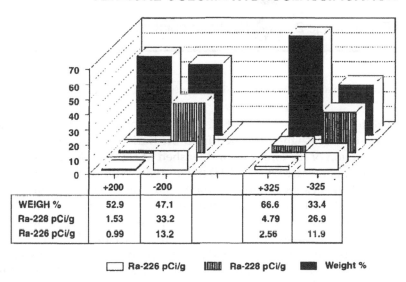

	+200	-200		+325	-325
WEIGH %	52.9	47.1		66.6	33.4
Ra-228 pCi/g	1.53	33.2		4.79	26.9
Ra-226 pCi/g	0.99	13.2		2.56	11.9

☐ Ra-226 pCi/g   ▥ Ra-228 pCi/g   ■ Weight %

FIGURE 7.20
Mass distribution and radioactivity of fractions of soil after wet sieving and hydroclassification.

NAREL (Williford, 1991a). The latter system used parallel hydrocyclones, shown in Figure 7.24, to produce a –200 mesh cut enriched in radioactive particles. The system worked well, producing a 70 to 90% volume reduction.

7.5.4.6 Mineral Table

Historically, mineral (or shaking) tables have been used to separate high density from low density minerals. The following discussion draws on background and experimental results

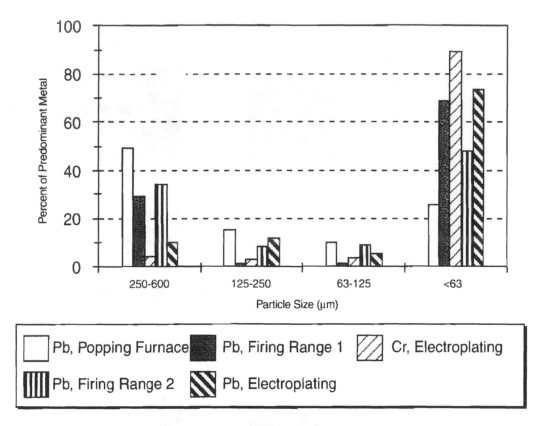

FIGURE 7.21
Metals percent distribution after hydroclassification of four soils.

from remediation research by Marino et al. (1997) and others cited below. Shaking tables consist of plane surfaces, slightly inclined from the horizontal, shaken with a differential motion in the direction of the long axis and washed at right angles to the direction of the motion with a stream of water (Figure 7.25) (Silva, 1986; Wills, 1984; Marino et al., 1997).

The wet shaker table is a rectangular or rhomboid-shaped surface with raised strips of rigid material placed longitudinally along the surface to form "riffles." It is operated essentially in a horizontal plane. A drive mechanism imparts an oscillating motion to the deck along its long axis as water flows perpendicular across the table. The separation produced on a wet shaker table is the result of several mineral processing principles simultaneously acting on the table feed. These principles include flowing film concentration, hindered settling, consolidation trickling, and asymmetrical acceleration (Weiss, 1985).

Operationally, a solid slurry is fed to the upper edge of the sloping table. As the slurry moves across the table, "longitudinal riffles" slow it, creating pools behind the riffles. The oscillating action of the table causes size classification and specific gravity stratification. Particles having similar specific gravities arrange themselves vertically according to size. The continuous addition of slurry and the action of the flow of cross water shears off the top layers of the stratified particles. The depth of the riffles decreases toward the left or discharge (concentrate) end of the table, allowing the lighter (lower specific gravity), coarser particles to pass over the riffles toward the downslope side of the table while the increasingly finer, denser particles move longitudinally along the table. The drive mechanism imparts a faster return stroke than forward stroke, forcing the particles to move toward the discharge end of the table. The heavy particles will discharge at the left side of the table,

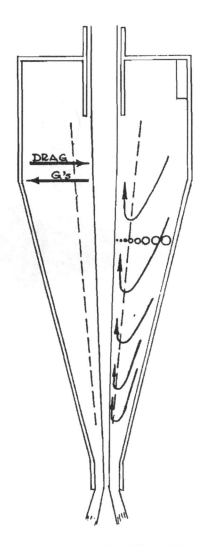

FIGURE 7.22
Graphic representation of forces and particle movement in a
hydrocyclone.

while the lighter particles will discharge toward the bottom of the table (Silva, 1986;
Pryor, 1965). Three fractions are collected and are referred to as tails (the least dense), mids
(denser), and cons (the densest fraction).

The effectiveness and speed of the separation is dependent on size and shape of the particles, difference in the particles specific gravity, the particle size range of the feed, the cross
water and feed flow rates, and the mechanical settings of the shaker table, which include
the speed, stroke, and slope of the table (Weiss, 1985).

Wet tabling can separate a wide range of particle sizes. If mineral processing is considered a starting point for soil remediation design, applicable particle sizing ranges from
6 mesh (3.36 mm) to 150 mesh (105 mm) (Green, 1984).

Tabling for separation from a soil mixture of contaminants remains under development.
Tabling is probably most appropriate for heavy metal-contaminated soils, especially firing
ranges, where operations are analogous, in part, to mineral beneficiation.

7.5.5 Attrition Scrubbing

Attrition scrubbing employs a high energy mixer to impart a mechanical scrubbing action on
a slurried soil. This mixing results in vigorous particle-to-particle scrubbing in a high solids

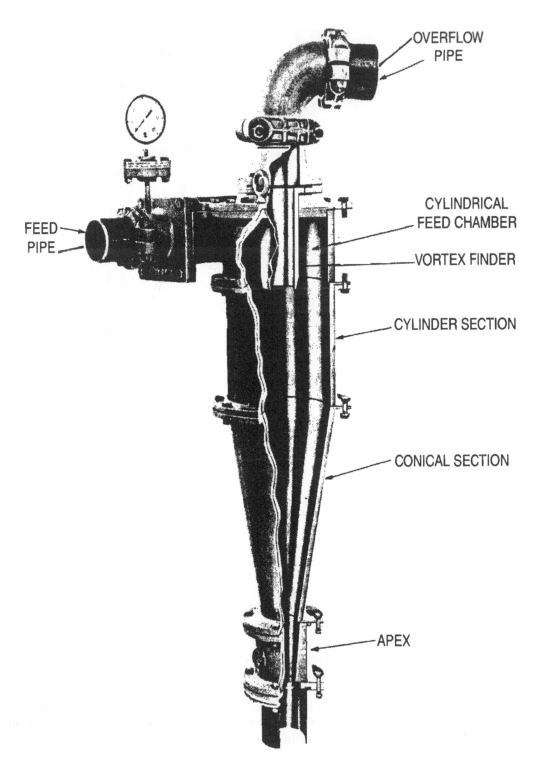

FIGURE 7.23
Cutaway view of hydrocyclone.

FIGURE 7.24
Hydrocyclones in trailer-mounted process train.

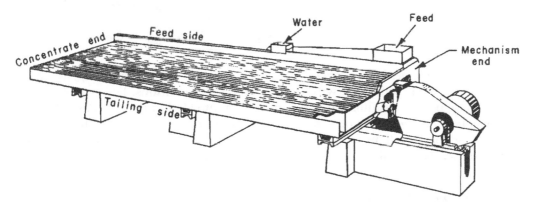

FIGURE 7.25
Mineral (shaking) table. (From Marino et al., 1997. Reproduced with permission of AIChe.)

environment, producing scouring, disintegration, and dispersion. Scouring removes coatings of films from individual soil grains and produces clean soil surfaces. Disintegration and dispersion form a "slime" (ultra fine particles), or a dispersed slurry resulting from the breakup of agglomerated particles. Figure 7.27 shows a diagram of a lab-scale attrition scrubber (Marino et al., 1997). Note the intense opposing flows generated by the twin impellers.

Industrial-scale attrition scrubbing (Figure 7.28) (Denver, 1991) can be utilized in a soil washing process after the large, oversize material has been removed by a grizzly and trommel. Attriting a heavy metal-contaminated soil can either concentrate the contaminants into a particular soil fraction or separate the soil particle from the metal surface and

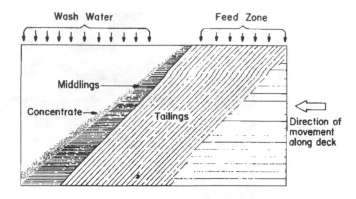

FIGURE 7.26
Idealized mineral separation on a shaking table.

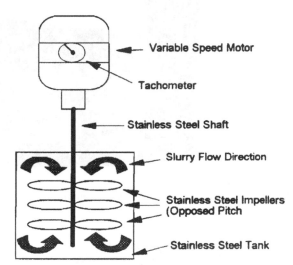

FIGURE 7.27
Schematic of a laboratory attrition scrubber. (From Marino et al., 1997. Reproduced with permission of AIChE.)

increase the effectiveness of a particle density separation. Attrition scrubbing, as a conditioning treatment, is combined with separations unit operations. For example, in the NAREL Phase II system (Williford, 1991a) and in Toronto Harbor Project (U.S. EPA, 1993), hydrocyclones followed attrition scrubbing. The section on process trains further elaborates on these configurations. The following section provides specific results from combining attrition scrubbing with hydroclassification or mineral tabling.

7.5.5.1 Sample Results for Attrition Scrubbing with Wet Tabling

The U.S. Army Corps of Engineers investigated mineral tabling with and without attrition scrubbing (Marino et al., 1997). The work was performed at the Waterways Experiment Station Environmental Laboratory. The soil came from the impact berm of the Army small arms training range and had a lead concentration of approximately 40,000 mg/kg. Separation was evaluated using a WEMCO Laboratory Attrition Scrubber in conjunction with a Wilfley Laboratory Wet Shaking Table. Comparing results in Figures 7.29 and 7.30 reveals that the vast majority of the lead contamination is concentrated on the cons fraction, which is a relatively small portion of the initial bulk soil mass. It was also seen that attriting the bulk soil prior to tabling produces a smaller, more concentrated fraction (cons), while simultaneously producing a larger, cleaner fraction (mids), which is the goal of this treatment.

These results indicate that attrition scrubbing enhances the separation achieved on a wet shaker table. Attrition scrubbing appears to liberate lead from the soil, producing a more

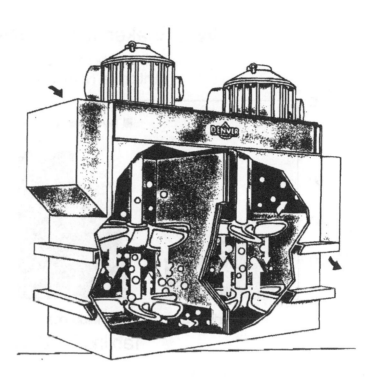

FIGURE 28
Graphic of industrial-scale attrition
scrubber.

discrete lead constituent within the bulk soil, and facilitating physical separation. Separating individual lead particles from the soil particles with attrition allowed a more distinct separation between the mids and the cons on the shaker table. This helped produce the smaller, more concentrated cons, and larger, less contaminated mids, as illustrated in Figures 7.29 and 7.30.

7.5.5.2   *Sample Results for Attrition Scrubbing with Hydroclassification*

In the hydroclassification study at the University of Mississippi (Williford et al., 1997) described previously, attrition scrubbing was applied to two soils, followed by hydroclassification. The soils were a firing range (no. 2) soil with a loamy sandy texture (48 wt%
<63 μm) and an electroplating facility soil with a sandy texture (18 wt% <63 μm). Figure 7.31 shows that attrition scrubbing of the electroplating facility soil produced a small (about 3%) increase in the distribution of chromium to the <63-μm fraction. The effect was clearer for lead, increasing the distribution to the <63-μm fraction from 73.3 to 87.5%. For the firing range soil, attrition scrubbing produced a more dramatic shift of 14 to 15% of the lead from the 500- to 2000-μm fraction to the <63-μm fraction.

7.5.6   Flotation

Flotation is a process in which fine particles are removed from suspension by attachment to air bubbles rising through the slurry. The theoretical basis for flotation rests on hydrophilic and hydrophobic interaction. In water, the bubble provides a hydrophobic interface. The resulting froth is carried out in the overflow. Suspended particles can be chemically conditioned to cause them to be air-avid and water-repellent (Klimpel, 1992; Averett, et al., 1990). As for its application to soil remediation, the soil matrix is slurried with water and chemically prepared to support the adherence of the contaminant with a bubble rising through the

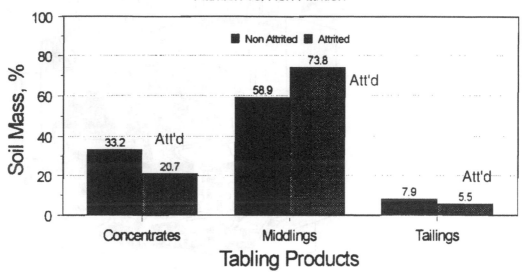

FIGURE 7.29
Weight distribution from wet shaker table separation of firing range soil.

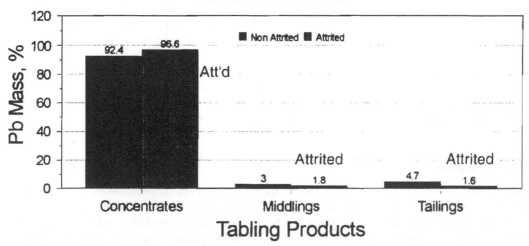

FIGURE 7.30
Lead distribution from wet shaker table separation of firing range soil.

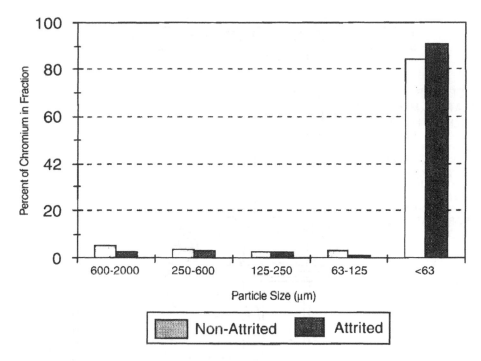

FIGURE 7.31
Chromium distribution from hydroclassification of electroplating facility soil. (From Williford et al., 1997. Reprinted with permission from Elsevier Science.)

water column and exiting the vessel. In mining applications, the process may be more indirect, with a soil fraction removed having the contaminant affixed. Once the contaminant adheres to the rising bubble, a stable froth must form that can be removed from the vessel. The soil remaining in the vessel has a reduced contaminant concentration. The efficacy of the separation of the contaminant from the soil matrix is defined in the mining industry as the "selectivity" of the process. As a result, the first goal of environmental flotation is to confirm that the system will be selective for the contaminant. In the mining industry, the tailings or the materials that do not leave the vessel or cell are the materials that have little value. For environmental applications, clean tailings are the desired product. Particle size is an important factor in the application of flotation. For mining operations, the applicable particle size range is from several hundred microns to approximately 20 μm (Klimpel, 1992). Laboratory evaluation for application of this technology to contaminated sediments suggests that particles finer than roughly 50 to 65 mesh can potentially be removed by this process. Conceptually, treatment of a sludge would require agitation of the sludge to separate and suspend the fine particles (attrition) followed by the flotation process.

Figure 7.32 shows a diagram for a bench-scale flotation cell (Denver, 1991) used for characterization/treatability studies. The mechanical agitator entrains air down into the cell, which then bubbles up through the slurry, attaching to hydrophobic particles and creating a low density froth that is scraped off the top surface. Typically, this froth would contain the highest concentration of contaminant.

Column flotation cells are becoming more widely used in commercial milling operations. In flotation columns, as in conventional flotation cells, hydrophobic mineral particles attach to rising bubbles and are removed as concentrate from the cell overflow while hydrophilic particles pass through the process without attaching to a bubble and are removed with the cell underflow. Unlike conventional flotation cells, columns use no mechanical

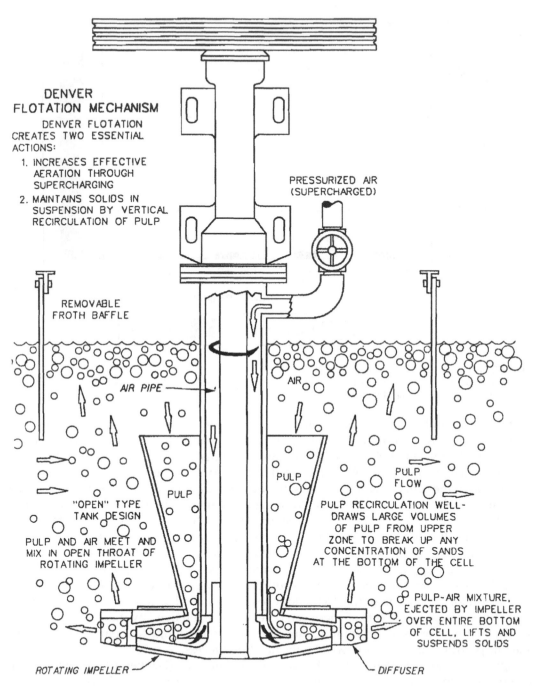

FIGURE 7.32
Bench-scale Denver flotation cell.

agitation, but an input of air-saturated water to suspend particles and disperse air. This makes them more energy efficient and less costly to maintain.

The simplicity, economics, and functional advantages of this external bubble generation system make it an attractive alternative to a conventional porous internal column sparger for commercial column flotation plants and laboratory column test facilities (Hill, 1988).

7.5.6.1   Sample Results for Application of Flotation to Contaminated Soil

Hazen Research, Inc. applied flotation at a western U.S. mine tailings site. Hazen investigated the application of froth flotation to remove sulfide minerals from the tailings to reduce the volume of sulfide material requiring fixation and produce clean non-sulfide soil for site remediation. Using bulk xanthate flotation, they successfully produced a soil that passed the EP toxicity test and a reduced volume of sulfides for fixation (Hansen, 1991).

The Dutch remediation firm, Heidemij Restsoffendiensten BV of Arnhem, the Netherlands, has also successfully applied froth flotation. At a plant in Moerdijk, near Rotterdam (November 1991), they first used vibrating screens to separate oversized material. The undersized material was slurried with water and sent to a hydrocyclone. The underflow was washed with surfactants and mechanically aerated. They observed that both organic and inorganic contaminants were enriched in the froth floating to the top. They reported that the underflow from the flotation was generally nonhazardous, while the froth was treated as a hazardous material (Mann and Besch, 1992).

In the early 1990s, the Great Lakes National Program Office of the U.S. EPA requested the U.S. Bureau of Mines to evaluate a number of mineral processing operations for the separation of metals-contaminated sediment (Allen, 1993). Results revealed the challenges to the practical applications of flotation. In this case, the flotation separations produced only small enrichment of contaminants. In the case of sediment from the Saginaw River (sediment #1) froth flotation was evaluated with xanthate and oleic acid at a range of pH. The best results (oleic acid at pH 7) produced a concentration of 31 to 73% of metallic contaminants in 18% of the original sediment. However, this enrichment was similar to that achieved with a size separation at 150 mesh. Real-world challenges were demonstrated with Indiana Harbor sediment. Froth flotation produced "voluminous, persistent froth," preventing selective separation of contaminants.

7.5.7   Other Technologies

Other properties may be exploited for separations. One promising approach uses high gradient magnetic separation (HGMS) to remove metal-containing particles. This is coupled with sulfate reducing bacteria which generates high surface area metal sulfides concentrating metals and organics. HGMS then removes and concentrates the metal-contaminated biomass from the original matrix, e.g., sediment (Allen, 1993; Watson et al. 1991).

7.6   Integrated Process Trains

Physical separations remediation generally adapts processes from the minerals processing unit operations. These processes exploit differences in particle size, density, surface, and other properties to effect a separation. Here, we shall present several examples of integrated process trains. Common features, criteria for planning, and economics will then be presented.

Working in concert with the U.S. Navy Civil Engineering Lab, the U.S. Bureau of Mines Salt Lake Research Center investigated physical separation of firing range soil. Based on mass and lead distribution relative to particle size, a conceptual process train was proposed (U.S. Bureau of Mines, 1991). The process (Figure 7.33) begins with a scrubbing trommel in which the soil flows into a rotating drum fitted with interior baffles and water spray. The rolling motion and the water condition, scrub, and declump the soil. The soil moves to the outlet where smaller material falls through a cyclindrical screen mounted around the

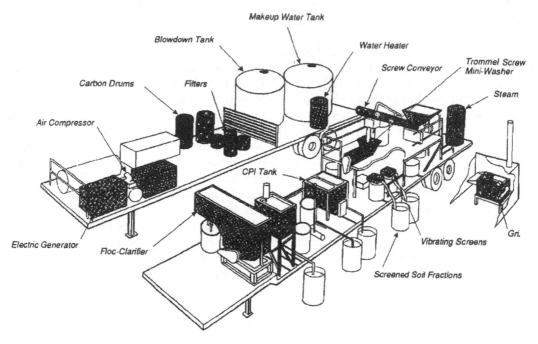

FIGURE 7.33
U.S. EPA Mobile Volume Reduction Unit (VRU).

mouth of the drum. The oversized material (e.g., gravel) rides to the edges of the screen and falls into a chute. Finer material moves on to other unit operations.

First-stage products (oversized and tailings) go on to secondary separation. Tailings might go to a "cleaning" or "concentrating" stage to concentrate contaminants into an even smaller volume. This approach can be taken if the contamination is preferentially associated with a distinct soil density fraction. A spiral concentrator is frequently used for this stage. As a soil/water slurry spirals down, the heavier soil fractions accumulate toward the inner radius and the less dense frac-tion moves toward the outer radius. The concentrate stream passes through the takeout ports. By the end of this stage, the soil has passed through separations based first on size and then on density. Further separations based on density difference may employ centrifuges or shaking tables. Differences in surface effects may be exploited with a flotation cell. A field demonstration based on these concepts was carried out at Camp Pendleton, CA (U.S. Navy, 1993).

7.6.1   Volume Reduction Unit (VRU)

A physical separation system, the mobile Volume Reduction Unit (VRU), was developed by the U.S. EPA Risk Reduction Engineering Laboratory (RREL). A demonstration test and an evaluation of the VRU technology were performed by the EPA under the Superfund Innovative Technology Evaluation (SITE) Program. This demonstration focused on the remediation of pentachlorophenol (PCP) and creosote at a wood products site. The physical condition of the wash water was modified to investigate a matrix of surfactant addition, pH adjustment, and temperature adjustment. Removal efficiencies of over 90% were achieved for the contaminants. Although this demonstration did not target metals, the system configuration and operational paradigm presented in the project report (U.S. EPA, 1993d) are highly relevant here.

Important features of this system were the following four stages:

- Soil preparation; screen out large objects
- Emission control (vapor) using activated carbon
- Soil washing process: mixed with water and extraction agents; sorted by size; large material cleaned for return to site; smaller material removed as sludge using flocculation with polymer or other chemical, followed by settling or by gravity alone
- Wastewater treatment

Other significant characteristics of the system included a capacity of 100 lb/h; trailer-mounted components providing mobility; connections among components allowing bypassing; steam heating for VOC stripping; complete wastewater treatment system consisting of clarification, sedimentation, flocculation, filtration, and carbon adsorption. Other unit operations, e.g., ion exchange can be added and can run at a wide range of soil-to-water ratios and with water at ambient to 150°F.

The VRU soil washing system successfully separated the contaminated soil into two unique streams: washed soil and fines slurry. The washed soil (86 to 90% of feed) was safely returned to the site following treatment. The fines slurry, which carried the majority of the pollutants from the feed soil, underwent additional treatment to separate the fines and contaminants from the water.

Figure 7.34 shows details of the VRU system. Note several key physical separation unit operations: grizzly, trommel screen, and vibrating screen. Also, very importantly, the system includes a flocclarifier, filters, and carbon drums for cleanup of effluent water and air streams.

7.6.2   Toronto Harbor Soil Recycle Treatment Train

A demonstration of the Toronto Harbor Commissioners (THC) Soil Recycle Treatment Train was performed under the Superfund Innovative Technology Evaluation (SITE) Program at a pilot plant facility in Toronto, Ontario, Canada (U.S. EPA, 1993c). The Soil Recycle Treatment Train, which consists of soil washing, biological treatment, and metals chelation, was designed to treat inorganic and organic contaminants in soil.

Soil from a site used for metals finishing and refinery and petroleum storage was processed in the pilot plant. The primary developer's claim to produce gravel and sand that met the THC target criteria for medium to fine soil suitable for industrial/commercial sites was achieved for the sand and gravel products. The fine soil from the biological treatment process exhibited anomalous oil and grease behavior and, although exhibiting a significant reduction in polynuclear aromatic hydrocarbon (PAH) compounds, did not meet the target level of 2.4 ppm for benzo(a)pyrene.

The soil washing process is diagrammed in Figure 7.34. The upper half illustrates the physical separations part of the system; the lower half, the metals removal part. The rotary trommel washer removes particles larger than 0.24 in. as a gravel fraction. The contaminated soil particles less than 0.24 in. and the wash water pass through the screen in the trommel washer into a holding tank where belt-type oil skimmers remove free oil from the water. The remaining soil and wash water are pumped through a separation hydrocyclone where the contaminated fines (less than 0.0025 in.) are separated from the coarser soil particles. Larger sand particles are easily separated from the fines, where the contaminants are concentrated. The fines are pumped to a lamellar separator and then to a gravity thickener, while the coarse sand is pumped to the attrition scrubbers.

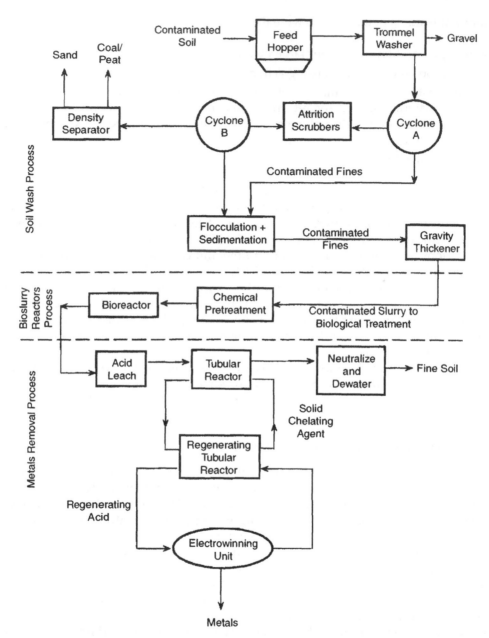

FIGURE 7.34
Toronto Harbor Commissioners (THC) process train for contaminated soil remediation.

Results showed that this process met targets for reducing total hydrocarbons for the gravel (1.97 to 0.024 in.) and the sand fractions (0.24 to 0.0025 in.). Although the metals contaminants in this soil were not high enough in concentration to require remediation, physical separation was observed as indicated in Table 7.4. These data are consistent with the often observed concentration of metals in the small diameter (high surface area) fractions of soil. In addition, a minor stream of coal and peat (only 1.6% of product), dramatically concentrated metal species. The feed contained 8200 mg/kg of oil and grease, while the coal/peat product stream had a concentration of 38,000 mg/kg. Although our focus here is metals removal, one must consider the negative impact of other contaminants on

TABLE 7.4

Toronto Harbor Attrition Soil Washing: Metals Concentrations in Feed and Product Streams

Contaminant/Stream	Feed Soil	Gravel	Coal/Peat	Sand	Fines
Copper (mg/kg)	18.3	6.4	32.9	13.8	83.1
Lead (mg/kg)	115	45.3	406	46	522
Zinc (mg/kg)	82.5	46	210	34.1	344

Adapted from U.S. Environmental Protection Agency, Remediation Technologies Screening Matrix and Reference Guide, EPA 542-B-93-005, prepared for the Office of Solid Waste and Emergency Response, U.S. Environmental Protection Agency and the Environics Directorate, U.S. Air Force, 1993b.

performance. In this demonstration, the metals removal process using acid and chelation appeared to succeed. However, this component of the test also experienced fouling from oil and grease and had to be cut short (U.S. EPA, 1993c).

7.6.3 Volume Reduction and Chemical Extraction System (VORCE)

The National Atmospheric Radiation and Engineering Lab in Montgomery, Alabama, used the finding of the bench-scale hydroclassification tests to design a trailer-mounted treatment system. The following photographs (Williford, 1991a) of the volume reduction and chemical extraction system (VORCE) provided a good view of several major components likely to be found on similar systems. These include the grizzly (Figures 7.7 and 7.8) for scalping large debris, the barrel trommel for disaggregation and screening of gravel (Figure 7.35), the settling tank (Figure 7.36) for removal of suspended fines, and the filter press (Figure 7.37) for concentration of these fines. Finally, Figure 7.38 shows an overall view, from left to right, of the system: grizzly with hopper, conveyor, trommel, hydrocyclones (closeup in Figure 7.24), and treatment water storage. This was a view of the system before addition of several improvements, including the enhancements in mass flow control, an enclosed screw conveyor, attrition scrubbers, and vibratory screens. After these improvements, the system comprised three trailer-mounted sections and had a capacity to treat from 1 to 2 tons of soil per hour.

7.6.4 Application of Physical Separations Systems

Successful physical separations depend on a number of factors including site and soil/contaminant characteristics, operational requirements, economics, and product residuals. Soil and contaminant characteristics and their implications have been previously addressed. Site and operational requirements will be similar to those for soil washing systems (U.S. EPA, 1990):

- Access roads
- Four acres for 20 ton/h system
- Utilities: water (130,000–800,000 gal/1000 tons); electricity; steam, and compressed air
- Site safety plan and personnel protecting and waste handling measures
- Control water infiltration at excavation to maintain consistent moisture content; use covers, dikes, and other runoff control; protect equipment from freezing
- Once chemical reagents are dissolved into the wash water, there should be no significant fire or explosion hazards

FIGURE 7.35
U.S. EPA Volume Reduction and Chemical Extraction System (VORCE) closeup of barrel trommel.

- Control noise and emissions to avoid nuisance as well as safety problems affecting nearby populations (places of work and residences)
- Storage capacity for cleaned material that is to be redeposited at the site (as well as for material to be removed from the site)
- Analytical capability, either with rapid turnover or onsite (beneficial) to verify attainment of required cleanup

Process residuals include contaminated soils, wastewater, wastewater treatment sludges, and air emissions. Disposition of treated product streams has a significant effect on process economics. A major rationale for physical separations is the simplified disposal, preferably onsite, of "cleaned" or depleted residuals with little or no further treatment. Requirements for transport offsite and for additional treatment, e.g., stabilization will, to the extent applied, negate the advantages of separation.

Cost estimates vary widely. A bulletin (U.S. EPA, 1990) cites vendor-supplied costs ranging from $50 to $205/ton of feed soil for a soil washing process. The upper end estimate includes costs for soil residue disposal. The costs are related to soil washing, because almost all the equipment and operational requirements are in common, except for those directly related to the chemical agents used for extraction in washing. A more thorough, and probably more realistic, analysis was performed in connection with U.S. EPA Volume Reduction Unit (VRU). In this case the cost to remediate 20,000 tons of contaminated soils using a 10-ton-per-hour (tph) soil washer was estimated at $136.67 per ton when the system was online 90% of the time. A breakdown of project component costs for this scenario

FIGURE 7.36
U.S. EPA Volume Reduction and Chemical Extraction System (VORCE) closeup of settling tank.

is available (U.S. EPA 1993d). A matrix of characteristics affecting treatment cost or performance vs. technologies is provided in Appendix D of the Remediation Technologies Screening Matrix and Reference Guide 2.0s (U.S. EPA, 1998d).

7.7   Summary

Heavy metals contamination is pervasive and represents a significant threat to human and ecological health. Unlike organic contaminants, metals do not degrade but remain onsite until transported away, either in an uncontrolled manner or as a result of a planned remediation. The major options for site remediation are to stabilize the metal contaminants in place, excavate and dispose of in a secure landfill, or separate the contaminants. The choice depends on site and contaminant characteristics, acceptable risks, and finally costs.

Physical separation is an adaptation of mineral processing equipment and methods used to deplete soil fractions of contaminant, simplifying or eliminating the need for treatment. This is designed to improve remedial operations and economics. Physical separation should not be seen as a distinct choice but as an option that should be integrated into soil remediation systems to the extent required for risk reduction and improved economics. For

FIGURE 7.37
U.S. EPA Volume Reduction and Chemical Extraction System (VORCE) closeup of filter press.

FIGURE 7.38
U.S. EPA Volume Reduction and Chemical Extraction System (VORCE) overall view of process train.

example, lead slugs have been separated from firing range soil, thus removing the majority of the mass of the lead. Residual soil was stabilized to prevent leaching.

Key elements for successful application of physical separation involved the determination of soil and contaminant characteristics, particularly the distribution into size and density fractions and liberation (debonding) from coatings. Soil size distribution is a decisive parameter. Clay-sized materials present problems for handling, mixing, and removal of suspended solids from residual streams. Small bench- and pilot-scale treatability studies are required to determine the above characteristics and to support design and configuration of a process train for a specific remediation project. In a research context, these treatability studies are needed to build the empirical knowledge base on the separation of soil and contaminant combinations.

References

Allen, J.P., Mineral Processing of Contaminated Sediment — Part 1, Draft Report, prepared by U.S. Bureau of Mines, Salt Lake Research Center, Salt Lake City, UT, for Great Lakes Program Office, Chicago, IL, 1993.

Amdur, M.O., J. Doull, and C.D. Klassen, *Casarett and Doull's Toxicology: The Basic Science of Poisons,* Pergamon Press, New York, 1991.

Averett, D.E., B.D. Perry, E.J. Torrey, and J.A. Miller, Review of Removal, Containment, and Treatment Technologies for Remediation of Contaminated Sediment in the Great Lakes, Miscellaneous paper EL-90-25, USACE Waterways Experiment Station, Vicksburg, MS, 1990.

Bricka, R., C. Williford, and L. Jones, Technology Assessment of Currently Available and Developmental Techniques for Heavy-Metals-Contaminated Soils Treatment, Technical Report IRRP-93-4, U.S. Army Corps of Engineers, Waterways Experiment Station, Vicksburg, MS, 1993.

Bricka, R.M., C.W. Williford, and L.W. Jones, Heavy Metal Soil Contamination at U.S. Army Installations: Proposed Research and Strategy for Technology Development, Technical Report IRRP-94-1, U.S. Army Engineer Waterways Experiment Station, Vicksburg, MS, 1994.

Briggs, D.J., *Soils,* Butterworths, London, 1977.

Buser, W. and F. Graf, Differenzierung von Mangan (II)-Manganit and delta-MnO_2 Durch Oberflachenmessung Nach Brunauer-Emmet-Teller, *Helv. Chim. Acta,* 38, 830, 1955.

Calmano, W. and U. Forstner, Chemical extraction of heavy metals in polluted river sediments in Central Europe, *Sci. Total Environ.,* 28, 77, 1983.

Denver Equipment Catalog, Colorado Springs, CO, 1991.

Forstner, U. and G. Wittmann, *Metal Pollution in the Aquatic Environment,* 2nd ed., Springer-Verlag, New York, 1981.

Fripiat, J. and M. Gastuche, Etude Physiochimique des Surfaces des Argiles – Les Combinaisons de la Kaolinite avec des Oxides de fer Trivalent, *Publ. Inst. Natl. Etude Agron. Congo, Belge,* 54, 7, 1952.

Goldberg, E., Marine geochemistry. I. Chemical scavengers of the sea, *J. Geol.,* 62, 249, 1954.

Green, D.W., Ed., *Perry's Chemical Engineers Handbook,* 6th ed., McGraw-Hill, New York, 1984.

Grim, R., *Clay Mineralogy,* 2nd ed., McGraw-Hill, New York, 1968.

Hanson, B.J., What the Hazardous Waste Industry Can Learn from the Mineral Processing Industry, presented at Hazpac '91, April 1991, Cairns Queensland, by Randol International Ltd., Golden, CO, 1991.

Hay, S., W.S. Richardson, and C. Cox, Wayne Interim Storage Site: Particle Size Fractionation and Radionuclide Characterization of Soil, a preliminary report for the Environmental Protection Agency and New Jersey Department of Environmental Protection, prepared by the U.S. EPA National Air and Radiation Environmental Lab, Montgomery, AL, 1991.

Hill, S.D., Column Flotation, U.S. Bureau of Mines, Salt Lake City Research Center, Salt Lake, UT, 1988.

Hirner, A., K. Kritsotakis, and H. Tobschall, Metal-organic associations in sediments. I. Comparison of unpolluted recent and ancient sediments affected by anthropogenic pollution, *Appl. Geochem.*, 5, 491, 1990.

Horowitz, A.J., *A Primer on Sediment-Trace Element Chemistry*, 2nd ed., Lewis Publishers, Chelsea, MI, 1991.

Horowitz, A.J. and K. Elrick, The Relation of stream sediment surface area, grain size, and composition to trace element chemistry, *Appl. Geochem.*, 2, 437, 1987.

Jones, B. and C. Bowser, *The Mineralogy and Related Chemistry of Lake Sediments, Lakes: Chemistry, Geology, Physics*, Lerman, A., Ed., Springer-Verlag, New York, 1978, 179.

Klimpel, R.R., Froth flotation, in *Encyclopedia of Physical Science and Technology*, 6, Academic Press, New York, 1992.

Krauskopf, K., Factors controlling the concentration of thirteen rare metals in sea water, *Geochim. Cosmochim. Acta*, 9, 1, 1956.

Lieser, K.H., Sorption and Filtration Methods for Gas and Water Purification, Series E, Bonnevie-Svendsen, Ed., 13, NATO Advanced Study Institute, 1975.

Manahan, S.E., *Hazardous Waste Chemistry, Toxicology, and Treatment*, Lewis Publishers, Boca Raton, FL, 1990.

Mann, M. and J. Besch, Divide and conquer, *Soils*, March, 20, 1992.

MARCOR Remediation, Inc., Discussions with Derek Rhodes of MARCOR, Hunt Valley, MA , 1997.

Marino, M.A., R.M. Bricka, and C.N. Neale, Heavy metal soil remediation: the effects of attrition scrubbing on a wet gravity process, *Environ. Progress*, 16(3), 208, 1997.

McCabe, W.L., J.C. Smith, and P. Harriott, *Unit Operations of Chemical Engineering*, 5th ed., McGraw-Hill, New York, 1993.

McCarter, W.A., Placer recovery, in *Yukon Placer Mining Industry 1978–1982*, R.L. Debicki, Ed., Exploraton and Geological Services, Northern Affairs Program, Indian and Northern Affairs Canada, presented at the D.I.A.N.D.-K.P.M.A. Placer Mining Short Course, Whitehorse, Yukon, 1982.

Mular, A.L. and R.B. Bhappu, *Mineral Processing Plant Design*, 2nd ed., Books on Demand, Ann Arbor, MI, 1980.

Pryor, E.J., *Mineral Processing*, 3rd ed., American Alsevier Publishing Co., Inc., New York, 1965.

Silva, M., Placer Mining Gold Recovery Methods, Special Publication 87, California Department of Conversation, Division of Mines and Geology, Sacramento, CA, 1986.

U.S. Bureau of Mines, Heavy Metal Removed from Small Arms Firing Ranges, R. McDonald, Ed., prepared for U.S. Naval Civil Engineering Research Laboratory, Salt Lake City Research Center, Salt Lake City, UT, 1991.

U.S. Department of Agriculture, *Soil Conservation Service, Soil Taxonomy: A Basic System of Soil Classification for Making and Interpreting Soil Surveys*, Robert E. Krieger Publishing Company, Malabar, FL, 1988.

U.S. Department of Defense, The DoD Environmental Scholarships/Fellowships and Grants Program, Application Package, Woodbridge, VA, 1993.

U.S. Department of Defense, Cleanup Pillar, http://iridium.nttc.edu/env/dod/grnbkclp.html, 1997.

U.S. EPA, Superfund Treatability Study Protocol: Bench-Scale Level of Soils Washing for Contaminated Soils, Interim Report, 1989a.

U.S. EPA, Innovative Technology: Soil Washing, OSWER Directive 9200.5-250FS, 1989b.

U.S. EPA, Soil Washing Treatment, Eng. Bull., Office of Emergency and Remedial Response, EPA/540/2-90/017, 1990.

U.S. EPA, Characterization Protocol for Radioactive Contaminated Soils, publication 9380.1-10FS, Office of Solid Waste and Emergency Response, 1991.

U.S. EPA, Remediation Technologies Screening Matrix and Reference Guide, EPA 542-B-93-005, prepared for the Office of Solid Waste and Emergency Response, U.S. Environmental Protection Agency and the Environics Directorate, U.S. Air Force, 1993b.

U.S. EPA, Innovative Treatment Technologies: Annual Status Report 5th ed., EPA 542-R-93-003, No. 5, prepared for the Office of Solid Waste and Emergency Response, U.S. Environmental Protection Agency, Washington, D.C., 1993b.

U.S. EPA, Toronto Harbor Commissioners (THC), Soil Recycle Treatment-Technology Demonstration Summary, EPA/540/SR-93/517, 1993c.

U.S. EPA, EPA RREL's Mobile Volume Reduction Unit, Applications Analysis Report, EPA/540/AR-93/508, Office of Research and Development, Washington, D.C., 1993d.

U.S. EPA, Innovative Treatment Technologies Annual Status Report, 6th ed., http://www.epa.gov/bbsnrmrl/attic/dldocs/ittasr6n.html, 1994.

U.S. EPA, Federal Facilities Restoration & Reuse Office, http://www.epa.gov/swerffrr/sitemaps.htm, 1998a.

U.S. EPA, Remediation Technologies Screening Matrix and Reference Guide Version 2.0, Section 2.6.1.1, Metals, http://www.frtr.gov/matrix/section2/2_6_1_1.html, 1998b.

U.S. EPA, Remediation Technologies Screening Matrix and Reference Guide Version 2.0, Section 2.6.1.2, Radionuclides, http://www.frtr.gov/matrix/section2/2_6_1_2.html, 1998c.

U.S. EPA, Remediation Technologies Screening Matrix and Reference Guide Version 2.0, Table of Contents, http://www.frtr.gov/matrix/section1/toc.html, 1998d.

U.S. Navy Civil Engineering Laboratory and U.S. Bureau of Mines, Heavy Metal Removal from Small Arms Ranges: A Pilot-Scale Demonstration at Marine Corps Base, Camp Pendleton, CA, 1993.

Watson, J.H.P., D.C. Ellwood, E.I. Hamilton, and J.B. Mills, The removal of heavy metals and organic compounds from anaerobic sludges, in *Proc. Congress and Treatment of Sludge*, Gen, Belgium: Technological Institute of the Royal Flemish Society of Engineers, 1991.

Webster's New World Dictionary, 2nd College ed., Simon and Schuster, New York, 1980.

Weiss, N.L., ed., *SME Mineral Processing Handbook*, Society of Mining Engineers, New York, 1985.

Williford, C.W., Trip report (and photographs) of Volume Reduction and Chemical Extraction System (VORCE), National Air and Radiation Environmental Laboratory, Montgomery, AL, prepared for U.S. Army Corps of Engineers (WES-EE-R: R.M. Bricka), Vicksburg, MS, 1991a.

Williford, C.W., Trip report (and photographs) of AWC Lockheed TRUClean process, prepared for Restoration Branch, U.S. Army Corps of Engineers (WES-EE-R: R.M. Bricka), Vicksburg, MS, 1991b.

Williford, Jr., C.W., Z. Li, Z. Wang, and R.M. Bricka, Vertical column hydroclassification of metal-contaminated soils, *J. Hazardous Mater.*, 66, 1, 1999.

Wills, B.A., Gravity separation. 1 and 2, *Mining Mag.*, October, 325, 1984.

8

Heavy Metals Extraction by Electric Fields

Akram N. Alshawabkeh and R. Mark Bricka

CONTENTS

8.1   Introduction

In situ remediation of heavy metal-contaminated fine-grained soils, such as silt and clay, is often hindered by low hydraulic conductivities. The resistance of such soils to hydraulic flow and their high sorption potential limit the success of *in situ* techniques that use hydraulic gradients. However, effective electroosmotic flow in clays and silts provides an option for *in situ* extraction of heavy metals using electric fields. Heavy metals transport by electroosmosis, enhanced by their migration to the opposite polarity electrode, is the basis of electrokinetic remediation, an innovative *in situ* cleanup technology. Electrodes are inserted in fully or partially saturated soil and a direct electric current is applied to produce an electric field. Ambient or introduced solutes move in response to the imposed electric field by electroosmosis and ionic migration. Electroosmosis mobilizes the pore fluid to flush solutes, usually from the anode (positive electrode) toward the cathode (negative electrode), while ionic migration effectively separates anionic (negative ions) and cationic (positive ions) species, drawing them to the anode and cathode, respectively. Because the process requires the presence of solutes, geochemical reactions including sorption, precipitation, complexation, and dissolution play a significant role in enhancing or retarding electrokinetic remediation.

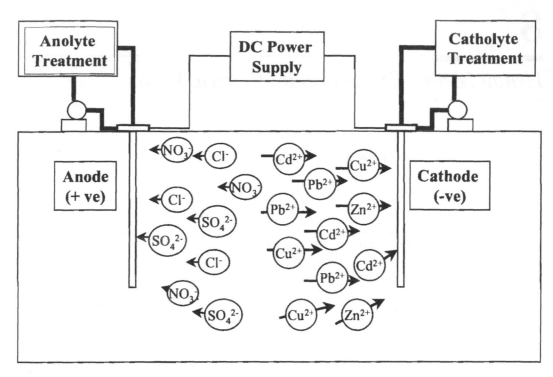

FIGURE 8.1
Schematic of field implementation of electrokinetic remediation.

Electrokinetic remediation can clean up sites contaminated with heavy metals as well as organics. Extraction of heavy metals is accomplished by pumping catholyte (cathode electrolyte) and anolyte (anode electrolyte), electroplating, precipitation/co-precipitation, or ion exchange either at the electrodes or in an external extraction system. The major advantages of the technology are that (1) it can be implemented *in situ* with minimal disruption, (2) it is well suited for fine-grained, heterogeneous media, where other techniques such as pump-and-treat may be ineffective, and (3) accelerated rates of contaminant transport and extraction can be obtained. A schematic of field implementation of the technique is displayed in Figure 8.1. The topics that are discussed in this chapter include principles of heavy metals transport under electric fields, electrolysis and geochemical reactions, process enhancement and conditioning, a review of recent findings and implementations of the technique, and a discussion of design requirements for *in situ* implementation.

8.2   Heavy Metals Transport under Electric Fields

Two major heavy metal transport mechanisms occur in soft soils (silt and clay) under electric fields: electroosmosis and ion migration. Electroosmosis is one of several electrokinetic phenomena that develop because of the presence of particle surface charge and the diffuse double layer. Discrete clay particles usually have a negative surface charge that influences and controls the particle environment. The net negative charge on the clay particle surfaces requires an excess positive charge (or exchangeable cations) distributed in the fluid zone adjacent to the clay surface forming the double layer. The quantity of these exchangeable

Anode (+)　　　　　　　　　　　　　　　　　　**Cathode (−)**

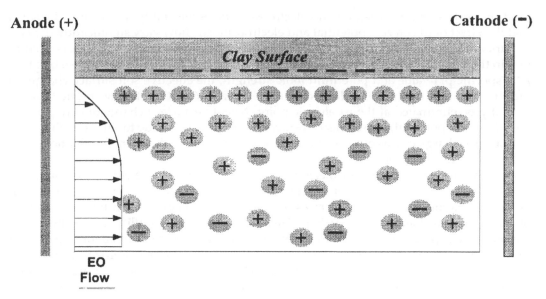

EO Flow

FIGURE 8.2
Electroosmosis — fluid movement with respect to clay particle surface.

cations required to balance the charge deficiency of clay is termed the cation exchange capacity (CEC), and is expressed in milliequivalents per 100 g of dry clay. Several theories have been proposed for modeling charge distribution adjacent to clay surface. The Gouy-Chapman diffuse double layer theory has been widely accepted and applied to describe clay behavior. A detailed description of the diffuse double layer theories for a single flat plate is found in Hunter (1981), Stumm (1992), and Mitchell (1993).

Electroosmosis is fluid movement with respect to clay particle surface as a result of applied electric potential gradients (Figure 8.2). The role of electroosmosis is significant in electrokinetic soil remediation, particularly under high water content and low ionic strength conditions. Several theories describe and evaluate water flow by electroosmosis, including Helmholtz-Smoluchowski theory, Schmid theory, Spiegler friction model, and ion hydration theory. Descriptions of these theories are given in Gray and Mitchell (1967) and Mitchell (1993). Helmholtz-Smoluchowski model is the most common theoretical description of electroosmosis and is based on the assumption of fluid transport in the soil pores because of transport of the excess positive charge in the diffuse double layer toward the cathode. The rate of electroosmotic flow is controlled by the coefficient of electroosmotic permeability of the soil, k_e ($L^2 T^{-1} V^{-1}$), which is a measure of the fluid flux per unit area of the soil (all formulations are provided based on a unit area of the soil, not the pore space) per unit electric gradient, where L is length, T is time, and V is electric voltage. The advective component of contaminant transport due to electroosmosis is given by

$$J_i^c = c_i k_e i_e \qquad (1)$$

where J_i^c is the rate of mass transport of contaminant (or species) i by electroosmosis per unit area (M $L^{-2} T^{-1}$); c_i is the concentration of species i (M L^{-3}); i_e is the electric gradient (V L^{-1}); and M is mass. The value of k_e is assumed to be a function of the zeta potential of the soil-pore fluid interface (which describes the electrostatic potential resulting from the soil surface charge), the viscosity of the pore fluid, soil porosity, and soil electrical permittivity. West and Stewart (1995) and Vane and Zang (1997) investigated the effect of pore fluid properties on zeta potential and electroosmostic permeability. The results displayed that the effect of pH on zeta potential and electroosmostic flow vary significantly depending upon the mineral type. Lockhart (1983) demonstrated that high electrolyte concentration in the pore fluid causes strong electrolyte polarization that limits electroosmotic flow. At a specific pH value and pore fluid ionic strength, the effective soil surface charge can drop to zero and reach the isoelectric point (Lorenz, 1969). The electroosmotic flow can virtually be eliminated at the isoelectric point. Negative surface charge of clay particles (negative zeta potential) causes electroosmosis to occur from anode to cathode while positive surface charge causes electroosmosis to occur from cathode to anode (Eykholt, 1992; Eykholt and Daniel, 1994).

The other important transport mechanism in soil under electric fields is ion migration, which is the transport of charged ions in the pore fluid toward the electrode opposite in polarity. Ions migrate at different rates in an electrolyte because of differences in their physicochemical characteristics such as size and charge. Ionic mobility defines the rate of migration of a specific ion under a unit electric field. The term is modified for migration in soils to "effective" ionic mobility in order to account for effective soil porosity and tortuosity. Rates of contaminant extraction and removal from soils by electric fields are dependent upon the values of the effective ionic mobilities of contaminants, and are given by

$$J_i^m = c_i \, u_i^* \, i_e \tag{2}$$

where J_i^m is the rate of mass transport of species i by ion migration per unit area ($M \, L^{-2} \, T^{-1}$), and u_i^* is the effective ionic mobility of species i ($L^2 \, T^{-1} \, V^{-1}$). Heavy metal ionic mobilities at infinite dilution are in the range of $10^{-4} \, cm^2 \, V^{-1} \, s^{-1}$. Accounting for soil porosity and tortuosity, the effective ionic mobilities are in the range of 10^{-4} to $10^{-5} \, cm^2 \, V^{-1} \, s^{-1}$, which cause heavy metals transport in clays at a rate of few centimeters per day under a unit electric gradient ($1 \, V \, cm^{-1}$).

Contaminant transport under electric fields can also be enhanced by hydraulic gradients. In heterogeneous soils, combined electric and hydraulic gradients can be used to produce uniform transport. While electroosmosis carries contaminants through silt and clay layers, an equivalent flow under hydraulic gradient carries contaminants through sand layers. Mass transport due to hydraulic gradients is simply calculated by

$$J_i^h = c_i \, k_h \, i_h \tag{3}$$

where J_i^h is the advective component of species i mass flux ($M \, L^{-2} \, T^{-1}$), k_h is the hydraulic conductivity of the soil ($L \, T^{-1}$), and i_h is the hydraulic gradient (dimensionless). Transport processes will also be affected, to a lesser extent, by hydrodynamic dispersion (mechanical dispersion and molecular diffusion).

A schematic of mass transport profiles of cationic and anionic species is provided in Figure 8.3. Transport profiles in Figure 8.3 are based on the assumptions that water advection components (electroosmosis and hydraulic) act from the anode to the cathode. The advective flow enhances transport of cations, which migrate from anode to cathode, and retards transport of anions, which migrate from cathode to anode. For a given time period (ΔT), cations will travel a net distance (X_{net}) given by

$$X_{net} = X_h + X_e + X_m \tag{4}$$

where X_h is distance traveled due to the hydraulic gradient ($X_h = k_h \, i_h \, \Delta T$), X_e is distance traveled due to electroosmosis ($X_e = k_e \, i_e \, \Delta T$), and X_m is distance traveled due to the migration ($X_m = u^* \, i_e \, \Delta T$). On the other hand, anions will travel a net distance given by

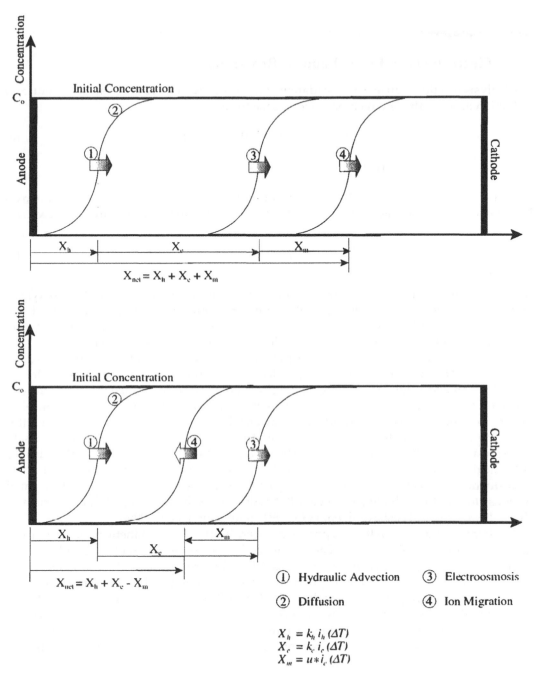

FIGURE 8.3
Schematic of mass transport profiles of cationic and anionic species.

$$X_{net} = X_h + X_e - X_m \qquad (5)$$

The difference between cations and anions transport is that the migrational components act in opposite directions.

8.3   Electrolysis and Geochemical Reactions

Electrolysis reactions cause water oxidation at the anode which produces an acid front, and reduction at the cathode which produces a base front:

$$2H_2O - 4e^- \Rightarrow O_2 \uparrow + 4H^+ \quad \text{(anode)} \tag{6}$$

$$4H_2O + 4e^- \Rightarrow 2H_2 \uparrow + 4OH^- \quad \text{(cathode)}$$

Rates of acid and base production depend upon the current density. Based on Faraday's law of equivalence of mass and charge, rate of ions production at the electrodes is given by

$$J_i = \frac{I_d}{z_i F} \tag{7}$$

where J_i is the mass flux per unit area of ion i (hydrogen ion at the anode and hydroxyl ion at the cathode), $M\,L^{-2}\,T^{-1}$; I_d is the current per unit area or (current density amp L^{-2}); z_i is the charge of ion i, and F is Faraday's constant (96,485 C mol^{-1}). Current densities used in electrokinetic remediation are usually in the order of few amps per square meter.

Within a few hours of processing, anode pH drops to around two and cathode pH increases to above ten. The rate of pH change is dependent upon the electric current and electrode volume. If no amendments (or enhancement agents) are used to neutralize water electrolysis reactions, the acid advances through the soil toward the cathode by ionic migration and electroosmosis, and the base initially advances toward the anode by ionic migration and diffusion. The counterflow due to electroosmosis (from anode to cathode) retards the back-diffusion and migration of the base front. The advance of this front is slower than the advance of the acid front also because the ionic mobility of H^+ is about 1.76 times that of OH^-. As a consequence, the acid front dominates the chemistry across the specimen except for small sections close to the cathode (Acar et al., 1990; Alshawabkeh and Acar, 1992; Probstein and Hicks, 1993; Acar and Alshawabkeh, 1993, 1994; Yeung and Datla, 1995).

Geochemical reactions in the soil pores significantly affect electrokinetic remediation and can enhance or retard the process. These geochemical reactions are highly dependent upon the pH condition generated by the process. The advance of the acid front from anode toward the cathode assists in desorption and dissolution of metal precipitates. However, formation of the high pH zone near the cathode results in immobilization to precipitation of metal hydroxides. Complexation can reverse the charge of the ion and reverse direction of migration. Limitations of electrokinetic remediation caused by high catholyte pH require innovative methods to enhance the technique and control immobilization and complexation of metals close to the cathode.

8.4   Enhancement Conditions

Catholyte pH can be controlled by neutralizing hydroxyl ions produced by electrolysis using weak acids or catholyte rinsing. The advantages of using weak acids are that (1) they form soluble metal salts, (2) their low solubility and migration rates will not cause a significant

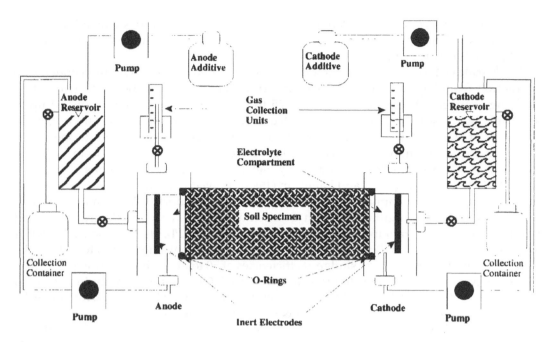

FIGURE 8.4
Typical bench-scale setup.

(orders of magnitude) increase in electric conductivity of the soil, and (3) they are biodegradable and, if properly selected, environmentally safe. However, improper selection of some acids may pose a health hazard. For example, the use of hydrochloric acid may pose a health hazard because (1) it may increase the chloride concentration in the groundwater, (2) it may promote the formation of some insoluble chloride salts, e.g., lead chloride, and (3) if it reaches the anode compartment, chlorine gas may be generated by electrolysis. Another procedure to control hydroxyl ions and enhance metals transport toward the cathode is the use of membranes. Ion selective membranes, which are impermeable to hydroxyl ions, can be used to separate the catholyte from the soil and thus prevent or minimize the transport of hydroxyl ions into the soil. These membranes are insoluble in most solvents and chemically resistant to strong oxidizing agents and strong bases.

Under certain circumstances, such as soils with high buffering capacity, the use of enhancement agents to solubilize the contaminants without acidification is necessary for cost-effective implementation. Chelating or complexing agents, such as citric acid and EDTA, have been demonstrated to be feasible for the extraction of different types of metal contaminants from soils. The enhancement agents should form charged soluble complexes with the metal contaminants.

8.5 Recent Developments

Several bench-scale studies during the late 1980s and early 1990s showed the potential of using electric fields for extraction of heavy metals from soils. Figure 8.4 shows a typical bench-scale setup. The setup usually holds a small soil sample in the range of 10 cm in diameter and 10 to 40 cm in length. Inert electrodes are placed in compartments filled with water (or electrolytes) and separated from the soil using filters or fabrics. Amendment solutions are usually supplied to the electrode compartments using pumps (when enhancement procedures are used).

Bench-scale tests conducted by Hamed (1990) and Hamed et al. (1991) demonstrated lead extraction from kaolinite at various concentrations below and above the soil cation exchange capacity. The process removed 75 to 95% of lead at concentrations of up to 1500 mg/kg across test specimens at reported energy expenditure of 29 to 60 kWh/m^3 of soil. Acar et al. (1994) demonstrated 90 to 95% removal of Cd^{2+} from kaolinite specimens with initial concentration of 99 to 114 mg/kg. However, because no enhancement procedure was used, these studies showed heavy metals accumulation at sections close to the cathode. Lageman et al. (1989) and Lageman (1993) showed that the process can migrate a mixture of different contaminants in soil. Lageman (1993) reported 73% removal of Pb at 9000 mg/kg from fine argillaceous sand, 90% removal of As at 300 mg/kg from clay, and varying removal rates ranging between 50 and 91% of Cr, Ni, Pb, Hg, Cu, and Zn from fine argillaceous sand. Cd, Cu, Pb, Ni, Zn, Cr, Hg, and As at concentrations of 10 to 173 mg/kg also were removed from a river sludge at efficiencies of 50 to 71%. The energy expenditures ranged between 60 and 220 kWh/m^3 of soil processed. Other laboratory studies reported by Runnels and Larson (1986), Eykholt (1992), and Acar et al. (1993) further substantiate the applicability of the technique to a wide range of heavy metals in soils.

Pamukcu and Wittle (1992) and Wittle and Pamukcu (1993) demonstrated removal of Cd^{2+}, Co^{2+}, Ni^{2+}, and Sr^{2+} from different soil types at variable efficiencies. The results showed that kaolinite, among different types of soils, had the highest removal efficiency followed by sand with 10% Na-montmorillonite, while Na-montmorillonite showed the lowest removal efficiency. The results indicated that soils of high water content, high degree of saturation, low ionic strength, and low activity (soil activity describes soil plasticity and equals plasticity index divided by % clay by dry weight) provide the most favorable conditions for transport of contaminants by electroosmotic advection and ionic migration. Highly plastic soils such as illite, montmorillonite, or soils that exhibit high acid/base buffer capacity require excessive acid and/or enhancement agents to desorb and solubilize contaminants before they can be transported through the subsurface and removed (Alshawabkeh et al., 1997), thus requiring excessive energy.

Runnells and Wahli (1993) showed the use of ion migration combined with soil washing for removal of Cu^{2+} and SO_4^{2-} from fine sand. A field study reported by Banerjee et al. (1990) also investigated the feasibility to use electrokinetics in conjunction with pumping to decontaminate a site from chromium. Although soil chromium profiles were not evaluated in this study, the results showed an increase in effluent chromium concentrations.

Hicks and Tondorf (1994) indicated that development of a pH front could cause isoelectric focusing, which retards ion transport under electric fields. They showed that this problem can be prevented simply by rinsing away the hydroxyl ions generated at the cathode. They demonstrated 95% zinc removal from kaolinite samples by using the catholyte rinsing procedure. Acar and Alshawabkeh (1996) showed extraction of lead at 5300 mg/kg from pilot-scale kaolinite samples. Alshawabkeh et al. (1997) studied electrokinetic extraction of heavy metals from clay samples retrieved from a contaminated army ammunition site. The soil contained calcium at 19,670 mg/kg; iron at 11,840 mg/kg; copper at 10,940 mg/kg; chromium at 9,930 mg/kg; zinc at 6,330 mg/kg; and lead at 1990 mg/kg. High calcium concentration hindered extraction of the metals. However, the results further showed that metals with higher initial concentration, less sorption affinities, higher solubilities, and higher ionic mobilities are transported and extracted faster than other metals. Rødsand et al. (1995) and Puppala et al. (1997) demonstrated that neutralization of the cathode reaction by acetic acid can enhance electrokinetic extraction of lead. Rødsand et al. (1995) and Puppala et al.

(1997) also showed that using membranes at the cathode has limited success in enhancing electrokinetic remediation. The reason is that heavy metals accumulate and precipitate on these membranes, resulting in a significant increase in the electrical resistivity of membrane. Unless these membranes are continuously rinsed and cleaned, the energy cost of this technique will substantially increase. Cox et al. (1996) demonstrated the feasibility of using iodine/iodide lixivant to remediate mercury-contaminated soil. The use of EDTA as an enhancement agent has also been demonstrated for the removal of lead from kaolinite (Yeung et al., 1996) and lead from sand (Wong et al., 1997). Reddy et al. (1997) showed that soils that contain high carbonate buffers, such as glacial till, hinder the development and advance of the acid front. Reddy et al. (1997) also demonstrated that presence of iron oxides in glacial till creates complex geochemical conditions that retard Cr(VI) transport. On the other hand, the study showed that presence of iron oxides in kaolinite and Na-montmorillonite did not seem to significantly impact Cr(VI) extraction.

With regard to radionuclides contamination, Ugaz et al. (1994) displayed that uranium at 1000 pCi/g of activity is efficiently removed from bench-scale kaolinite samples. A yellow uranium hydroxide precipitate was found in sections close to the cathode. Enhanced electrokinetic processing showed that 0.05 M acetic acid was enough to neutralize the cathode reaction and overcome uranium precipitation in the soil. Other radionuclides such as thorium and radium showed limited removal (Acar et al., 1992a). In the case of thorium, it was postulated that precipitation of these radionuclides at their hydroxide solubility limits at the cathode region formed a gel that prevented their transport and extraction. Limited removal of radium is believed to be either due to precipitation of radium sulfate or because radium strongly binds to the soil minerals causing its immobilization (Acar et al., 1992a).

It should be mentioned that electric fields are also effective for the removal of organic pollutants such as phenol, gasoline hydrocarbons, and TCE from contaminated soils. Successful application of the process has been demonstrated for extraction of the BTEX (benzene, toluene, ethylene, and m-xylene) compounds and trichloroethylene from kaolinite specimens at concentrations below the solubility limit of these compounds (Bruell et al., 1992; Segall and Bruell, 1992). High removal efficiencies of phenol and acetic acid (up to 94%) were also achieved by the process (Shapiro et al., 1989; Shapiro and Probstein, 1993). Acar et al. (1992b) reported removal of phenol from saturated kaolinite by the technique. Two pore volumes were sufficient to remove 85 to 95% of phenol at an energy expenditure of 19 to 39 kWh/m^3. Wittle and Pamukcu (1993) investigated the feasibility of removal of organics from kaolinite, Na-montmorillonite, and sand samples. Their results showed the transport of acetic acid and acetone toward the cathode. Samples mixed with hexachlorobenzene and phenol showed accumulation at the center of each samples. The results of some of these experiments were inconclusive, either because contaminant concentrations were below detection limits or because the samples were processed for only 24 h, which might not be sufficient to demonstrate any feasibility in electrokinetic soil remediation. Recently, the Department of Energy (DOE), Environmental Protection Agency (EPA), Monsanto, General Electric, and Dupont have also applied electric fields for electroosmotic extraction using layered horizontal electrodes or the Lasagna process (DOE, 1996). Ho et al. (1997) reported 98% removal efficiency of p-nitrophenol, as a model organic compound, from soil in a pilot-scale study using the Lasagna process. Although removal of free phase nonpolar organics is questionable, Mitchell (1991) stated that this could be possible if they would be present as small bubbles (emulsions) that could be swept along with the water moving by electroosmosis. Acar et al. (1993) stated that unenhanced electrokinetic remediation of kaolinite samples loaded up to 1000 mg/kg hexachlorobutadiene has been unsuccessful. However, Acar et al. (1993) reported that hexachlorobutadiene transport was detected only when surfactants were used.

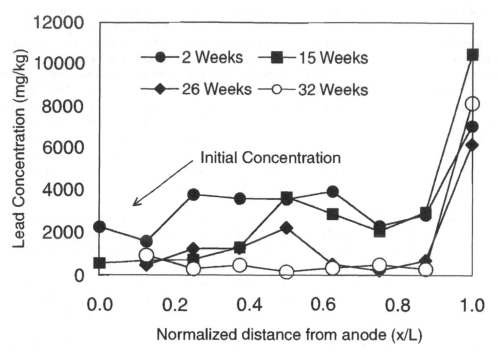

FIGURE 8.5
Lead profiles in one of the pilot tests for electrokinetic remediation.

8.6 Field Demonstrations

Several field demonstrations of electrokinetic remediation are being conducted by Electro-kinetics, Inc. (EK Inc., Baton Rouge, LA) with collaboration and support from the Environ-mental Laboratory (EL) U.S. Army Corps of Engineers Waterways Experiment Station (Vicksburg, MS). A pilot-scale study was conducted on enhanced removal of lead from firing range soil. The study treated 1.5-ton samples of clayey sandy soil contaminated with lead at concentrations in the range of 3500 mg/kg. Electrode spacings of 90 and 180 cm were used. Figure 8.5 shows lead profiles in one of the pilot tests after 2, 15, 26, and 32 weeks of processing. The figure shows lead transport front that moves at a rate in the range of 0.4 to 1.4 cm/day. Final analysis demonstrated lead reduction to less than 400 mg/kg. EK Inc. and WES followed the pilot-scale study by a field demonstration of the technology at an army firing range site. Electrode spacings of 150 cm are being used at a current density of 3.0 amp/m². WES is also involved in an Environmental Security Technology Certification Program (ESTCP) project to demonstrate electrokinetic extraction of chromium (up to 14,000 mg/kg) and cadmium (up to 1,900 mg/kg) from one half acre, tidal marsh site containing two waste pits at Naval Air Weapons Station, Point Mugu, CA. Although the study is not completed, over 80% or treated soil sections now have chromium and cadmium concentrations below detection limits.

Sandia National Laboratory (SNL) in Albuquerque, NM, reported successful field dem-onstration of removal of chromium (VI) from unsaturated soil (moisture content in the range of 2 to 12% by weight) beneath the SNL Chemical Waste Landfill (CWL) (Lindgren et al., 1998). The study reported removal of 600 g of Cr(VI) after 2700 h of processing. Other field demonstrations include extraction of uranium from Oak Ridge K-25 Facility (Oak Ridge, TN) being conducted by Isotron Corporation (New Orleans, LA) and supported by the Department of Energy Office of Technology Development. Isotron Corporation and Westinghouse Savannah River Company reported difficulties in mercury extraction, but good transport of lead and chrome in a field demonstration at Old TNX Basin, Savannah River Site (South Carolina). Lageman (1993) also reported successful demonstrations by Geokinetics International, Inc., in Europe for *in situ* and *ex situ* electrokinetic extraction of metals and organics. Field studies have also been conducted for extraction and treatment of soils contaminated with organics. The Lasagna™ process was used to treat an area of 14 m² up to a depth of 5 m at the Paducah Gaseous Diffusion Plant (PGDP), Paducah, KY. The process reduced trichloroethylene (TCE) concentration in the soil (tight clay) from the 100 to 500 parts per million (ppm) range to an average concentration of 1 ppm (DOE, 1996).

8.7 Theoretical Modeling

Theoretical formulation of the process result in a system of partial differential equations for transport of solutes coupled with nonlinear algebraic equations for geochemical reactions. The system is quite complex and numerical stability becomes a critical issue when dealing with large number of species (e.g., more than 20) in two-dimensional and three-dimensional problems. There are attempts to model contaminant transport under electrical gradients in limited one-dimensional and two-dimensional conditions. In most of these models, the number of species is limited to 10. Shapiro et al. (1989) and Shapiro and Probstein (1993) described a one-dimensional model for species transport under electric fields. The model accounted for ion diffusion, migration, and electroosmotic advection in predicting species transport rate. The model assumed incompressible soil medium, constant hydraulic head distribution, and steady-state electroosmostic flow. Water electrolysis reactions were used to calculate constant flux boundary conditions for hydrogen ion at the anode and hydroxyl ion at the cathode. The results were compared with the experiments for the case of acetic acid extraction with constant voltage at the boundaries. Geochemical reactions included were first-order sorption and water and acetic acid dissociation. Comparisons showed good agreement in one case of acetic acid removal from 0.4-m-length kaolinite sample. Jacobs et al. (1994) enhanced the Shapiro and Probstein (1993) model to predict one-dimensional transport of zinc under electric fields. The model accounted for zinc precipitation and dissolution reactions and demonstrated the role of background ion concentrations on the process. Jacobs and Probstein (1996) further modified the code to model two-dimensional species transport under electric fields. They applied the two-dimensional code for the case of electroosmotic extraction of phenol from kaolinite. The model solved three PDEs for transport of phenol, sodium ion, and chloride ion. Hydrogen and hydroxyl ion concentrations were calculated using zero net charge equation and water equilibrium equation. Limited number of species and geochemical reactions were incorporated (water and phenol dissociation) because of the complex two-dimensional simulation of the process.

Mitchell and Yeung (1991) proposed a model in a study of the feasibility of using electrical gradients to retard or stop migration of contaminants across earthen barriers. Principles of irreversible thermodynamics were employed and a one-dimensional model was developed for transport of contaminants across the liner. The model reasonably predicted the transport of sodium and chloride ions across the liner. Geochemical reactions were not incorporated in this model. Eykholt (1992) presented an attempt to model the pH distribution during the process using mass conservation equation accompanied by empirical relations to account for the nonlinearity in the parameters controlling the process. One transport differential equation was formed assuming that hydrogen and hydroxyl ions have the same diffusion coefficients and ionic mobilities. In this model, the development of negative pore water pressure was modeled using a modified Smoluchoweski equation, as described by Anderson and Idol (1986). The complexity in electrical potential distribution was modeled using proposed empirical relations. Haran et al. (1997) presented a one-dimensional model for extraction of hexavalent chromium from soils using electric fields. The model accounted for transport of H^+, OH^-, CrO_4^{-2}, K^+, Na^+, and SO_4^{-2}. Geochemical reactions included sorption (described by a retardation coefficient) and water equilibrium.

Acar et al. (1988, 1989) presented a one-dimensional model to estimate pH distribution during electrokinetic soil processing. The model demonstrated the impact of electrolysis reactions on pH distribution during electrokinetic remediation. Alshawabkeh and Acar (1992) described a modified formulation and presented a system of differential/algebraic equations for the process that accounted for adsorption/desorption, precipitation/dissolution, and acid/base reactions. Acar and Alshawabkeh (1994) modeled the change in soil and effluent pH during electrokinetic soil processing. This attempt assumed linear electric and hydraulic gradients throughout the process and disregarded the coupling of these components. Alshawabkeh and Acar (1996) and Acar and Alshawabkeh (1996) enhanced the model and modified the code for stimulating reactive extraction of heavy metals by electric fields. The model predicted reactive transport of hydrogen, lead, hydroxyl, and nitrate ions. The model accounted for lead hydroxide precipitation-dissolution, lead sorption (assuming linear pH-dependent isotherm), and water equilibrium reactions.

8.8 Practical Considerations

Studies on practical aspects of electrokinetic remediation are rare. Schultz (1997) provided an economic modeling and calculations of optimum spacings, time and energy requirements of one-dimensional field applications based on electroosmotic transport. Alshawabkeh et al. (1999) discussed practical aspects of one-dimensional full-scale *in situ* applications. These studies also address optimum conditions for one-dimensional applications.

8.8.1 Electrode Requirements

Electrodes could be placed in one-dimensional or two-dimensional configurations, which affect the total number of electrodes, time, energy, and extent of remediation. One-dimensional configurations differ depending upon spacing between same-polarity electrodes (Figure 8.6). Decreasing spacing between same-polarity electrodes minimizes the area of inactive electric field but increases the cost of the process. In two-dimensional configurations, the goal is to achieve axisymmetrical (radial) flow toward a center electrode. For extraction of positively charged heavy metals, it is recommended that the cathode be the center electrode, which allows accumulation of the cationic contaminants in a smaller zone around the cathode. Outer electrodes (anodes) are placed at specific distances from the center cathode to achieve a relatively radial flow. The electrodes can be placed in a hexagonal, square, or triangular configuration (Figure 8.7). Hexagonal (honeycomb) electrode configuration consists of cells, each containing a cathode surrounded by six anodes. The square configuration consists of a cathode and four (or possibly eight) anodes surrounding the cathode. Similarly, a triangular arrangement consists of one cathode surrounded by three anodes.

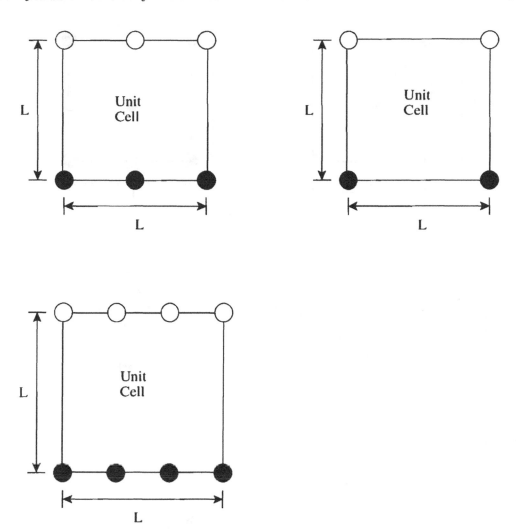

○ Anode   ● Cathode

FIGURE 8.6
One-dimensional configurations.

One way of calculating electrode requirements is to evaluate the number of electrodes based on a unit surface (plane) area. For each configuration, the number of electrodes per unit area is calculated considering a unit cell,

$$N = \left[\frac{F_1}{L_E^2}\right]_{1D} = \left[\frac{F_1}{\pi R_E^2}\right]_{radial} \tag{8}$$

where N is the number of electrodes per unit surface area of the site (L^{-2}); L_E is the anode-cathode spacing in one-dimensional applications (L); R_E is the radial distance between the electrodes in two-dimensional applications (L); and F_1 is a factor depending on electrode configuration. One-dimensional configurations with same-polarity electrode spacing of half and one third anode-cathode spacing require 100 and 200% increase in number of electrodes,

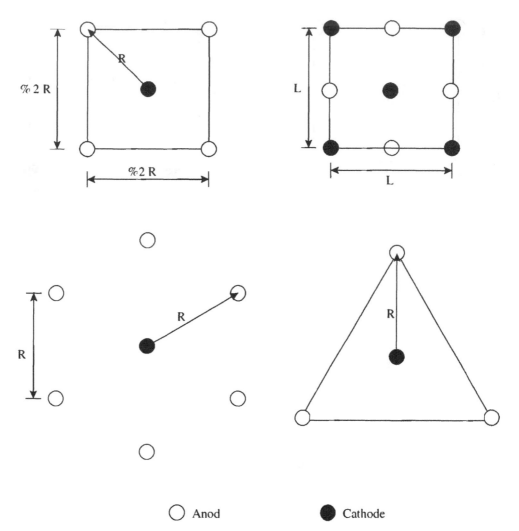

FIGURE 8.7
Square, hexagonal, and triangular configurations.

respectively, when compared to the one-dimensional case of equal electrode spacings.
Hexagonal configuration requires 15% increase in number of electrodes when compared
to two-dimensional square configuration. Configurations with more electrodes produce
more uniform electric fields and minimize the areas with ineffective electric fields.

8.8.2   Electric Field Distribution

Mathematical models for contaminant transport under electric fields assume nonlinear
one-dimensional electric field distributions (due to nonuniform ionic strength) for predicting species transport (Alshawabkeh and Acar, 1992, 1996; Shapiro and Probstein, 1993;
Jacobs et al., 1996). Similar procedure can be used for two-dimensional applications. One
could also assume uniform steady state conditions and use the Laplace equation to
describe the two-dimensional electric field distribution. Numerical methods, such as finite
element and finite difference, can be used for solving Laplace equation for different
boundary conditions (or electrode configuration). Theory of functions is another option

for solving the Laplace equation. However, these solutions provide electric field distribution but do not provide a mechanism for comparing the effectiveness of different configurations.

8.8.3 Remediation Time Requirements

An important aspect for design of *in situ* electrokinetic remediation is the time required for cleanup. The duration of the remediation process is a function of contaminant transport rate and electrode spacing. As electroosmotic advection and ionic migration are the prominent transport mechanisms, hydrodynamic dispersion can be neglected to simplify the analysis. Assuming linear one-dimensional electric field distribution and a homogeneous soil medium, the rate of species transport can be calculated by

$$v = \frac{(u_i^* + k_e)i_e}{R_{dc}} \tag{9}$$

where v = rate of species transport (or velocity, $L\,T^{-1}$). R_{dc} is a delaying factor (dimensionless) to account for the time required for contaminant desorption and dissolution (R_{dc} is similar to the retardation factor in advection-dispersion contaminant transport, but accounts also for other chemical reaction such as precipitation). The value of R_{dc} depends on soil type, pH, and type of contaminant. Sorption retardation factor can be used as an estimate of R_{dc} ($R_{dc} = 1$ for nonreactive contaminants). If enhancement agents are used to solubilize heavy metals, this factor should be modified accordingly.

If the spacing between electrodes of opposite polarity is chosen to be L_E, the time (T_E) required for remediation can be estimated simply by L_E/v or

$$T_I = \frac{1}{\beta} \frac{L_E}{\sigma^* i_e} \tag{10}$$

where T_E is the time required for cleanup (T), σ^* is the effective electric conductivity of the soil medium (siemens L^{-1}), and β is a soil parameter to calculate the reactive transport rate of a species relative to the electric conductivity of a medium, which is given by

$$\beta = \frac{(u_i^* + k_e)/R_{dc}}{\sigma^*} \tag{11}$$

β is a lumped property of the contaminant and the soil ($L^3\,C^{-1}$). β is similar to transference number in electrochemistry but also account for electroosmosis, soil conditions, and retardation caused by geochemical reactions. Typical values of β for contaminated fine-grained soils are estimated to be in the range of 1×10^{-9} to 1×10^{-6} m^3/C.

If time is to be calculated using current density, Equation 10 becomes

$$T_I = \frac{1}{\beta} \frac{L_E}{I_d} \tag{12}$$

where I_d is the electric current density $= I/A$ (amp L^{-2}), I is the total current (amp), and A is the cross-sectional area of the soil treated (L^2).

8.8.4   Cost

The total costs for full-scale *in situ* implementation of electrokinetic remediation can be divided into five major components (Alshawabkeh et al., 1999): (1) costs for fabrication and installation of electrodes, (2) cost of electric energy, (3) cost of enhancement agents, if necessary, (4) costs of any post-treatment, if necessary, and (5) fixed costs. Impacts of electrode configuration and spacing on these cost components are addressed separately.

Cost of electrodes depends upon the number of electrodes per unit surface area and is given by

$$C_{electrode} = C_1 N \tag{13}$$

where $C_{electrode}$ is electrode costs per unit volume of soil to be treated ($ L^{-3}); N is number of electrodes per unit surface area of soil (L^{-2}), which is given by Equation 8; and C_1 is cost of electrode per unit length ($ L^{-1}). C_1 generally includes the unit costs for material and fabrication of the electrode, drilling and preparation of the borehole, placement of the electrode, and any other related materials, membranes, or fabrics required for electrode installations. Increasing electrode spacings decreases the value of N and hence decreases total electrode costs.

Factors that impact energy requirements and cost for electrokinetic remediation at a specific site include soil and contaminant properties, electrode configuration, and processing time. Energy consumption changes during processing because of changes in electric conductivity. However, energy calculations could by simplified by averaging soil electrical conductivity throughout the process. Accordingly, energy expenditure per unit volume of contaminated soil is given by the following equation:

$$W = \frac{\phi_{max} I_d T_E}{L_E} \tag{14}$$

where W is energy expenditure per unit volume of soil (J L^{-3}) and ϕ_{max} is the maximum voltage applied across the electrodes (V). Substituting Equation 10 into Equation 14 results in the following equation:

$$W = \frac{\phi_{max}}{\beta}. \tag{15}$$

Equation 15 indicates that energy requirement could be considered independent of electrode spacings if energy source (maximum voltage) is the controlling factor. In other words, two one-dimensional schemes (for one specific site) with different spacings should result in same energy expenditure if same total voltage was used in both schemes. The difference between the two schemes would be in time requirements. However, electrode configuration is a design factor if the energy source is not the limiting factor.

Based on energy expenditure calculation, energy cost can be estimated by

$$C_{energy} = \frac{C_2 \phi_{max}}{3,600,000\beta} \tag{16}$$

where C_{energy} is electric energy cost per unit volume of soil treated ($ L^{-3}) and C_2 is electric energy cost ($ kWh^{-1}). It is evident from Equation 13 that energy cost is highly dependent upon both the soil and contaminant characteristics. A high coefficient of electroosmotic permeability, k_e, or a high ionic mobility of the contaminant, u_i^*, will increase the value of β and reduce the energy expenditure and cost. High contaminant concentrations or high ionic strength of the pore fluid will increase the electrical conductivity of the soil, reduce the value of β, and thus increase the energy expenditure. An increase in the delaying factor R_{dc} due to complex geochemical reactions will also increase energy expenditure.

Cost of enhancement agents and chemicals that are used to improve the efficiency of electrokinetic remediation is a significant component of the total cost. Chemicals are used for either neutralizing pH conditions or enhancing solubility of target contaminants or both. Therefore, chemicals cost is divided into two components. The cost of chemicals required for pH neutralizing will depend upon electric current applied and is given by the following equation:

$$C_{n-chemical} = C_3 \frac{I_d}{L} \frac{M_W}{\alpha F} T_E$$ (17)

where $C_{n-chemical}$ is the cost of chemicals required to neutralize electrolytes per unit soil volume (\$ L^{-3}), C_3 is the cost of the chemical agent (\$ M^{-1}), M_W is the molecular weight of the neutralizing chemical, and α is a factor depending upon the stoichiometry of the neutralizing reaction (dimensionless). Substituting the time required for remediation results in the following equation:

$$C_{n-chemical} = \frac{C_3 M_w}{\beta} \frac{}{\alpha F}$$ (18)

Equation 18 shows that chemicals cost is independent of electric current or spacing and is dependent on soil characteristics. This is because electric current and electrode spacings affect time requirements. For example, increasing the current will decrease the time required for remediation, such that the same total charge is introduced for any electric current value.

Post-treatment costs should also be considered if effluent treatment is required. These costs are highly site- and contaminant-specific. An estimate of effluent treatment costs could be evaluated per unit volume of the soil as follows:

$$C_{post-treat} = \frac{C_4 k_e i_1}{(L_E/n)} T_E$$ (19)

where $C_{post-treat}$ is the post-treatment cost per unit volume of the soil (\$ L^{-3}) and C_4 is the cost of treatment per unit volume of the electrolyte (effluent) collected (\$ L^{-3}). Substituting for the value of T_E (time required for remediation), effluent treatment cost is given by

$$C_{post-treat} = C_4 \frac{n k_e}{\beta \sigma^*} = \frac{C_4 n R_{dc} k_e}{\sigma^* + k_e}.$$ (20)

Volume and cost of effluent treatment depends on the ratio of transport under electro-osmosis relative to total transport rate. In order to minimize the volume collected, it is necessary to maximize transport by ionic migration and minimize transport by electro-osmosis. If contaminant transport occurs only because of migration, then this cost component will be zero and one needs only to treat electrolyte in electrode well. However, if electroosmosis is the only mechanism used for contaminant transport (e.g., for noncharged contaminants) then the cost of treatment will be equal to ($C_4 n R_{dc}$), indicating that the cost depends upon the number of pore volumes required for remediation. If the contaminant is readily available for transport, then $R_{dc} = 1$ and one pore volume is enough for remediation. However, if extraction is retarded by geochemical reactions, then it is obvious that the pore volumes required will increase depending upon the value of R_{dc}. Sometimes catholyte recycling is used, thus adding another component that should be considered for evaluation of total volume of water collected.

Other costs for full-scale implementation include mobilization and demobilization costs of various equipment, site preparation, security, progress monitoring, insurance, labor, contingency, and miscellaneous expenses. These cost components are divided into fixed (e.g., mobilization and demobilization) and variable (e.g., monitoring, insurance, rentals) components. Variable costs are simply evaluated by multiply cost rate by the total time required for remediation, i.e.,

$$C_{variable} = \frac{C_5}{\beta} \frac{L_E}{\sigma^* i_e} \qquad (21)$$

where $C_{variable}$ is the total variable cost per unit soil volume ($\$ L^{-3}$) and C_5 is the variable cost rate per unit soil volume ($\$ L^{-3} T^{-1}$). C_5 is evaluated by estimating the variable daily cost (for monitoring, insurance, rentals, etc.) and dividing by the total volume of site. C_5 is highly dependent upon the size of the site and decreases as volume of contaminated soil increases.

The total cost per unit volume of soil to be treated is thus given by

$$C_{total} = C_{electrode} + C_{energy} + C_{chemical} + C_{post-treat} + C_{fixed} + C_{variable}$$

where C_{total} is the total cost per unit volume of soil to be treated ($\$ L^{-3}$) and C_{fixed} is the fixed cost per unit volume of soil to be treated ($\$ L^{-3}$). Cost evaluation indicates that electrode configuration will impact electrode, energy, and variable costs. Other costs (chemicals, fixed, and post-treatment) are independent of electrode configuration and spacing.

References

Acar, Y.B. and A.N. Alshawabkeh, Principles of electrokinetic remediation, *Environ. Sci. Technol.*, 27(13), 2638, 1993.

Acar, Y.B. and A.N. Alshawabkeh, Modeling conduction phenomena in soils under an electric current, in *Proc. XIII Int. Conf. Soil Mechanics and Foundation Eng.* (ICSMFE), New Delhi, India, 1994.

Acar, Y.B. and A.N. Alshawabkeh, Electrokinetic remediation: I. Pilot-scale tests with lead-spiked kaolinite, *ASCE J. Geotechnical Eng.*, 122(3), 173, 1996.

Acar, Y.B., R.J. Gale, G. Putnam, J.T. Hamed, and I. Juran, Determination of pH Gradients in Electrochemical Processing of Soils, report presented to the Board of Regents of Louisiana, Civil Engineering Department, Louisiana State University, 1988.

Acar, Y.B., R.J. Gale, G. Putnam, and J.T. Hamed, Electrochemical processing of soils: its potential use in environmental geotechnology and significance of pH gradients, in *2nd Int. Symp. Environ. Geotechnology*, Shanghai, China, May 14–17, Envo Publishing, Bethlehem, PA, 1, 25, 1989.

Acar, Y.B., R.J. Gale, G.A. Putnam, J.T. Hamed, and R.L. Wong, Electrochemical Processing of soils: theory of pH gradient development by diffusion, migration, and linear convection, *J. Environ. Sci. and Health*, A25(6), 687, 1990.

Acar Y.B., R.J. Gale, A. Ugaz, and S. Puppala, Feasibility of Removing Uranium, Thorium, and Radium from Kaolinite by Electrochemical Soil Processing, report prepared for the Office of Research and Development, Risk Reduction Engineering Laboratory, USEPA, Report No. EK-BR-009-092, Electrokinetics, Inc., Baton Rouge, LA, 1992a.

Acar, Y.B., H. Li, and R.J. Gale, Phenol removal from kaolinite by electrokinetics, *ASCE J. Geotech. Eng.*, 118(11), 1837, 1992b.

Acar, Y.B., A.N. Alshawabkeh, and R.J. Gale, Fundamentals of extracting species from soils by electrokinetics, *Waste Manage.*, 13(2), 141, 1993.

Acar, Y.B., J.T. Hamed, A.N. Alshawabkeh, and R.J. Gale, Cd(II) removal from saturated kaolinite by application of electrical current, *Géotechnique*, 44(3), 239, 1994.

Alshawabkeh, A.N. and Y.B. Acar, Removal of contaminants from soils by electrokinetics: a theoretical treatise, *J. Environ. Sci. Health*, A27(7), 1835, 1992.

Alshawabkeh, A.N. and Y.B. Acar, Electrokinetic remediation: II. Theory, *ASCE J. Geotechnical Eng.*, 122(3), 186, 1996.

Alshawabkeh, A.N., S.K. Puppala, Y.B. Acar, R.J. Gale, and R.M. Bricka, Effect of solubility on enhanced electrokinetic extraction of metals, in *Proc. of In Situ Remediation of the Geoenvironment* (In Situ Remediation '97), Minneapolis, MN, October 5–8, 1997.

Alshawabkeh, A.N., A. Yeung, and R.M. Bricka, Practical aspects of in situ electrokinetic remediation, *ASCE J. Environ. Eng.*, 125(1), 1999.

Anderson, J.L. and W.K. Idol, Electroosmosis through pores with nonuniformly charged walls, *Chem. Eng. Commun.*, 38, 93, 1986.

Banarjee, S., J. Horng, J. Ferguson, and P. Nelson, Field-Scale Feasibility of Electrokinetic Remediation, report presented to USEPA, Land Pollution Control Division, PREL, CR 811762-01, 1990, 122p.

Bruell, C.J., B.A. Segall and M.T. Walsh, Electroosmotic removal of gasoline hydrocarbons and TCE from clay, *ASCE J. Environ. Eng.*, 118(1), 68, 1992.

Cox, C.D., M.A. Shoesmith, and M.M. Ghosh, Electrokinetic remediation of mercury-contaminated soils using iodine/iodide lixivant, *Environ. Sci. Technol.*, 30(6), 1933, 1996.

DOE, Lasagna™ Soil Remediation: Innovative Technology Summary Report, Innovative Technology Summary Report prepared for U.S. Department of Energy, Office of Environmental Management, Office of Science and Technology, 1996. http://www.em.doe.gov/plumesfa/intech/lasagna

Eykholt, G.R., Driving and Complicating Features of the Electrokinetic Treatment of Contaminated Soils, Ph.D. dissertation, Department of Civil Engineering, University of Texas at Austin, 1992.

Eykholt, G.R. and D.E. Daniel, Impact of system chemistry on electroosmosis in contaminated soil, *ASCE J. Geotech. Eng.*, 120(5), 797, 1994.

Gray, D.H. and J.K. Mitchell, Fundamental aspects of electroosmosis in soils, *ASCE J. Soil Mechanics and Foundation Eng. Div.*, 93(6), 209, 1967.

Hamed, J., Decontamination of Soil Using Electroosmosis, Ph.D. dissertation, Louisiana State University, Baton Rouge, 1990.

Hamed, J., Y.B. Acar, and R.J. Gale, Pb(II) removal from kaolinite using electrokinetics, *ASCE J. Geotech. Eng.*, 117(2), 241, 1991.

Haran, B.S., B.N. Popov, G. Zheng, and R.E. White, Mathematical modeling of hexavalent chromium decontamination from low surface charged soil, electrochemical decontamination of soil and water, Special Issue of *J. Hazardous Mater.*, 55 (1-3), 93, 1997.

Hicks, R.E. and S. Tondorf, Electrorestoration of metal contaminated soils, *Environ. Sci. Technol.*, 28(12), 2203, 1994.

Ho, S.V., C.J. Athmer, P.W. Sheridan, and A.P. Shapiro, Scale-up aspects of the Lasagna™ process for in situ soil decontamination, electrochemical decontamination of soil and water, Special Issue of *J. Hazardous Mater.*, 55(1–3), 39, 1997.

Hunter, R.J., *Zeta Potential in Colloid Science*, Academic Press, New York, N.Y., 1981.

Jacobs, R.A. and R.F. Probstein, Two-dimensional modeling of electromigration, *AICHE J.*, 42(6), 1685, 1996.

Jacobs, R.A., M.Z. Sengun, R.E. Hicks, and R.F. Probstein, Model and experiments on soil remediation by electric fields, *J. Environ. Sci. Health*, A29(9), 1933, 1994.

Lageman, R., Electro-reclamation: application in the Netherlands, *Environ. Sci. Technol.*, 27(13), 2638, 1993.

Lageman, R., W. Pool, and G. Seffinga, Electro-reclamation: theory and practice, *Chem. Ind.*, 18, 585, 1989.

Lindgren, E.R., M.G. Hankins, E.D. Mattson, and P.M. Duda, Electrokinetic Demonstration at the Unlined Chromic Acid Pit, Report Abstract, SAND97-2592, Sandia National Laboratory, 1998.

Lockhart, N.C., Electroosmotic dewatering of clays: I, II, and II, *Colloids and Surfaces*, 6, 238, 1983.

Lorenz, P.B., Surface conductance and electrokinetic properties of kaolinite beds, *Clays and Clay Minerals*, 17, 223, 1969.

Mitchell, J.K., Conduction phenomena: from theory to geotechnical practice, *Géotechnique*, 41(3), 299, 1991.

Mitchell, J.K., *Fundamentals of Soil Behavior*, 2nd ed., John Wiley & Sons, New York, 1993.

Mitchell, J.K. and T.C. Yeung, Electrokinetic flow barriers in compacted clay, *Transportation Research Record*, No. 1288, Soils Geology and Foundations, 1991.

Pamukcu, S. and J.K. Wittle, Electrokinetic removal of selected heavy metals from soil, *Environ. Progress AIChE*, 11(4), 241, 1992.

Probstein, R.F. and R.E. Hicks, Removal of contaminants from soil by electric fields, *Science*, 260, 498, 1993.

Puppala, S.K., A.N. Alshawabkeh, Y.B. Acar, R.J. Gale, and R.M. Bricka, Enhanced electrokinetic remediation of high sorption capacity soils, electrochemical decontamination of soil and water, Special Issue of *J. Hazardous Mater.*, 55(1–3), 203, 1997.

Reddy, K.R., U.S. Parupudi, S.N. Devulapalli, and C.Y. Xu, Effect of soil composition on removal of chromium by electrokinetics, electrochemical decontamination of soil and water, Special Issue of *J. Hazardous Mater.*, 55(1–3), 135, 1997.

Rødsand, T., Y.B. Acar, and G. Breedveld, Electrokinetic extraction of lead from spiked Norwegian marine clay, in *Characterization, Containment, Remediation, and Performance in Environmental Geotechnics*, Geotechnical Special Publications No. 46, ASCE, New York, 2, 1518, 1995.

Runnels, D.D. and J.L. Larson, A Laboratory study of electromigration as a possible field technique for the removal of contaminants from ground water, *Ground Water Monitoring Rev.*, Summer, 81, 1986.

Runnells, D.D. and C. Wahli, In situ electromigration as a method for removing sulfate, metals, and other contaminants from ground water, *Ground Water Monitoring & Remediation*, 13(1), 121, 1993.

Schultz, D.S., Electroosmosis technology for soil remediation: laboratory results, field trial and economic modeling. electrochemical decontamination of soil and water, Special Issue of *J. Hazardous Mater.*, 55(1–3), 81, 1997.

Segall, B.A. and C.J. Bruell, Electroosmotic contaminant removal processes, *ASCE J. Environ. Eng.*, 118(1), 84, 1992.

Shapiro, A.P. and R.F. Probstein, Removal of contaminants from saturated clay by electroosmosis, *Environ. Sci. Technol.*, 27(2), 283, 1993.

Shapiro, A.P., P.C. Renaud, and R.F. Probstein, Preliminary studies on the removal of chemical species from saturated porous media by electroosmosis, *PhysicoChem. Hydrodyn.*, 11(5/6), 785, 1989.

Stumm, W., *Chemistry of the Solid-Water Interface, Processes at the Mineral-Water and Particle-Water Interface in Natural Systems*, John Wiley & Sons, New York, 1992.

Ugaz, A., S.K. Puppala, R.J. Gale, and Y.B. Acar, Electrokinetic soil processing: complicating features of electrokinetic remediation of soils and slurries: saturation effects and the role of the cathode electrolysis, *Chem. Eng. Commun.*, 129, 183, 1994.

Vane, M.L. and G.M. Zang, Effect of aqueous phase properties on clay particle zeta potential and electroosmostic permeability: implications for electrokinetic remediation processes, electrochemical decontamination of soil and water, Special Issue of *J. Hazardous Mater.*, 55(1–3), 1, 1997.

West, L.J. and D.I. Stewart, Effect of zeta potential on soil electrokinesis, in *Characterization, Containment, Remediation, and Performance in Environmental Geotechnics*, Geotechnical Special Publication No. 46, ASCE, New York, 2, 1535, 1995.

Wittle, J.K. and S. Pamukcu, Electrokinetic Treatment of Contaminated Soils, Sludges, and Lagoons, Final Report, DOE/CH-9206, Argonne National Laboratory, Chicago, IL.

Wong, J.S., R.E. Hicks, and R.F. Probstein, EDTA-enhanced electroremediation of metal contaminated soils, electrochemical decontamination of soil and water, Special Issue of *J. Hazardous Mater.*, 55(1–3), 61, 1997.

Yeung, A.T. and S. Datla, Fundamental formulation of electrokinetic extraction of contaminants from soil, *Can. Geotech. J.*, 32(4), 569, 1995.

Yeung, A.T., C. Hsu, and R.M. Menon, EDTA-enhanced electrokinetic extraction of lead, *ASCE J. Geotech. Eng.*, 122(8), 666, 1996.

Section II

Biological Methods and Processes

9

The Relationships between the Phytoavailability and the Extractability of Heavy Metals in Contaminated Soils

Lenom J. Cajuste and Reggie J. Laird

CONTENTS

9.1   Introduction

Since the beginning of this century, with the era of intensive Mexican industrialization there has been considerable interest in applying the wastewater collected from the metropolitan area of Mexico City to croplands of the Mezquital Valley. This municipal wastewater has been recognized as a valuable source of plant nutrients such as N, P, and minor elements that favor adequate crop growth (Cajuste et al., 1991; Juste and Mench, 1992). In spite of these beneficial effects, there is concern about the potential hazard associated with the consumption of the edible portion of crops grown in this area because of metal phytotoxicity or transfer of the metal into the food chain (Chang, 1984; Adriano, 1986; Rappaport et al., 1988; Miller et al., 1995). Various chemical extractants such as the EDTA method (Haq et al., 1980), the DTPA soil test of Lindsay and Norvell (1978), and the AB-DTPA procedure (Soltanpour, 1985) have been proposed to evaluate the phytoavailability of heavy metals from contaminated soils.

TABLE 9.1

Some Heavy Metals Regarded as Environmental Pollutants

Element	Sources	Concentration Range (mg kg⁻¹)
Cd	Burning of coal ash	1–200
	Water pipes	—
	Smelting	6,000–80,000
Ni	Oil (residual/diesel)	—
	Coal ash	3–1,300
	Gasoline additives	—
Pb	Coal ash	10–7,000
	Motor vehicle, leaded gasoline	—
	Battery reclamation	2.16–43,700 (sediment)
		2–135,000 (soil)
		0–140 mg L⁻¹ (waters)
	Battery recycling	210–75,850 (soil)
Cr	Wood preserving, tanneries, mining	—
	Chromium production	500–70,000

Adapted from Peters, R.W. and L. Shem, Treatment of soils contaminated with heavy metals, in *Metal Speciation and Contamination of Soils*, H.A. Allen, C.P. Huang, G.W. Bailey, and A.R. Bowers, Eds., Lewis Publishers, Boca Raton, FL, 1995.

9.2 Background

9.2.1 Sources of Soil Contamination

The contamination of soil can derive from distinct activities such as (1) industrial operations, (2) agricultural activities, and (3) agricultural domestic and urban activities. Some environmental contaminants and their associated sources are presented in Table 9.1.

As a consequence of the adverse effects of the technical civilization of the world, soil as a component of the biosphere has been viewed as a natural buffer system that controls the fate of chemical elements in the environment.

Heavy metals are common sources of soil contaminants; they are also present in natural waters because of natural processes or man's activities. In this aqueous phase they do not stay under the soluble form for a long time; they are rather present as suspended colloids or they are fixed by organic material and mineral substances (Kabata-Pendias and Pendias 1992). The anthropogenic sources of trace elements in waters are associated with mining of coal and mineral ores, and manufacturing and municipal wastewater operations (Förstner, 1995).

Most of the regional contamination of the soil originates from industrial regions and urban areas where factories, motor vehicles, and municipal waste are the common sources of heavy metals. Furthermore, fertilizers, pesticides, and sewage sludge constitute other important sources of heavy metals in the soil (Table 9.2). Effects of sewage sludge application on soil composition are of great environmental concern because of their heavy cumulative load, which tends to increase the soils' levels of Ni, Cr, Pb, Zn, Cu, and Hg because of long-term use (Chaney, 1978; Henry and Harrison, 1992).

TABLE 9.2

Sources of Some Trace Elements in Agricultural Soils

Element	Sewage Sludge	Phosphate Fertilizers	Nitrogen Fertilizers	Pesticides (%)
	mg kg^{-1} dry weight			
As	2–26	2–1200	2.2–120	22–60
B	15–1000	5–115	—	—
Cd	2–2500	0.1–170	0.05–8.5	—
Cr	20–40,600	66–245	3.2–19.0	—
Cu	50–3300	1–300	<1–15	12–50
Hg	0.1–55	0.02–1.2	0.3–2.9	0.8–42
Mo	1–40	0.1–60	1–7	—
Ni	16–5300	7–38	7–34	—
Pb	50–3000	7–2225	2–27	60
Se	2–9	0.5–25	—	—
Zn	700–49,000	50–1450	1–42	1.3–25

Adapted from Kabata-Pendias, A. and H. Pendias, *Trace Elements in Soil and Plants*, CRC Press, Boca Raton, FL, 1992.

In addition to sewage sources of trace pollutants, heavy metals also can derive from other soil contaminant sources. For example, lead originates from motor vehicle exhaust, especially in the vicinity of highways, or from smelters and mining operations. Copper and chromium as soil contaminants are found in discharges from tanneries and wood preserving plants. Another common soil contaminant is arsenic which has found wide use in the past as a preservative of hides in tanning (Connell, 1997). All of these elements, including Se, B, Mo, are of primary concern to plant, animal, and human health. Of these, in terms of uptake and accumulation by food chain crops, cadmium poses the greatest long-term threat to human health. A report from CAST (Council for Agricultural Science and Technology, 1976) summarizes the impact of these metals on plant and animal health; for example, Cd, Cu, Ni, Zn, and Mo are the elements that plants can accumulate and then become problems throughout the food chain. Copper, nickel, and zinc can cause phytotoxicity at elevated levels in acid soils. Boron, molybdenum, and arsenic are of concern for both plant and animal health, and Se is of concern for animal health. Because they form sparingly soluble compounds in the soil, mercury and lead are not taken by plants in amounts harmful to consumers and have not posed a direct food chain problem for land application of sewage sludge. However, through other pathways of the food chain, such as direct ingestion from soil or the edible portion of crops, they can be harmful to the consumers.

A wide range of organic substances also occurs as contaminants in soils; of particular interest are petroleum products, paints, oils, waxes, solvents, plasticizers, pigments, and so forth. These organic substances can be divided into several groups including pesticides, polychlorinated biphenyls (PCBs), polycyclic aromatic hydrocarbons (PAHs), halogenated aliphatics, ethers, phtalate esters, phenols, monocyclic aromatics, and nitrosamines (Sommers and Barbarick, 1988; Jorgensen and Johnsen, 1989).

All of these substances would be expected to be present in urban areas and agricultural land. The concentrations of most tend to decrease with time because of environmental transformation and degradation processes; however, some, will tend to persist in the environment for many years, or even many decades.

9.2.2 Chelating Agents as Soil Tests for Heavy Metals

Many heavy metals (essential and nonessential) are taken up by plants at a level often without any cause for concern; however, when crops are cultivated in contaminated soils, metal plant concentrations may reach phytotoxic levels. There is, therefore, a need to monitor heavy metal content of soils. Considerable research has been done on the extraction of heavy metals from contaminated soils by chelating agents, primarily the EDTA (ethylenediamine-tetraacetic acid) and the DTPA (diethylenetriaminepentaacetic acid). These chelates decrease the activity of metal ions in soil solution by giving rise to the formation of soluble metal chelate complexes.

Initial efforts in selecting a chelate for studying the fate of metals in the environment have centered around EDTA. In an experiment carried out on treatment of a soil contaminated with Cd, Cr, Cu, Pb, and Ni, Ellis et al. (1986) reported that the EDTA chelated and solubilized all of the metals to some degree. On the other hand, EDTA was seen to offer good potential as soil washing additive for the removal of Pb from soils (Peters and Shem, 1995).

Elliot et al. (1989) in a series of batch experiments evaluated extractive decontamination of Pb-polluted soil using EDTA. Their results showed that increasing EDTA concentration resulted in a greater Pb release and that recovery of Pb was higher under acidic than alkaline conditions.

Many studies have indicated that soil-test-extractable heavy metals can be correlated with their associated levels in plants. Bowman et al. (1981) found that Ni was complexed and therefore was maintained in soluble form from different agricultural soils and potential waste disposal sites, in the presence of a high level of Ca ions and a small amount of chelating agent EDTA.

The Lindsay and Norvell (1978) procedure using DTPA is one of the most popular soil chelating extractants. Haq et al. (1980) in a greenhouse study compared nine extractants for plant available Zn, Cd, Ni, and Cu in 46 Canadian soils contaminated with heavy metals; they concluded that DTPA-extractable metals were best correlated with heavy metals uptake by plant.

Because the DTPA soil test is to some extent tedious and laborious to perform in laboratories, Soltanpour and Schwab (1977) proposed a simplification of the method by combining the DTPA test with ammonium bicarbonate and by making some adjustments relative to the solution pH and the extracting time. Barbarick and Workman (1987), by comparing DTPA and AB-DTPA-extractable Cd, Cu, Ni, Pb, and Zn with metal plant concentration from a greenhouse experiment, found a highly significant linear relationship between these parameters. They also indicated that either extractant could predict some accumulation of these metals in the field, and that the relationships between AB-DTPA-extractable metal and their associated levels in plant were markedly influenced by soil pH. Rappaport et al. (1988) evaluated the ability of DTPA to predict heavy metal (Cd, Cu, Ni, and Zn) availability to corn grown in three soils amended with high doses of sewage sludge. Corn, grain, and stover yields, and DTPA-extractable Cd, Cu, and Ni were closely related to sludge rate, although the critical phytotoxic level of these metals could not be established.

Intensive efforts have been directed (Wang 1997) at the development of sequential fractionation schemes that quantitatively partition the total amount of a heavy metal into soil pools that may be interpreted for predicting metal phytoavailability from contaminated soils. In this regard, greenhouse and field studies have been conducted to relate these chemical fractions to plant uptake and soil test extractable levels.

The objectives of this study were (1) to compare the EDTA extractant to DTPA and AB-DTPA for the removal of metals from contaminated soils; and (2) to determine the relationships between EDTA, DTPA, and AB-DTPA extractable metals of the wastewater irrigated soils and the metal concentrations in alfalfa (*Medicago sativa* L.) grown in the same fields.

TABLE 9.3

Sampling Sites as Related to Date of Wastewater Application

Sites	Location	Series	First Year under Irrigation	No. of Years Irrigated
2-1, S-2, S-3	Tlaxcoapan	Lagunilla	1912	87
2-4, S-5	Tlahualilpan	Tepatepec	1912	87
2-6, S-7, S-8	Mixquiahuala	Progreso	1920	79
S-9, S-10	Tepatepec	Tepatepec	1954	35
S-11	Actopan	Lagunilla	1972	27
S-12, S-13	Atitalaquia	Lagunilla	1976	23
S-14, S-15	Clavijero	Tepatepec	1976	23
S-16, S-17	Xicudo	Progreso	1979	20

TABLE 9.4

Range, Mean, and Standard Deviation of Soil Characteristics and Metal Concentrations in Alfalfa Tissue

Parameter	Range (n = 17)	Mean	Standard Deviation
pH	6.9–8.1	7.5	0.34
OM (g kg^{-1})	14–53	34	0.85
Clay (g kg^{-1})	240–700	412.4	13.7
Soil Cd (µg g^{-1})	17.4–39.1	30.8	1.17
Soil Ni (µg g^{-1})	11.5–35.1	21.6	6.05
Soil Pb (µg g^{-1})	7.7–63.0	41.7	1.52
Plant Cd (µg g^{-1})	0.48–1.97	1.28	0.47
Plant Ni (µg g^{-1})	3.18–9.09	6.35	1.49
Plant Pb (µg g^{-1})	1.26–7.70	3.30	1.68

9.3 Materials and Methods

9.3.1 Field Observations

Composite soil samples (Typic Calciorthids, 0 to 15 cm depth) were collected from 17 alfalfa plots located in the irrigation District 03 in the Mexican Valley of Mezquital. These plots were irrigated with wastewater from domestic and industrial effluents of the metropolitan area of Mexico City (Cajuste 1991), over different periods of time (Table 9.3).

9.3.2 Sample Preparation and Analysis

Soil samples were dried and ground for laboratory analyses. Soil characterization included the determination of pH, percent clay, percent organic matter, and total Cd, Ni, and Pb. Plant samples collected from the 17 plots were dried at 70°C for 48 h and ground in a Wiley mill to pass through a 1-mm sieve. They were then digested with a $HClO_4 \cdot H_2SO_4$ acid mixture (1:4) and analyzed for Cd, Ni, and Pb using an AA spectrophotometer (Table 9.4). A batch experiment was conducted to determine extractable Cd, Ni, and Pb from the soils by using the following conventional methods for plant micronutrients: 0.05 M EDTA, pH 7.0; 0.005 M DTPA, pH 7.3; 0.005 M AB-DTPA, pH 7.6; and the same solutions at pH 6.0 and 8.4. Additionally, duplicates of 1-g samples of soil were analyzed using the following

TABLE 9.5

Removal of Heavy Metals from the Soils (Average Values) by Chelates at Different Solution pH Values

	pH	Cd	Ni	Pb
		(µg g⁻¹)		
EDTA (0.05 *M*)	6.0	2.56	2.11	2.90
	7.0[a]	2.17	7.57	2.44
	8.4	1.99	8.26	2.49
DTPA (0.005 *M*)	6.0	1.02	1.24	2.80
	7.3[a]	1.14	1.86	0.66
	8.4	1.00	2.32	0.62
AB-DTPA (0.005 *M*)	6.0	1.07	1.26	2.18
	7.6[a]	2.26	4.74	1.89
	8.4	1.22	3.34	1.27

[a] pH of the solution of the conventional method.

extractants: 0.01 *M* and 0.1 *M* EDTA, DTPA, and AB-DTPA at pH 7.0, 7.3, and 7.6, respectively. Cd, Ni, and Pb were determined by the same procedure mentioned above.

Statistical analyses (correlation and regression procedures) were carried out to compare soil tests and to relate the amounts of Cd, Ni, and Pb extracted from the soils to their associated concentrations in alfalfa tissue.

9.4 Results and Discussion

9.4.1 Effect of Chelates on Metal Removal

Amounts of metal extracted from the soils, on the average, varied widely, according to solution pH and chelate concentrations. The conventional EDTA method was, in general, more effective in removing metals than the DTPA and AB-DTPA soil tests, presumably because of its stronger solution concentration than the other solutions, thus favoring greater metal solubility (Table 9.5). An increase in chelate concentration generally resulted in greater metal release by EDTA and DTPA, but this effect was not consistent in the extraction of metals by AB-DTPA (Table 9.6). Increased solubility of metal ions as influenced by chelate complexation has been reported by many researchers (Ellis et al., 1986; Peters and Shem, 1995; Elliott et al., 1989; Bowman et al., 1981); they suggest that complexing agents offer the potential to be effective extractants of heavy metals from contaminated soils because of the increase in metal solubility.

There was no consistent effect of solution pH on the extraction of metals; for instance, an increase in soil test pH showed a negative effect on the level of EDTA extractable Cd and the amounts of Pb removed by chelates (Table 9.5). The quantity of Ni extracted by the chelates, however, showed an opposite trend and DTPA-extractable Cd was insensitive to pH variation.

9.4.2 Extractable Metals and Phytoavailability Relationships

Correlation coefficients between soil tests were highly significant ($P = 0.01$) (data not presented). Correlations between the amounts of metals extracted by the conventional

TABLE 9.6

Averaged Amounts of Cd, Ni, and Pb Removed from the Soils as Influenced by Solution of Chelates

Chelates	pH	Solution	Cd	Ni	PB
			($\mu g\ g^{-1}$)		
EDTA	7.0	0.01	1.4	1.8	1.0
		0.05[a]	2.6	7.6	2.1
		0.1	2.2	10.4	2.4
DTPA	7.3	0.005[a]	1.1	1.9	0.7
		0.01	6.8	2.3	1.5
		0.1	12.3	7.7	7.7
AB-DTPA	7.6	0.005[a]	2.3	4.7	1.9
		0.01	10.4	3.2	1.3
		0.1	9.7	2.8	1.5

[a] Solution concentration of the conventional method.

TABLE 9.7

Correlation Coefficients (r) between Metals Extracted by the Chelates and Total Metal Content in Soil and Metal Concentration in Alfalfa Tissue

	ED1	ED2	ED3	DT1	DT2	DT3	AB1	AB2	AB3	pH	OM	Cl
Cd												
Cdtot	0.78***	0.77***	0.80***	0.80***	0.83***	0.50*	0.75***	0.67***	0.68***	−0.5*	0.52*	−0.10
CdPp	0.48*	0.55**	0.50*	0.52*	0.45	0.37	0.46	0.36	0.37	−0.19	0.22	0.37
Ni												
Nitot	0.54**	0.52*	0.67**	0.81***	0.78***	0.58**	0.80***	0.81***	0.74***	0.63**	0.20	−0.01
NiPp	0.72***	0.62**	0.64**	0.46	0.18	0.51*	0.35	0.30	0.39	0.62**	0.04	0.09
Pb												
Pbtot	0.78***	0.84***	0.89***	0.47*	0.88***	0.69***	0.71***	0.92***	0.86***	−0.43	0.30	0.06
PbPP	0.68**	0.72**	0.67**	0.54**	0.47*	0.61**	0.34	0.56**	0.58**	−0.28	0.27	−0.03

Note: ED1 to ED2, DT1 to DT3, and AB1 to AB3 mean 0.01, 0.05, and 0.1 *M* EDTA, 0.005, 0.01 and 0.1 *M* DTPA, and 0.005, and 0.01, and 0.1 *M* AB-DTPA, respectively.
   Cdtot, Nitot, and Pbtot = Total Cd, total Ni, and total Pb in soils, respectively.
   CdPp, NiPp, and PbPp = Cd, Ni, and Pb concentration in plant, respectively.

methods (0.05 *M* EDTA, pH 7.0; 0.005 *M* DTPA, pH 7.3; 0.005 *M* AB-DTPA, pH 7.6) as well as at 0.01 *M* and 0.1 *M* concentrations of the chelates, and their associated total levels in soils and plant tissue were carried out. As only the top portion of the alfalfa plant was sampled and inasmuch as dry matter was not correlated with metal concentration in alfalfa tissue, metal concentration in the plant was used instead of total metal uptake. In general, total Cd and Pb correlated very significantly with soil tests (Table 9.7). Although soil organic matter had a positive and significant effect on total Cd, percent clay was not similarly correlated (Table 9.7). This seems to indicate that Cd bound to organic matter was more important as a source of soil Cd than the Cd retained on the mineral exchange complex.

A significant correlation was found between plant Cd concentration and Cd removed by the three EDTA concentrations and the conventional DTPA procedure. Phytoavailable Cd, however, correlated poorly with AB-DTPA-Cd and the other soil properties (Table 9.7). Likewise, total Pb and Pb concentration in the plant were not correlated with soil pH, organic matter, or percent clay (Table 9.7).

Relationships between total levels of Ni in soil and plant and their amounts complexed by the chelates showed, to some extent, a different pattern. Total Ni was directly related to

Ni extracted by soil tests and plant Ni concentration was found to be strongly influenced by soil pH, and only by EDTA-Ni and 0.1 M DTPA-Ni (Table 9.7).

In the case of soil Ni, similar observations were made by Wang et al. (1997) who found that DTPA-extractable Ni correlated significantly with soil Ni fractions as a result of a sequential fractionation analysis from a pot experiment. They suggested that the DTPA-extractable Ni in terms of phytoavailability was related to the concentrations in soil fractions.

The amounts of metals removed from the soils by the conventional chelating methods over the whole period of irrigation management correlated poorly with plant metal concentrations; extractable metals, in general, accounted for less than 50% of the variability of plant metal concentrations. This indicates that data for metals extracted during the first period of irrigation management may not belong to the same population as the data from the sites irrigated during the last 20 years. The separation of the data on the accumulation of metals for sampling sites with relatively short- and long-term periods of irrigation management, corresponding to time intervals of 20 to 23 and 45 to 87 years, respectively, improved the prediction of metal phytoavailability dependence, solely on extractable metals or on these in combination with other soil variables.

Regression equations between plant-available metals and their associated extractable metals for Group I (20 to 23 years of irrigation management) and Group II (45 to 87 years of irrigation management) are shown below. (Data for one site in each group were deleted for each of the three metals, as metal concentrations in alfalfa plants grown in the two sites were much higher than the other values and would have biased the regression equations.)

GROUP I

Pb P_p = 0.912 + 1.114 EDTA-Pb	R^2 = 0.854**	(1)
Pb P_p = 2.918 + 1.910 EDTA-Pb + 1.191 OM	R^2 = 0.667*	(2)
Cd P_p = 4.191 + 0.048 EDTA-Cd – 0.117 pH	R^2 = 0.669*	(3)
Ni P_p = –3.960 + 0.295 EDTA-Ni – 1.636 pH	R^2 = 0.873**	(4)

GROUP II

Pb P_p = 0.078 + 0.998 EDTA-Pb + 3.357 OM	R^2 = 0.853**	(5)
Cd P_p = –2.017 + 0.380 EDTA-Cd – 0.332 pH	R^2 = 0.514*	(6)
Ni P_p = 9.351 + 0.391 EDTA-Ni + 0.109 OM	R^2 = 0.796**	(7)
Cd P_p = 8.298 – 0.615 pH + 0.665 OM	R^2 = 0.718**	(8)

In these equations, Pb P_p, Cd P_p, and Ni P_p mean Pb, Cd, and Ni concentration in the plant, respectively.

In Group I, the highest regression coefficient was observed when the plant concentration of Ni was regressed on the amount of Ni extracted with 0.05 M EDTA, pH 7.0, and soil pH. This regression accounted for about 87% of the variability of plant Ni concentrations. In the case of Group II, the combined effect of EDTA-extracted Pb and percent organic matter was responsible for 85% of the variability in plant Pb concentrations. On the other hand, the removal of Pb by EDTA alone could explain 85% of the variation in plant Pb concentrations (Equation 1).

A somewhat different situation was observed with Cd in Group II, where about 72% of the variability in plant concentration of Cd could be explained by differences in soil pH and percent organic matter (Equation 8). EDTA extractable Cd, on the other hand, had only a moderate influence on phytoavailable Cd as reflected in a value of R^2 = 0.514 (Equation 6).

These results are in agreement with earlier findings of some researchers (Haq et al., 1980; Ellis et al., 1986; Peters and Shem, 1995; Elliot, 1989; Bowman et al., 1981; Soltanpur and Schwab, 1977; Barbarick and Workman, 1987; Peters and Ku, 1987; Wang et al., 1997; Xiu et al., 1991; Jorgensen and Johnsen, 1989) who showed that these three extractants have proved to be reliable chemical methods for assessing the levels of toxic metals in soils.

9.5   Conclusions

The three chelating agents appeared to be effective extractants in removing metals from contaminated soils. The EDTA procedure was the best single extracting agent for most metals. Extractable metals were, to some extent, influenced by solution pH and chelate concentrations. Correlation coefficients between soil tests were highly significant, but extractable metals accounted for less than 50% of the variability of plant metal concentrations. The separation of the soils into two groups, based on the number of years of irrigation management, improved the prediction of metal phytoavailability, solely in terms of the amounts of extractable metals or on these in combination with other soil variables.

9.6   Summary

Chelating agents are known to be effective extractants of heavy metals from contaminated soils. This study was aimed to compare the DTPA procedure with the EDTA and AB-DTPA soil tests for extraction of Cd, Ni, and Pb, and to determine the relationships between soil extractable metals and associated alfalfa (*Medicago sativa* L.) plant concentrations.

Soil samples (Typic Calciorthids) were collected from 17 unfertilized plots irrigated with wastewater. They were analyzed for metal contents using varying chelate concentrations and at three pH values. Plant samples were obtained from the sites and were analyzed for metal content. Results indicated that EDTA, in general, extracted greater amounts of metals than DTPA, regardless of solution pH. The amounts of metals removed were influenced by extractant concentrations. Correlation coefficients between soil tests were highly significant ($P = 0.01$), but regression analysis between plant and soil extractant concentration accounted for less than 50% of the variation; however, the separation of the soil sites by amounts of time under irrigation with wastewater improved the relationship between these two parameters.

References

Adriano, D.C., Nickel, in *Trace Elements in the Terrestrial Environment*, Springer-Verlag, New York, 1986, 362.

Barbarick, K.A. and S.M. Workman, Ammonium bicarbonate-DTPA and DTPA extractions of sludge-amended soils, *J. Environ. Quality*, 16, 125, 1987.

Bowman, R.S., M.S. Essington, and G.A. O'Connor, Soil sorption of nickel: influence and composition, *Soil Sci. Soc. Am. J.*, 45, 860, 1981.

Cajuste, L.J., R. Carrillo-G, E. Cota-G, and R.J. Laird, The distribution of metals from wastewater in the Mexican Valley of Mezquital, *Water, Air, Soil Pollut.*, 57, 673, 1991.

Chaney, R.L., Plant acummulation of heavy metals and phytotoxicity resulting from utilization of sewage sludge and sludge compost on cropland, in *Proc. 1977 Conf. Nat. Composting of Municipal Residues and Sludges*, Rockville, MD, 1978, 86.

Chang, A.C., J.E. Warneke, A.L. Page, and L.J. Lund, Accumulation of heavy metals in sewage sludge-treated soils, *J. Environ. Quality*, 13, 287, 1984.

Connell, D.W., *Environmental Chemistry*, Lewis Publishers, Boca Raton, FL, 1997, 289.

Council for Agricultural Science and Technology, Application of Sewage Sludge to Cropland: Appraisal of Potential Hazards of the Heavy Metals to Plants and Animals, CAST, Ames, IA, 1976, 64.

Elliott, H.A., G.A. Brown, G.A. Schields, and J.H. Lynn, Restoration of Pb polluted soils by EDTA extraction, in *7th Int. Conf. Heavy Metals in the Environment*, Vol. 2, J.-P. Vernet, Ed., Geneva, Switzerland, 1989, 64.

Ellis, W.D., T.R. Fogg, and A.N. Tafuri, Treatment of soil contaminated with heavy metals, in *Land Disposal, Remedial Action, Incineration and Treatment of Hazardous Waste*, 12 Ann. Res. Symp. EPA 600/9–86/022, Cincinnatti, OH, 1986, 201.

Förstner, U., Land contamination by metals: global scope and magnitude of problem, in *Metal Speciation and Contamination of Soil*, H.E. Allen, C.P. Huang, J.W. Bailey, and A.R. Bowers, Eds., Lewis Publishers, Boca Raton, FL, 1995.

Haq, A.V., T.E. Bates, and Y.K. Soon, Comparison of extractants for plant-available Zn, Cd, Ni and Cu in contaminated soils, *Soil Sci. Soc. Am. J.*, 44, 772, 1980.

Henry, C.L. and R.B. Harrison, Fate of trace metals in sewage sludge compost, in *Biogeochemistry of Trace Metals*, D.C. Adriano, Ed., Lewis Publishers, Boca Raton, FL, 1992.

Jorgensen, S.E. and I. Johnsen, *Principles of Environmental Science and Technology*, Elsevier, Amsterdam, 1989, 270.

Juste, C. and M. Mench, Long-term application of sewage sludge and its effect on metal uptake by crops, in *Biogeochemistry of Trace Metals*, D.C. Adriano, Ed., Lewis Publishers, Boca Raton, FL, 1992.

Kabata-Pendias, A. and H. Pendias, *Trace Elements in Soil and Plants*, CRC Press, Boca Raton, FL, 1984.

Lindsay, W.L. and W.A. Norvell, Development of DTPA test for zinc, iron, manganese and copper, *Soil Sci. Soc. Am. J.*, 42, 421, 1978.

Miller, R.W., A.S. Azzari, and D.T. Gardiner, Heavy metals in crops as affected by soil types and sewage sludge rates, *Commun. Soil Sci. Plant Anal.*, 26, 703, 1995.

Peters, R.W. and L. Shem, Treatment of soils contaminated with heavy metals, in *Metal Speciation and Contamination of Soils*, H.A. Allen, C.P. Huang, G.W. Bailey, and A.R. Bowers, Eds., Lewis Publishers, Boca Raton, FL, 1995.

Peters, R.W. and Y. Ku, The effect of citrate, a weak complexing agent, on the removal of heavy metals by sulfide precipitation, in *Metals Speciation, Separation and Recovery*, J.W. Patterson and R. Passino, Eds., Lewis Publishers, Boca Raton, FL, 1987.

Rappaport, B.D., D.C. Martens, R.B. Renaud, Jr., and T.W. Simpson, Metal availability in sludge-amended soils with elevated metal levels, *J. Environ. Quality*, 17, 42, 1988.

Soltanpour, P.N., Use of ammonium bicarbonate DTPA to evaluate elemental availability and toxicity, *Commun. Soil Sci. Plant Anal.*, 16, 323, 1985.

Soltanpour, P.N. and A.P. Schwab, A new soil test for simultaneous extraction of macro and micro-nutrients in soils, *Commun. Soil Sci. Plant Anal.*, 3, 195, 1977.

Sommers, L.E. and K.A. Barbarick, Constraint to land application of sewage sludge, in *Utilization, Treatment and Disposal of Waste on Land*, SSSA, Madison, WI, 1988.

Wang, P., E.Qu, Z. Li, and L.M. Shuman, Fraction availability of nickel in loessial soil amended with sewage or sewage sludge, *J. Environ. Quality*, 26, 795, 1997.

Xiu, H., R.W. Taylor, J.W. Shuford, and W. Tadesse, Comparison of extractants for available sludge-born metals: a residual study, *Water, Air, Soil Pollut.*, 57, 913, 1991.

10

Restoration of Selenium-Contaminated Soils

K.S. Dhillon and S.K. Dhillon

CONTENTS

1-56670-457-X/00/$0.00+$.50
© 2000 by CRC Press LLC

10.1 Introduction

Selenium (Se), depending upon concentration, can be beneficial or toxic to plants, animals, and humans. Dietary intake below 0.04 mg/kg results in Se deficiency diseases, and when it exceeds 4 mg/kg toxicity diseases may appear (Lakin and Davidson, 1973). The Food and Nutrition Board (1980) of the U.S. National Academy of Sciences has accepted 5 mg Se/kg as the critical level between toxic and nontoxic feeds. Soils that supply sufficient Se to produce vegetation containing >5 mg Se/kg are referred to as seleniferous soils. Selenium toxicity problem is associated with sporadically distributed Se toxic soils throughout the Great Plains and Rocky Mountains regions of the United States, Prairie regions of Canada, Queensland in Australia (Rosenfeld and Beath, 1964), Sangliao, Weihe, and Hua Bei plains of China (Tan et al., 1994), and Haryana, Punjab, and West Bengal states in India (Arora et al., 1975; Dhillon and Dhillon, 1991a; Ghosh et al., 1993). Animal and human productivity is closely linked to the level of Se in plants and grains (Yang et al., 1983; Dhillon and Dhillon, 1991a, 1997a). *Cruciferae* spp. are capable of accumulating Se to several hundred micrograms per gram without showing Se phytotoxicity symptoms (Banuelos et al., 1990). Recent interest in the volatilization of Se is related to the buildup of excessive levels of Se in soils. Biological volatilization of Se may be carried out by microorganisms as well as by plants. Ross (1984) estimated that as much as 10,000 tonnes of Se may be emitted to the atmosphere annually in the northern hemisphere alone and more than 1/4 of it originates from soils and plants. In spite of well known toxic effects of Se, it was not acknowledged as a pollutant for a long time. With its inclusion in the list of inorganic carcinogenic agents (Shubik et al., 1970), a large number of papers have been appearing from different corners of world determining the status of Se in every material composing the environment. In 1985 the United States Environmental Protection Agency (U.S. EPA) postulated that Se should receive closer scrutiny as a potential contaminant of the food chain.

Until the mid-1970s, parent material was considered as an important factor controlling the level of Se in geoecosystem in the juvenile landscapes (Moxon and Rhian, 1943; Anderson et al., 1961; Rosenfeld and Beath, 1964; Brown and Shrift, 1982). Human activities contribute substantially to the redistribution and cycling of Se on a global scale. Anthropogenic activities, which include disposal of coal generated fly ash, mine tailings, and agricultural drainage water, use of fertilizers and underground water for crop production, and domestic household sources such as dandruff shampoo, have been linked to Se toxicity problem (Thomson and Heggen, 1982; Nriagu and Pacyna, 1988; Jacobs, 1989; Dhillon and Dhillon, 1990; Frankenberger and Benson, 1994). Total worldwide input of Se into soils from anthropogenic activities has been estimated to be 6,000 to 76,000 t/yr (Nriagu and Pacyna, 1988). The atmosphere is playing an important role in the mass balance of Se in grassland ecosystems, and total input from atmospheric deposition is calculated to be typically in the range 0.2 to 0.7 mg/m²·yr (Haygarth et al., 1991).

The most effective strategies for remediation of a contaminated site should protect all components of the biosphere, i.e., land, air, surface water, and groundwater as well as health of the general public (McNeil and Waring, 1992). In recent years, a large number of papers have appeared on restoration of Se-contaminated soils. Particularly after the mid-1980s, when Se was shown to bioaccumulate and was positively identified as the cause of death and deformities of waterfowl in the Kesterson Reservoir, many research efforts were made to restore seleniferous soils and waters. Research strategies on restoration of seleniferous soils have generally followed on-site management. Some researchers have even attempted to work out strategies to live with seleniferous soils with no harmful effects of Se on fauna

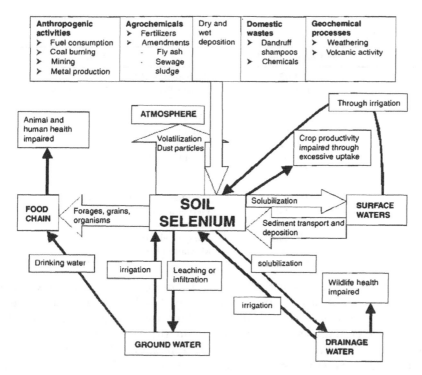

FIGURE 10.1
Schematic diagram of selenium inputs/outputs in the soil and possible impact on the environment.

and flora. This chapter reviews research carried out in different parts of the globe in terms of Se accumulation in soils due to natural and anthropogenic sources, and it suggests various options to restore the Se contaminated soils or to manage these soils in such a manner that entry of Se into the food chain is restricted to permissible levels.

10.2 Source and Nature of Contamination

Enrichment of soil with Se is governed by the type of parent material, process of soil genesis, and anthropogenic activities related to inadvertent use of Se-rich materials for increasing soil productivity. The natural fluxes of Se are small compared with emissions from industrial activities, implying that mankind has become the key agent in the global atmospheric cycle of Se in soil-plant system (Figure 10.1). Total emission of Se into the atmosphere ranged from 2.5 to 24 thousand t/yr, which included 42% from anthropogenic sources (Nriagu, 1989).

10.2.1 Parent Material

With the association of Se with alkali disease since the early thirties, researchers have continued to characterize the sources of Se in soil. A detailed account of geological distribution of Se in relation to the development of seleniferous soils of the United States has been given by Anderson et al. (1961) and Rosenfeld and Beath (1964). It has been estimated that 0.1 to 1.8 thousand t Se/yr is emitted into the atmosphere through volcanic activity (Nriagu, 1989)

TABLE 10.1

Selenium Content of Rocks

Rock Type	Se Content (mg/kg)	Ref.
Meteorites	3–15	Rosenfeld and Beach (1964)
Igneous rocks	0.01–0.05	Kabata-Pendias and Pendias (1984)
Sedimentary rocks		
Marine shales	2–24	Web et al. (1966)
Black pyritic shales	0.2–6.5	Web et al. (1966)
Carbonaceous shales	2.3–52.0	Rosenfeld and Beath (1964)
Phosphate rocks	1–300	Rosenfeld and Beath (1964)
Sandstones	0.2–46	Rosenfeld and Beath (1964)
	2	Web et al. (1966)
Limestones	0.1–6.0	Rosenfeld and Beath (1964)
	0.2–1.0	Web et al. (1966)
Uranium deposits	526–2630	Rosenfeld and Beath (1964)
Coal	0.46–10.60	Pillay et al. (1969)
	1–20	Mayland (1989)

and this leaves the igneous rocks poor in Se (Table 10.1). Among sedimentary rocks, Se concentration is higher in shales, due to its association with clay, than limestones and sandstones.

Cretaceous sedimentary rocks like shale, sandstone, limestone, conglomerates, etc. form the parent material of seleniferous soils in arid and semiarid parts of the western United States. Selenium content of sedimentary rocks ranged from 2.3 to 52.0 mg/kg. Exceptionally high concentrations of Se (156 mg/kg) in sedimentary rocks have been reported in Pierre shales of Cretaceous age; 680 mg/kg in phosphate rocks of Permian age, and 890 mg/kg of tuffs of Eocene age. Fleming and Walsh (1957) assumed the source of Se in Irish lacustrine soils containing 30 to 1200 mg of Se/kg to be pyritic shale of early Carboniferous age, with as much as 28.5 mg Se/kg. Shales are also considered the principal source of Se in toxic soils of Israel (Abu-Erreish and Lahham, 1987).

In northwestern India, transportation of Se-rich material from the nearby Shivalik Range through flood water and its deposition in depressions has resulted in the development of seleniferous soils (Dhillon and Dhillon, 1991a). The toxic sites are located at the dead end of seasonal rivulets coming from upper ranges of the Shivalik Hills.

Total Se concentration of parent material of a particular soil can influence the Se concentration in plants. Doyle and Fletcher (1977) reported that average total Se concentration in whole wheat plants was highest (2.18 mg/kg) when grown over lacustrine clay followed by that on glacial till (1.50 mg/kg), lacustrine silt (1.08 mg/kg), and aeolian sand (0.64 mg/kg). They suggested that soil parent material maps could form a suitable sampling base for designing rapid plant sampling programs to outline areas where Se excess or deficiency problems are most likely to occur.

10.2.2   Fertilizers

Fertilizers have become an integral part of modern agriculture, as 50% of the world's agricultural production is being attributed to fertilizer use. Use of fertilizers also implies incidental addition of toxic elements such as Cd, F, and Se to soils. These elements are present as impurities in fertilizer raw materials. The Se content of fertilizers differs widely depending upon the choice of raw materials and manufacturing procedures (Table 10.2). Normal superphosphate is expected to contain about 60%, and concentrated superphosphate about 40%, as much as the phosphate rock from which it is made. The decrease in Se concentration results from volatilization and during processes such as smelting. Concentrated superphosphate

TABLE 10.2

Total Se Contents of Fertilizers and Raw Materials

Fertilizer/Raw Materials	Se Content (mg/kg)	Ref.
Rock phosphate	0.77–178	Robbins and Carter (1970)
Pyrite	1–300	Rosenfeld and Beath (1964)
	3.1–25	Gissel-Nielsen (1971)
	25–41.6	Gissel-Nielsen (1971)
Sulphuric acid	0.25–10.1	Gissel-Nielsen (1971)
Phosphoric acid	9.3	Gissel-Nielsen (1971)
	0.01–0.40	Robbins and Carter (1970)
Superphosphate	4.2–8.0	Gissel-Nielsen (1971)
	10	Senesi et al. (1979)
Concentrated superphosphate	0.54–3.88	Robbins and Carter (1970)
PK	3.6–5.5	Gissel-Nielsen (1971)
NPK	0.02–4.0	Gissel-Nielsen (1971)
Phosphatic fertilizers	0.5–25.0	Kabata-Pendias and Pendias (1984)
Ammonium nitrate	13.25	Senesi et al. (1979)
Natural sulphur	<1–68.2	Steudel et al. (1984)

and single superphosphate contained 70 and 105 mg Se/kg, respectively (Robbins and Carter, 1970). With the application of 300 kg ammonium nitrate/ha (containing 10 mg Se/kg), 3 g Se/ha would enter the soil and application of 800 kg superphosphate/ha (containing 13.25 mg Se/kg) resulted in an input of 10.6 g Se/ha (Senesi et al., 1979). The estimated worldwide emissions of Se applied through fertilizers into the soil range from 20 to 100 t/yr (Nriagu and Pacyna, 1988). The contribution to total Se content of the plants from Se in the fertilizers is negligible, unless high seleniferous raw materials are employed (Gissel-Nielsen, 1971).

In Se-deficient regions, addition of Se to the soil either directly or through super-phosphate is recommended for raising the Se level of vegetation. In New Zealand and Finland, application of 10 g Se/ha with carrier fertilizer has been recommended to raise the level of Se in feedstuffs (Korkman, 1985). Although there does not exist any report linking Se toxicity in soils and the use of fertilizers, continuous use of Se-rich fertilizers should substantially contribute to total load of Se in soils. For instance, buildup of Cd to toxic levels in agricultural soils has been traced to the use of phosphatic fertilizers in many countries in the Asia-Pacific region (Bramley, 1990; McLaughlin et al., 1966).

10.2.3 Fly Ash

Finely divided residue resulting from combustion of bituminous or subbituminous coal in the furnace of thermal power generation plants is termed as fly ash (FA). Of the residue left after combustion of coal, about 40% occurs as bottom ash or slag, 60% as fly ash, and, where emission control devices are employed, < 1% escapes to the atmosphere as aerosol (Eisenberg et al., 1986). Incineration of municipal waste is another source of aerosol and FA. Release of Se into atmosphere through anthropogenic combustion can affect its temporal and geographical distribution in terrestrial vegetation (Haygarth et al., 1993a,b).

Fly ash generation in the United States was estimated to be 1.2×10^9 tonnes in 1987 (Pattishall, 1998), and particulate emissions from coal combustion may increase to 5×10^6 t/yr by 2000 AD. Selenium concentration in FA is inversely related to particle size. With decrease in diameter from 50 to 0.5 mm, the Se content of FA increased from 3.5 to 59 mg/kg (Campbell et al., 1978). The average total Se concentration of coal in the Powder River Basin is 5.8 mg/kg, with a range of 0.2 to 44 mg/kg (Boon and Smith, 1985). Fly ash from 21 states contained Se ranging from 1.2 to 16.5 mg/kg (Gutenmann et al., 1976).

Generation of FA is fast increasing in developing countries as well. For example, annual FA generation is expected to exceed 100 million tonnes by 2000 AD in India (Kumar and Sharma, 1998), which may contain as much as 27 mg Se/kg. Burning of coal contributes 1.5 to 2.5 times more Se to the environment compared to natural weathering. Worldwide emissions of Se into soils from coal-generated FA varies from 4.1 to 60 thousand t/yr (Nriagu and Pacyna, 1988).

In some countries 30 to 80% of the FA is being used for gainful applications such as manufacture of bricks, cement, etc. The Netherlands has achieved 100% utilization of FA since the beginning of 1990s (vom Berg, 1998). In many developing countries such as India, the FA utilization level is very low (3 to 5%) and a large proportion is dumped on wasteland (Kumar and Sharma, 1998). In fact, in spite of available technologies for gainful utilization of FA, large quantities of ash produced in thermal power plants are ending up in vast areas close to the power plants in these countries. From FA transported to the landfills as solid residues or flushed with water to ash ponds, Se and other toxic elements may easily enter the aquatic environments. Laboratory experiments have revealed that 5 to 30% of toxic elements in FA are leachable (Kumar et al., 1998), and hence FA holds the potential to contaminate underground waters.

Fly ash is also being used as a soil amendment to create physical conditions conducive for plant growth as well as to supply essential plant nutrients. With an application of 5 to 10% FA, significant increases in crop yields varying from 8 to 25% and in some cases even from 100 to 200% has been reported (Doran and Martens, 1972; Elseewi et al., 1978; Kansal et al., 1995; Kumar et al., 1998). Giedrojc et al. (1980) reported that optimum rate of FA was 200 to 400 t/ha for potato and rye, 800 t/ha for peas, 400 t/ha for oats, and beyond this reduction in yield was observed. Application of FA at 10% amounts to an addition of 224 tonnes of FA/ha, and if contained 20 mg Se/kg, it corresponds to an addition of 4.48 kg Se/ha. Compared to the recommended application of 10 g Se/ha for raising Se level of crops to meet the nutritional requirements of animals, as in New Zealand and Finland, this value is on the higher side.

Furr et al. (1978a) found that sweet clover voluntarily growing in deep layers of fly ash at a landfill accumulated as much as 205 mg Se/kg (dry wt). Studies on bioavailability of Se contained in FA (12 to 21.3 mg/kg) revealed that depending upon soil reaction, the application rate has to be carefully controlled to obviate the possible accumulation of toxic levels of Se (Furr et al., 1978 a,b). Experimental feeding of animals for 91 to 173 days on seleniferous diets (prepared from Se-rich materials grown on FA disposal sites or FA amended soils) did not result in any outward signs of selenosis (Furr et al., 1975; Stoewsand et al., 1978), but tissue Se concentration was elevated. Development of selenosis in animals is therefore likely if feeding on seleniferous diets is continued for longer periods. Thus, use of FA as soil amendment has every possibility leading to the development of seleniferous soils. The quality of soils receiving FA as an amendment, thus, needs to be continuously monitored. Establishment of long-term field experiments might reveal the pollution potential in terms of Se accumulation by plants as associated with these soils.

10.2.4   Sewage Sludge

Annual global discharge from urban refuse, municipal sewage sludge, and other organic wastes including excreta on land is estimated to be 670×10^9 tonnes, which leads to an addition of 0.05 to 4.06 thousand tonnes of Se/yr into the soil (Nriagu and Pacyna, 1988). Being a rich source of essential nutrients, raw sewage is preferred for use in crop production, especially for vegetables near the cities, and has become a source of income for municipal corporations in many developing countries. In developed countries, specifically treated sludge is commercially marketed for application on gardens and lawns. Typical Se concentration of sludges range from 1.7 to 17.2 mg/kg in the United States (Chaney, 1985) and from 1 to 10 mg/kg in the U.K. (Sauerbeck, 1987). Kabata-Pendias and Pendias (1984) cited a typical global range of 2 to 9 mg Se/kg in sewage sludge. The maximum permissible Se concentration in sewage sludge considered acceptable for application to agricultural land as suggested by Sauerbeck (1987) is 25 mg/kg.

Application of sludge containing Se to soil does not always lead to immediate transfer of Se into plants. Furr et al. (1976) did not observe any significant increase in Se levels in the edible portion of some crops grown in pots in which soil was amended with commercially marketed sewage sludge containing 1.8 mg Se/kg. Application of 3050 m³/ha of sewage sludge to a silty loam soil resulted only in slight increase in its Se content (El-Bassam et al., 1977). In a long-term experiment, composted sewage sludge containing 1.74 ± 0.45 to 9.59 ± 1.26 mg Se/kg was applied to different crops for 10 years, but there was no significant increase in the Se content of different crops even after maximum cumulative sludge application of 1,800 t/ha (Logan et al., 1987). Cumulative Se applied came out to be 8.34 kg/ha, which is 834 times the recommended level of Se to be applied for raising Se levels of crops in Se-deficient areas of Finland or New Zealand (Korkman, 1985). Although sludge application increased the level of Se in soil from 0.1 to 1.2 mg/kg, it was not reflected in the Se uptake by crop plants. Possibly, Se is lost as H_2Se or $(CH_3)_2Se$ under aerobic conditions, especially in the presence of organic matter (Adriano, 1986). Heavy organic matter addition to the soil as compost favors the formation of volatile Se compounds resulting in losses of Se in the gaseous form (Kabata-Pendias and Pendias, 1984). Most of the Se in forest soils is associated with hydrophobic fulvates, which are very mobile and can easily leach down to lower horizons and ultimately contaminate the water bodies (Gustafsson and Johnsson, 1992). Frankenberger and Karlson (1994) reported that alkylselenide production in soil is often carbon limited, and it is possible to achieve >tenfold increase in volatile Se evolution with the addition of organic amendments to soil. Srikanth et al. (1992) studied the distribution of Se in both soil and perennial forage grass *Panicum maximum* (Guinea grass) cultivated in the sludge containing 4.6 to 9.4 mg Se/kg along the bank of River Musi, Hyderabad (India). They, however, found that the mean concentration of Se in guinea grass grown in sewage sludge ranged from 3.24 to 9.26 (mean 5.35) mg/kg, which was two to four times more than that of the control.

10.2.5 Groundwater

Besides through soil, Se can easily enter the food chain through water. The U.S. EPA has prescribed the upper limit of Se in water used for drinking purposes as 10 µg/L and that used for irrigation of crops as 20 µg/L. The Se content of groundwater is the lowest from Sweden and the highest from France (Table 10.3). Water from wells drilled into any of the geologic formations of the Cretaceous Colorado group in Central Montana (U.S.) may contain as much as 1000 µg Se/L (Donovan et al., 1987). The recommended dietary allowance for adults is 50 to 70 µg/day with correspondingly lower intake for younger age groups (McDowell, 1992). In most studies published on daily intake of Se, contribution of drinking water is neglected. Daily consumption of drinking water containing the EPA's upper limit of Se would be responsible for a significant fraction of total intake by human beings. At a water consumption of 2 L/day, drinking water constitutes about 1 to 6% of Se intake by humans in England (Commins, 1981).

In northwestern India, typical symptoms of Se toxicity, i.e., hair loss, deformation of nails, and nervous breakdown, are observed in human beings living in seleniferous regions

TABLE 10.3

Selenium Content (µg/L) of Ground Water Used for Irrigation of Crops and Drinking Purposes

Country	Irrigation Water	Drinking Water	Ref.
France	2.36–200	<2–10	Robberecht and Grieken (1982)
Israel	0.90–27	26–1800	Robberecht and Grieken (1982)
Italy	<0.02–1.94	—	Robberecht and Grieken (1982)
Sweden	0.11–0.15	0.061	Robberecht and Grieken (1982)
United States	<1–480	<0.2–3.5	Robberecht and Grieken (1982)
Australia	0.008–33	<1	Robberecht and Grieken (1982)
Argentina	48–67	—	Robberecht and Grieken (1982)
Belgium	<0.05–1.33	<0.05–0.842	Robberecht et al. (1983)
India	2.5–69.5	<0.05–0.843	Dhilon and Dhillon (1990)
Finland	—	0.013–1.034	Wang et al. (1991)

TABLE 10.4

Selenium Content (mg/kg) of Wheat and Soil as Influenced by Cropping Sequences

Cropping Sequence	Amount of Irrigation Water Applied per ha (cm)	Wheat (45-60 days old shoots)	Soil	
			Total	Available
Rice-wheat (n = 31)	200	162.5 ± 115.8	1.87 ± 0.92	0.047 + 0.018
Corn-wheat (n = 37)	60	8.2 ± 11.8	0.44 ± 0.28	0.022 ± 0.022

(Dhillon and Dhillon, 1997a). Selenium content of groundwater frequently used for drinking purposes particularly by field workers at the toxic sites varies from 2.5 to 69.5 µg/L. Daily intake of groundwater by field workers in tropical/subtropical countries may range from 5 to 7 L/day and it must be a substantial contribution to total Se intake.

Presence of large amount of Se in groundwater has accentuated the problem of Se toxicity in India (Dhillon and Dhillon, 1990). The rice-wheat sequence requires 3.3 times more irrigation water than the corn-wheat sequence. Wheat following rice, therefore, accumulated 20 times more Se than wheat following corn (Table 10.4). Toxicity symptoms of Se, i.e., snow-white chlorosis, appeared in wheat that followed rice continuously for 8 to 10 years.

In the San Joaquin Valley of California, irrigated farmland gave rise to highly saline shallow groundwater which was collected through subsurface drainage and delivered to Kesterson Reservoir for storage and reuse for irrigation purposes. The drainage water, essentially a soil leachate, commonly contained Se in the range of 250 to 350 µg/L (Presser and Barnes, 1985). Even concentrations up to 4200 µg/L have been reported in subsurface irrigation drainage water. Accumulating this drainage water just for 4 to 5 years resulted in Se levels beyond toxic limits and caused chronic and acute selenosis of the aquatic wildlife (Ohlendorf et al., 1986).

In different geographic regions, Se content in rainwater varied from <0.001 to 2.5 µg/L (Robberecht et al., 1983). Selenium originates in the atmosphere either from volatilization of Se through biological activity in aquatic (Chau et al., 1976) and terrestrial ecosystems (Doran and Alexander, 1977), or through burning of coal at high temperature (Campbell et al., 1978), incineration of refuge (Wagde et al., 1986), or fine particles generated through volcanic eruptions are washed down to the earth through rainwater. The total input of Se from wet, dry, vapor, and particulate deposition to the soil-herbage system varies from 0.2 to 0.7 mg/m^2·yr (Haygarth et al., 1991).

10.3 Selenium Content of Seleniferous Soils

Early research on Se content of seleniferous soils in the Great Plains of the United States was compiled by Anderson et al. (1961). In a monograph by Rosenfeld and Beath (1964), Se status of seleniferous soils from several other countries was described. More recently, Jacobs (1989) and Frankenberger and Benson (1994) have contributed state-of-the art chapters on Se in the soil-plant-animal system.

Most of the seleniferous soils in the United States seem to have originated from cretaceous sedimentary deposits consisting of shales, limestone, sandstone, and coal. Shales also form the principal source of Se in toxic soils of Ireland, Australia, and Israel. Distribution of Se in surface and subsurface soils is not uniform. In highly seleniferous areas of the Great Plains, Se content of surface soils ranged from 1.5 to 20 mg/kg and that of subsurface soil varied from 0.7 to 16 mg/kg. A maximum of 98 mg Se/kg has been recorded in the toxic region (Rosenfeld and Beath, 1964). Only recently, Se toxicity problems have developed as a result of disposal of Se-rich drainage water from irrigated farmland in San Joaquin Valley of California. Average Se content in soils from where drainage water is being collected ranged from 0.28 to 2.32 mg/kg (Seversen and Gaugh, 1992). In upper 20 cm soil, the Se content ranged from 4 to 25 and 0.7 to 1.5 mg/kg at Kesterson Reservoir and Lahontan Valley, respectively (Tokunaga et al., 1994).

In China, soils with elevated levels of Se exist in some large accumulation plains such as Sangliao, Weihe, and Hua Bei Plains. Soils containing total Se ≥3.0 mg/kg and water-soluble Se ≥ 0.02 mg/kg are associated with Se poisoning. In typical seleniferous soils of China, the water-soluble Se concentration was as high as 42.9 µg/kg (Tan et al., 1994).

Total and water-soluble Se in soils from the toxic region of northwestern India ranged from 0.23 to 4.55 and 0.02 to 0.16 mg/kg (Dhillon et al., 1992). Soils with as high as 10 mg Se/kg have been reported (Singh and Kumar, 1976), but no cases of Se poisoning in animals and human beings have been reported so far from this region.

Acute poisoning and chronic selenosis has been reported from the regions where total Se content in surface soils ranged from 0.3 to 0.7 mg/kg in Canada, 0.3 to 20 mg/kg in Mexico, 1 to 14 mg/kg in Columbia, 1.2 to 324.0 mg/kg in Ireland, and up to 6.0 mg/kg in Israel (Rosenfeld and Beath, 1964).

Forms of Se in soils and the conditions governing their solubility are discussed in detail by Zingaro and Cooper (1974), Vokal-Borek (1979), and Elrashidi et al. (1987). Haygarth et al. (1991) have critically reviewed the available information. Redox potential and pH are the most important parameters controlling solubility and chemical speciation of Se in cultivated soils. Identification of the chemical forms of Se in soils is very difficult because of the presence of Se in small amounts and complex matrix of soils. But recent innovations in analytical chemistry have allowed the scientists to trace out the forms of Se in minute details. Selenates and selenites are the major form of Se in agricultural soils. Soluble selenates are the form of Se in alkaline soils, whereas a large fraction of Se is present as selenite in acidic soils. Selenites and selenates can be reduced to elemental Se either through mildly reducing agents in acidic environments or by microorganisms. Insoluble selenides and elemental Se constitute the highly immobile forms of Se in poorly aerated reducing environments. Oxidation of elemental Se to selenite and trace amounts of selenate by certain microorganisms has also been reported by Sarathchandra and Watkinson (1987). Organic forms of Se such as seleno-amino acids represent an important source of plant available Se and selenomethionine is more bioavailable than selenocystine. In some Californian soils, nearly 50% of the Se may even be in the organic forms, i.e., as analogues of S-amino acids

(Abrams et al., 1990). Production of methylated derivatives of Se such as dimethyl selenide or dissolved organic selenide compounds through microbial processes has been noticed by Ganje and Whitehead (1958).

10.4 Restoration of Selenium-Toxic Soils

There are two main options available in restoration of soils contaminated with toxic metals:

1. On-site management of contaminants in order to reduce exposure risk
2. Excavation of the contaminated soil and transport off-site

The use of the second option is dictated by the size of contaminated site and availability of suitable landfill site. At present, off-site burial of contaminated soil is extensively being used in Australia. However, it should be regarded as a last resort treatment as it merely shifts the contamination problem elsewhere (Smith, 1993). On-site containment may provide an inexpensive and rapid solution in contrast to the problem associated with off-site transport of contaminated material (Ellis, 1992).

According to Pierzynski et al. (1994), the first option can be split into three categories:

1. Reduction of inorganic contaminant to an acceptable level
2. Isolation of contaminant to prevent any further reaction with the environment
3. Reducing the biological availability

Research efforts on restoration of seleniferous soils have been progressing on the lines as discussed above. Although Se-toxic soils have been known to exist in different parts of the world since the early 1930s, emphasis on restoration of Se-contaminated soils has greatly increased since Se contamination came into light at Kesterson Reservoir — a large shallow marsh (1200 acres) in California's San Joaquin Valley created to store and dispose-off agricultural drainage water.

Until the 1960s, when high Se areas were located predominantly in dry and nonagricultural regions, the management of toxic soils was limited to the mapping of seleniferous soils, withdrawing from cultivation of all food plants and maintaining as fenced farm, selection of safe routes for trailing of livestock, eradication of Se-accumulating plants, etc. (Rosenfeld and Beath, 1964). During the following decades, research efforts were increasingly aimed to identify the source and distribution of Se in the environment and to understand the mechanisms controlling its transfer and accumulation in soil-plant-animal-human system. More recently, when Se contamination is being associated with anthropogenic activities such as metal refining (Nriagu and Wong, 1983), fly ash waste (Adriano et al., 1980), agricultural drainage waters (Presser and Barnes, 1985), and irrigation practices (Dhillon and Dhillon, 1990), research efforts have shifted toward finding the practical means of complete removal or immobilization of Se in the contaminated system.

10.4.1 Bioremediation

Bioremediation is a well established technology for the removal of organic contaminants. Use of microorganisms to transform inorganic contaminants such as Se is now increasingly being considered to restore contaminated soil. Bioremediation results in the change in the oxidation state of Se, leading to forms which are less available to plants or which lead to volatilization/precipitation.

Selenium has long been known to undergo various oxidation and reduction reactions mediated through microorganisms that directly affect its oxidation state and behavior in the environment (Doran, 1982). The nature of Se reduction can be either dissimilatory, i.e., reduction of Se compounds as terminal electron acceptors in energy metabolism, or assimilatory, i.e., when Se compounds are reduced and used as a nutrient source (Brock and Madigan, 1991). Perhaps McCready et al. (1966) were the first to propose that the reduction of selenite to elemental Se via dissimilatory reduction can be a detoxification mechanism, as it enabled *Salmonella* to tolerate higher concentrations of selenite than other microorganisms. Kovalski et al. (1968) reported that mechanism of adaptation and resistance to high Se concentration of microorganisms isolated from high Se soils was the ability of these organisms to reduce Se to the elemental state. Among the microorganisms isolated from silty clay loam soil, 11% fungi, 48% actinomycetes, and 17% bacteria were capable of reducing selenate, and 3% fungi, 71% actinomycetes, and 43% bacteria could reduce selenite to elemental Se (Bautista and Alexander, 1972). Reduction of Se compounds as a result of microbial action was stimulated by the addition of available C source and no activity was noticed in steam sterilized soils (Doran and Alexander, 1977).

Due to chemical similarity of Se to S, many biogeochemical transformations of Se were generally regarded as nonspecific reactions catalyzed by enzymes involved in S biogeochemistry (Heider and Bock, 1993). However, it is now clear that some microorganisms have evolved biochemical mechanisms unrelated to S metabolism for using selenate, the most predominant form of oxidized Se in the environment, as a terminal electron acceptor (Losi and Frankenberger, 1997a). Oremland et al. (1991) reported that selenate respiring bacteria are ubiquitous in nature, functioning even in highly saline soils and sediments and the reduction reactions are unaffected by sulfate concentration. However, if selenate is reduced by sulfate reduction pathways, the presence of sulfate inhibits selenate reduction (Zehr and Oremland, 1987). *Thauera selenatis* isolated from Se-contaminated drainage water in California's San Joaquin Valley has been the most intensively studied selenate-reducing microorganism (Macy et al., 1993). It conserves energy to support growth by coupling the oxidation of acetate to the reduction of selenate to primarily selenite. In the presence of selenate and nitrate, the selenite produced from selenate reduction is further reduced to elemental Se (DeMoll-Decker and Macy, 1993). Another Se respiring organism, *Enterobacter cloacae* strain SLD1a-1 (Losi and Frankenberger, 1997b), is a facultative anaerobe, respiring selenate when grown anaerobically and reducing selenate to elemental Se. Still another selenate-reducing microorganism, designated SES-3, grows in a specific medium with lactate as the electron donor and selenate as the electron acceptor (Oremland, 1994).

10.4.1.1 *Bioremediation Technologies Based on Dissimilatory Se Reduction*

Based on the results of investigations carried out during the last decade, several microbial treatment technologies for *ex situ* remediation of Se-contaminated water have been projected for practical utilization (Gerhardt et al., 1991; Macy et al., 1993; Oremland, 1994; Lortie et al., 1992; Owens, 1997). It is difficult to compare the effectiveness of different technologies because of the variable conditions used. However, a common feature is that contaminated water is treated before disposal and includes a pretreatment step to remove nitrogen oxyanions. Water is passed through a system containing selenate-reducing microorganism. After immobilization, the elemental Se is separated out. Using pilot studies, the technologies have been found to possess potential to be economically feasible.

The process of bioremediation as proposed by Macy (1994) offers considerable improvement over that of Squires et al. (1989) and consists of *Thauera selenatis* gen. nov. sp. nov., which is able to reduce nitrate and selenate simultaneously using a different terminal reductase. Under optimum conditions in a bioreactor (e.g., correct pH and ammonia level) in 286 days, *T. selenatis* reduced selenate and selenite in drainage waters from 350 to 450 μg of Se/L to an average of 5 μg of Se/L and that of nitrate was reduced from 260 to 380 mg of N/L to <1 mg of N/L. A three-step biological treatment process, called Algal-Bacterial Selenium Removal System (ABSRS), to remove Se and nitrate from drainage waters was proposed by Lundquist et al. (1994). The system is patented as the Oswald Process. Aerobic algal growth removes nitrates to <10 mg N/L. In an anoxic unit, denitrifying and selenate-respiring bacteria carry out reduction of selenate to selenite in the biomass suspension before finally adding $FeCl_3$ to precipitate out Se. Soluble Se levels in the drainage waters were reduced from 200 to 400 μg/L, to 7 to 12 μg/L.

Oremland (1991) has also patented another process, in the first stage of which algae depletes nitrate concentration in the contaminated water to <1 m*M* under aerobic conditions. Water is then fed to an anoxic bioreactor containing selenate-respiring bacteria where selenate is reduced to insoluble elemental Se. On an overall basis, Se levels of more than 50 mg/L as selenate were reduced to less than 0.2 mg Se/L in 7 days of incubation. The Owens Process (Owens, 1997) used a technology based on anaerobic reduction of selenate to elemental Se. Selenium reduction will not take place until nitrate is consumed. After the consumption of nitrate, Se reduction takes place stepwise: from selenate to selenite to elemental Se.

A bench-scale plug-flow bioreactor inoculated with mixed *Pseudomonas* cultures has been designed and tested by Altringer et al. (1989). The reduction of selenate into elemental Se is a two-step reaction in which selenate is reduced to selenite, and then possibly to selenide, and eventually to red amorphous granules of elemental Se. After over 1 year of operation, steady-state rates of Se removal from simulated San Louis drainwater averaged up to 86%. Lortie et al. (1992) characterized a *Pseudomonas stutzeri* isolate which is capable of rapidly reducing both selenite and selenate into elemental Se at initial concentration of both oxyanions of Se up to 48.1 m*M*. Optimal Se reduction occurred under aerobic conditions.

10.4.1.2   *Deselenification through Volatilization*

Assimilatory reduction leads to synthesis of selenoamino acids, which can be more toxic than Se oxyanions (Besser et al., 1989). However, process of microbial transformation of toxic Se species into less toxic methylated volatile Se compounds has been developed into an important mechanism responsible for reducing Se concentration in the toxic environments. Bacteria and fungi are the two major groups of Se methylating organisms isolated from soils and sediments (Abu-Erreish et al., 1968; Doran, 1982); in water, bacteria possibly play a more dominant role (Thompson-Eagle and Frankenberger, 1991). Dimethylselenide (DMSe) is found to be the predominant product of microbial methylation of Se, which is 500 to 700 times less toxic than selenite and selenate ions. The pathway for methylation of inorganic Se as proposed by Doran (1982) is given as

$$SeO_3^{2-} \quad \rightarrow \quad Se^0 \quad \rightarrow \quad HSeX \quad \rightarrow \quad CH_3SeH \quad \rightarrow \quad (CH_3)_2Se$$

Selenite	Elemental Se	Selenide	Methane Selenol	Dimethyl selenide

Deselenification of toxic Se species, including selenate and various organoselenium compounds into a less toxic volatile form (DMSe), is apparently a widespread transformation in seleniferous environments (Chau et al., 1976; Doran, 1982). Intensive investigations carried out by Frankenberger and associates at the University of California (Riverside), on the characterization of naturally occurring microbial Se transformations and the factors affecting them, has led to the development of a sound, economically feasible bioremediation program for seleniferous environments. The microorganisms responsible for methylation of Se into DMSe are naturally present in saline, alkaline drainage waters, and soils, and their activity can be dramatically accelerated by the addition of specific amendments.

Frankenberger and Karlson (1989) hold patents for a land treatment technology to detoxify seleniferous soil through volatilization of methylated Se compounds. In a field study conducted for over 2 years at contaminated areas at Kesterson Reservoir for soils, containing Se concentration ranging from 10 to 209 mg/kg (media 39 mg/kg), the highest emission rate recorded was 808 μg Se/m$^2 \cdot$ h when soil was treated with citrus peel, ammonium nitrate, and zinc sulphate. Of the initial Se inventory, 62% was reduced in surface layer. Volatilization of Se in the field is related to the carbon source, aeration, moisture, and temperature.

Among different C sources tested, Se methylation rate was found to be the highest with pectin, resulting in Se removal up to 51.4% in 118 days (Karlson and Frankenberger, 1988). In an other field study conducted for 22 months (Frankenberger and Karlson, 1995), the most effective amendment was found to be the cattle manure, as it could remove 59% of Se inventory from a sediment composed mainly of clay. In a long-term field study carried out by Flury et al. (1997), 68 to 88% of the total Se (0 to 15 cm) was volatilized within a period of over 100 months. Casein-amended soils resulted in the highest Se removal rates and the process of volatilization was more active in the warmer and drier months. Natural bioremediation by Se volatilization and precipitation processes in aquatic environments by a eurhaline green microalga has been reported by Fan and Higashi (1997). A species of *Chlorella* isolated from a saline evaporation pond was shown to transform Se aerobically into a variety of alkylselenides as well as elemental Se.

As soon as Se is methylated into volatile compounds, it escapes into the atmosphere and gets diluted and dispersed by air currents away from the contaminated site. The inhaled DMSe is found to be nontoxic to animals at concentration up to 8034 mg/kg or 34,000 mg/m^3 (Frankenberger and Karlson, 1994).

10.4.2 Phytoremediation

The use of plants to remove contaminants from the soil is termed as phytoremediation. These plants are called "hyperaccumulators," as these can tolerate about 10 to 100 times higher metal concentration in their shoots than agronomic species. Most hyperaccumulators are endemic metallophytes and are used for locating economic mine deposits (Brooks et al., 1977). Hyperaccumulator plants should exhibit hypertolerance to metals in soils and shoots; extreme uptake of metals from soils and hypertranslocation of metals from roots to shoots. Chaney (1983) visualized the hyperaccumulating process as a method to remove soil contaminants and introduced the concept of developing a "phytoremediation crop" to decontaminate polluted soils. The value of metals in the biomass might offset part or all of the cost of cleaning up the toxic site. The higher the biomass and the higher the concentration of a metal in the biomass ash, the higher the economic value. Attempts have been made to identify Se hyperaccumulators and use them for managing the Se toxic soils (Banuelos et al., 1990; Parker et al., 1991; Wu et al., 1988). Parker and Page (1994) reviewed the work done on remediation of Se toxic soils using hyperaccumulator plants. The concept of phytoremediation has been employed to get rid of excessive Se from previously contaminated soils and sediments, to prevent Se migration in irrigation drainage water by reducing soluble soil Se level, and to decontaminate Se-enriched drainage water prior to discharge.

10.4.2.1   Characteristics of Soils and Crops Suitable for Phytoremediation

Soils which can be decontaminated using phytoremediation technology should be able to provide a suitable environment for adequate plant growth. Selenium-toxic soils of the United States and in most other countries are alkaline in reaction, contain free $CaCO_3$ and lie in a region of low rainfall, and are sporadically distributed in highly productive areas of the world (Anderson et al., 1961). About 1000 acres of seleniferous soils located in north-western India are alkaline and calcareous in nature (Dhillon et al., 1992). Notwithstanding the high level of Se, the soils are highly productive. However, in San Joaquin Valley of California, the Se problem is not just a localized problem. Nearly 16,000 ha of farmland are affected by high levels of salinity and water table where excessive Se and boron are coexisting problems. Salinity and B levels of exposed evaporation ponds sediments range from 14.0 to 52.8 dS/m and 17 to 55 mg/L, respectively (Retana et al., 1993).

Soil characteristics of the respective areas will determine the suitability of a particular phytoremediation crop. An ideal phytoremediation crop should possess the following characteristics:

1. Rapidity and ease of establishment
2. Potential persistence of the crop
3. Management and harvesting using conventional equipment
4. Deep and extensive root system
5. Higher capacity to bioaccumulate Se and biomass

10.4.2.2   Classification of Selenium-Accumulating Plant Species

Plants can be classified into three groups on the basis of their ability to accumulate selenium when grown on seleniferous soils (Rosenfeld and Beath, 1964).

10.4.2.2.1   Primary Accumulators or Hyperaccumulators

Plants which are capable of accumulating Se in excess of 100 mg/kg dry weight. These prefer to grow on seleniferous soils and include many species of *Astragalus*, *Oonopsis*, and *Stanleya*.

10.4.2.2.2   Secondary Accumulators

Plants which may accumulate more than 50 to 100 mg Se/kg, e.g., *Aster* and some species of *Astragalus* and *Atriplex*.

10.4.2.2.3   Nonaccumulators

Plants which do not normally accumulate Se in excess of 50 mg/kg when grown on seleniferous soils, e.g., grasses and other cultivated plants. However, some members of so-called nonaccumulators (e.g., *Brassica* spp.) can accumulate large amounts of Se without showing phytotoxicity symptoms (Banuelos et al., 1990) and may be properly categorized as secondary accumulators.

The accumulator species possess a unique pathway wherein Se is incorporated in specialized and nontoxic amino acids, Se methylselenocysteine, and Se methylselenocystathionine, which are not found in nonaccumulating species (Brown and Shrift, 1981). Exclusion of Se from proteins of accumulators is thought to be the basis of Se tolerance. Selenium absorbed by nonaccumulating plants is converted into Se metabolites which are analogs of essential S compounds and interfere with cellular biochemical reactions resulting in disturbed protein metabolism. Studies with nonaccumulating species revealed a positive relationship between increase in overall plant tissue Se concentration and the protein Se concentration, and the increase in protein Se concentration was associated with the reduction of plant growth (Brown and Shrift, 1980). Thus, there exists a mechanism in nonaccumulating plants which restrict Se uptake by plants with a greater Se tolerance and consequently reducing the incorporation of Se into its proteins.

10.4.2.3 *Phytoremediation as a Technology*

The concept of phytoremediation came into being in early 1980s when a large tract of Se toxic soils in San Joaquin Valley was discovered. Within a short span of time, phytoremediation has now developed into a full-fledged technology, shown by the recently held symposium on Phytoremediation of Trace Element Contaminated Soils and Water at the University of California, Berkeley, in June 1997. Both hyperaccumulator and nonaccumulator plants have been tested for the possibility of their involvement in phytoremediation.

10.4.2.3.1 *Hyperaccumulators*

Prior to 1970s, accumulators were not used for remediation of Se-contaminated soils although these have been known ever since the problem of Se toxicity was recognized. Selenium accumulators are known to thrive very well on Se toxic soils, but so far no significant efforts have been made to include them in phytoremediation strategy. Recently, when it was realized that none of the known crop species can be established on highly saline seleniferous soils of San Joaquin Valley, attention was diverted to screen Se accumulators for their adaptation in these circumstances (Parker et al., 1991). Only two Se accumulating species, i.e., *Astragalus racemosus* and *A. bisulcatus*, exhibited EC_{50} values >20 dS/m and their growth was also unaffected by the B concentration up to 4 mM during seedling growth in solution/sand culture. These plants could accumulate 600 to 700 mg Se/kg in the shoots and could reduce Se inventory by 3 to 4 kg/ha in a greenhouse study using a column of soil collected from a Se toxic area of the Kesterson Reservoir (Retana et al., 1993). Highly saline drainage waters are also accompanied by high SO_4 levels that inhibit absorption of Se (Mikkelsen et al., 1988a). But Se accumulation by primary accumulators was not affected by the presence of sulfate ions in soil solution; rather these plants could maintain relative preference for Se absorption (Bell et al., 1992). Tolerance of Se accumulators to high soluble salts including boron and sulfate and high Se accumulation capacity warrants their inclusion in future studies on the suitability of vegetation for remediation of Se-enriched soils. Harvesting these species and their removal from the site could contribute significantly to Se dissipation strategies. Extraction of Se from harvested biomass may turn out to be a profitable proposition as prophesied by Chaney (1983). The practicality of including Se accumulators in remedial measures, however, may be limited, because they are (1) genetically poor, (2) susceptible to pests and diseases, (3) not responsive to fertilizer application, and (4) seed is not commercially available (Parker and Page, 1994). Thus, there is need to bring genetic improvement in these species so as to make their involvement possible in future strategies of remediation of Se-toxic soils.

10.4.2.3.2 *Nonaccumulating Species*

An important step toward remediation of a seleniferous area will be to identify crop plant species that can tolerate high levels of both Se and salinity for their possible use in bioextraction of Se from deteriorated agricultural soils. Among the nonaccumulating plant species, tall fescue (*Festuca arundinacea* Schred.) (Wu et al., 1988) and mustard (*Brassica juncea*) (Banuelos et al., 1990) have emerged as the possible choice for their use in phytoremediation strategies. Both of these species possess all the important characteristics of a phytoremediation crop and present a promising potential for their use on Se toxic soils even with high level of salinity.

Among the five crop species tested, tall fescue displayed the greatest tolerance to both high levels of Se and salinity in solution culture at 1 and 2 mg Se/L (Wu et al., 1988). It accumulated Se up to 200 mg/kg in the root and 400 mg/kg in shoot with very little reduction in plant dry matter production. The amount of Se accumulated by different plant species was inversely related to their Se tolerance. In a field study where tall fescue was grown for 1 year, Se level was reduced by 50% in the top 15 cm of soil. In another study, although tall wheatgrass [*Elytrigia pontica* (Podb.) Holb.], alkali sacaton [*Sporobolus airoides* (Torr.) Torr.] and weeping alkaligrass [*Puccinelia distans* (L.) Parlat] could tolerate salinity up to 20 dS/m and B up to 4 mM and accumulate large amounts of dry matter (9.5 to 14.5 t/ha); these plants did not qualify for use in the Se remediation process because of low Se tissue concentration (Parker et al., 1991).

In normal soils, sunflower (*Helianthus annus* L.) absorbed more than 400 mg Se/kg. Among cereals, wheat (*Triticum aestivum* L.) accumulated more than 80 mg Se/kg without showing toxicity symptoms (Hamilton and Beath, 1963; Dhillon and Dhillon, 1991a). Wild mustard, moderately tolerant to salinity, could accumulate several hundred milligrams Se/g without Se phytotoxicity (Banuelos et al., 1990). In a 2-year field study, Brassica species resulted in a 36% decrease in total Se concentration in surface soil (0 to 45 cm) and 48% in subsurface soil (45 to 90 cm) (Banuelos et al., 1997). Banuelos et al. (1993a) observed that growing of either wild mustard, canola, or tall fescue in a greenhouse reduced total Se from the soil by 41, 31, and 38%, respectively. Repeated irrigation with saline irrigation water rich in Se (154 µg Se/L) resulted in significant increase in Se content of wild mustard (Banuelos et al., 1993b). Thus, wild mustard allowed for dissipation of Se from contaminated soils and also for the disposal of potentially toxic drainage water. In fact, wild mustard was found to be better suited to short-term reclamation due to its higher total uptake and constant uptake rate. Tall fescue, on the other hand, may prove better for long-term reclamation as it absorbs Se slowly and is removed from the soil by repeated harvesting.

Agroforestory farming practices offer another novel phytoremediation technique to remove Se from the soil-plant system (Cervinka, 1994). By growing salt-tolerant trees like *Euclyptus* and halophytic plants, the volume of contaminated drainage water is substantially reduced. Obviously, removal of excess Se and other salts is facilitated from a smaller volume of water. More work needs to be carried out to determine the tolerance of *Euclyptus* and other trees and shrubs to the extreme salinities and their ability to selectively take up selenate.

10.4.2.4 *Phytovolatilization*

Besides accumulating Se in their tissues, plants can scavenge Se from contaminated soils and convert it to volatile forms such as dimethylselenide (DMSe) and dimethyldiselinide (DMDSe). This process termed as "phytovolatilization" when coupled with total Se uptake from the soil is being actively attempted as a practical methodology for the cleanup of the elevated levels of Se in soils. Working with Se accumulator plants (*A. racemosus*), the release of volatile Se compounds from intact higher plants was first shown by Lewis et al. (1966). Subsequently, this phenomenon was found to occur in nonaccumulator plants such as alfalfa. Phytovolatilization is widespread among various plant species. Terry et al. (1992) recorded that rice, broccoli, and cabbage are the best Se volatilizers among 15 crop plants studied and volatilized Se at 1500 to 2500 µg/kg · day on dry weight basis. The uptake and volatilization of Se by agricultural crops is dependent upon several environmental, chemical, and biological factors such as temperature, light, concentration of competitive ions, the concentration and chemical species of Se, plant age, and the presence of certain microbial species (Terry and Zayed, 1994). In spite of the smaller mass and lower Se concentration, the plant root rather than shoot is the primary site of Se volatilization. Interestingly, removal of shoots substantially enhanced the rate of Se volatilization by roots (Terry and Zayed, 1994).

Indian mustard has the highest rate of Se volatilization compared to other species when the data for root and shoot volatilization were considered together (Banuelos et al., 1995). With high Se uptake and volatilization rates, Brassica is emerging as one of the most valuable phytoremediation crop. More work needs to be carried out under actual field conditions to quantify total Se removal through uptake and volatilization by both cultivated and noncultivated plants. Transgenic plants are being generated taking Indian mustard as a model plant with an enhanced ability to volatilize Se and to phytoremediate Se-contaminated environments (Terry and Zayed, 1997). Biochemical pathways for the Se volatilization by both accumulators and nonaccumulators as described by Terry and Zayed (1994) has a number of key-limiting steps that must be targeted by molecular biologists for further improvements.

10.5 Other Remedial Measures

10.5.1 Covering Selenium-Contaminated Sites with Selenium-Free Soil

The simplest form of on-site containment is achieved by covering the soil with a clean material (Jefferis, 1992). This technique has been attempted for reclaiming seleniferous soils at Kesterson Reservoir. In the fall of 1988, when discharge of agricultural drainage water was terminated, the evaporation ponds at Kesterson Reservoir were covered with a clean soil layer of 15 cm thickness having water-soluble Se content between 15 to 20 $\mu g/kg$. Effectiveness of the clean cover on the contaminated soil was monitored for 2 years in comparison with the native soil (Wu, 1994). At the two monitoring points, water-extractable Se content of the soil below the covering layer was 273 and 233 $\mu g/kg$. For the first year, water-soluble Se content of the cover soil at one of the sites increased from 20 to 600 $\mu g/kg$ and, thereafter, it decreased. At the other site, Se content of cover soil remained almost unchanged. Because all the plant species established in the new environments had deep root system, large absorption of Se from the underlying contaminated layer was observed. At another site within Kesterson Reservoir covered with 0.53-m-deep imported nonseleniferous soil in 1988, a significant upward movement of Se^{6+} into nonseleniferous soil was observed (Tokunaga et al., 1994). Although no detrimental effect of Se on wildlife has been reported from the area covered with clean soil, it cannot be conceived as a long-term measure. Obviously, Se remains in place and is accessible to deep-rooted plants, and is prone to upward and downward movement with time.

10.5.2 Permanent Flooding

Flooding of soil environments results in reduced conditions where $H_2SeO_3^-$ or SeO_3^{2+} are the important Se species present in adsorbed form on the surface of hydrous sesquioxides (Cary et el., 1967). Thus, under anoxic conditions Se immobilizes into an insoluble fraction which is unavailable to aquatic biota. This technique will, however, be applicable only in situations where permanent flooding is feasible both in terms of availability of good quality water and soil use. Permanent flooding of Se toxic areas at Kesterson Reservoir with low Se water was proposed by Horne (1991) as another detoxification measure. Initially, storage of Se-contaminated drainage water has resulted in heavily contaminated biota containing Se varying from 100 to 300 mg/kg (on dry weight basis). The hypothesis was tested in a mesocosm at Kesterson Reservoir over 2 to 3 years by measuring the decline in Se content of plants and animals. A rapid initial decline in Se, lasting a few months, was followed by a slower, irregular but persistent decrease in Se content of biota. Selenium level decreased to a lower level (3 to 4 mg/kg) in vegetation than in animals (14 to 15 mg/kg). The microalga *Chara* was the most common submerged vegetation and reduced Se levels in it were considered safe for wildlife. This is important from the point of view of the detoxification process of the entire wetland. Permanent flooding seems to detoxify the aquatic environments faster than methods such as microbial volatilization and extraction via vegetation. A disadvantage of permanent flooding is that the toxicant remains in place and can be reactivated by drying and re-wetting. Thus, even the ecological advantage is temporary.

10.5.3   Chemical Immobilization

Bioavailability of a toxic element is governed by the chemical forms that will affect the life cycle of plant and other organisms. In fact, only bioavailable fractions of contaminants pose a toxicological or an environmental risk. Selenium exists in the soil in several chemical forms that differ in their solubility and availability to plants. Selenate is the most mobile form of Se and can be immobilized or made biologically unavailable by reduction to elemental Se or by formation of selenides or Se-sulphides. Major factors controlling the mobility and bioavailability of Se are pH and redox condition, adsorption to soil particles and organic matter, and the presence of competitive ions.

10.5.3.1   pH and Redox Conditions

Soil pH and redox potential affect plant availability by changing the oxidation state and reducing the mobility of Se (Masscheleyn et al., 1990). In acid and neutral soils, Se is commonly found as Se^{4+} complexes of oxides and oxyhydroxides of ferric iron with extremely low solubility and is largely unavailable to plants (Cary and Allaway, 1969; Hamdy and Gissel-Nielsen, 1977). In neutral and alkaline soils, the Se^{6+} oxidation state predominates, which is generally soluble, mobile, and readily available for plant uptake (Soltanpour and Workman, 1980; Ylaranta, 1983a). Leaching studies with selenate and selenite in a sandy loam soil adjusted to pH 2 to 9 revealed that selenate was mobile at all pH values and was completely leached from a 30-cm-long column with <3 pore volumes, whereas selenite was only slightly leached even with 50 pore volumes of solution (Ahlrichs and Hossner, 1987).

10.5.3.2   Adsorption of Selenium in Soil Environment

Reduced bioavailability of Se can be achieved through complexation resulting in the reduction of soil solution concentration of Se. Soluble selenate and selenite forms of Se may be rendered unavailable to plants due to adsorption on soil particles. Once adsorbed, such forms are poorly exchangeable (Neal, 1990). Ylaranta (1983b) did not observe any adsorption of selenate, but 77% of the added selenite was adsorbed onto clay soil, 34% in a fine sandy soil, and 39% in a carex peat. Christensen et al. (1989) found that just after 1 day, fixation on clay, silt, sand, and whole soil was 78–87, 67–79, 3–14, and 31–39%, respectively. Fixation of Se correlated positively with clay content, iron content, and surface area of the soils and negatively with sulphuric acid extractable P (Hamdy and Gissel-Nielsen, 1977; Rajan and Watkinson, 1976). Christensen et al. (1989) removed organic matter from clay using hydrogen peroxide and showed that fixation capacity was reduced by 50%, thus demonstrating the importance of organic matter in fixing Se added to the soil. Bisbjerg and Gissel-Nielsen (1969) reported that Se uptake by plants from a muck soil (14% organic mater) was ten times less than that from some mineral soils. Singh et al. (1981) observed that adsorption of selenate and selenite was higher in a soil with elevated levels of organic carbon (0.9%) than from a soil with lower organic carbon content (0.4%). Soil organic matter was negatively correlated with Se uptake by plants (Ylaranta, 1983b). After adding inorganic selenite and selenate in water corresponding to 500 mm rainfall, Ylaranta (1982) observed that only 0.2% of Se added to clay and sandy soil was leached, but selenite and selenate were leached from peat up to 7 and 84%, respectively. This indicates that although soil organic mater was negatively correlated to uptake of Se by plants, yet Se after complexing with organic matter is susceptible to leaching losses. Selenium is mainly associated with hydrophobic fulvates which are very mobile (Gustafsson and Johnsson, 1992). Thus, Se losses may not be high from clay and sandy soils, but will be significant from organic matter-rich soils.

Bioavailability of Se can also be substantially reduced due to complexation with Fe and Al oxides and oxyhydroxides. Selenium toxicosis has been often reported from areas where soils contained more than 1 mg Se/kg, but in some acidic Hawaiian soils containing large amounts of sesquioxides and Se ranging from 1 to 20 mg/kg, no Se toxicosis has been reported (Anderson et al., 1961). In acidic soils, Se is commonly found as selenium complexes of oxides and oxyhydroxides of ferric iron with extremely low solubility, and in such complexes Se is largely unavailable to plants (Cary and Allaway, 1969; Hamdy and Gissel-Nielsen, 1977). In seleniferous soils of Israel, barium chloride was found to be the most effective chemical in reducing Se absorption by alfalfa (Ravikovitch and Margolin, 1959). Even small application of barium chloride virtually eliminated uptake of selenate-Se by alfalfa. The mechanism responsible for decreased Se uptake seems to be the formation of sparingly soluble barium selenate. A major inherent problem associated with immobilization is that although Se becomes less bioavailable, its concentration in soil remains unchanged. On-site immobilization is technologically complex and expensive, and it has been suggested that it should be restricted to highly contaminated soils (Peters and Shem, 1992). Addition of organic matter and iron oxide can be easily recommended as a practical measure to counteract the Se toxicity problems.

A typical example of the role of alternatively reduced and oxidized condition in regulating Se availability in the soil has been reported by Dhillon and Dhillon (1991a). In northwestern India, typical Se toxicity symptoms of snow-white chlorosis have been observed in wheat crop which followed rice (*Oryza sativa* L.) after completion of 8 to 10 cycles of rice-wheat sequence at the same site. Rice is grown under flooded conditions and accordingly the Se content of rice straw and grain was, respectively, two to three and four to five times less than that of following crop of wheat grown under upland conditions. Wheat straw and grain contained 17.01 ± 14.8 and 33.1 ± 15.3 mg Se/kg, respectively.

10.5.4 Presence of Competitive Ions in Soil Solution

The selenium accumulation by plants is significantly influenced by the presence of other ions like sulphate, phosphate, and nitrate in the growth medium. The antagonistic interaction between sulfate and selenate for plant uptake was observed by Hurd-Karrer in as early as 1938 (Hurd-Karrer, 1938). In recent decades, this relationship has been confirmed in the greenhouse and field studies (Pratley and McFarlane, 1974; Mikkelsen et al., 1988b; Dhillon et al., 1977; Mikkelsen and Wan, 1990; Bawa et al., 1990). Reduction in Se absorption by 60 to 70% in a number of crops has been achieved by application of S through gypsum in alkaline calcareous seleniferous soils of northwestern India. Farmers of the region have adopted this practice as a practical measure for reducing transfer of Se from soil to food chain crops (Dhillon and Dhillon, 1991b, 1997b). Among S sources tested, $(NH_4)_2SO_4$ was the most effective in reducing Se uptake from seleniferous soils of the U.K. (Williams and Thornton, 1972). Probably, both NH_4^+ and SO_4^{2-} ions are involved in reducing Se uptake. Most of the soils in the United States that produce seleniferous vegetation are already naturally high in sulphate-S. Allaway (1970) suggested that had these soils contained little or no S, accumulation of Se by plants might have even been much higher than the present level. In California, the seleniferous drainage waters generally contain high levels of sulphate salinity. Mikkelsen et al. (1988a) observed that plant Se was reduced from 620 mg Se/kg to less than 7 mg Se/kg in the presence of sulphate salinity. Recently, in contrast to antagonistic effect, synergistic effect of Se and S has been reported at low levels of S in the soil solution (Mikkelsen and Wan, 1990).

Presence of phosphate ion in the soil may either decrease or increase Se uptake by plants (Carter et al., 1972; Levesque, 1974; Singh and Malhotra, 1976). Plants grown on single-superphosphate–amended soils generally had lower concentration than monocalcium phosphate-amended soils (Davies and Watkinson, 1966). Possibly, gypsum present in the superphosphate reduced the bioavailability of Se.

10.5.5   Selecting Plants with Low Selenium Absorption Capacity

As early as 1938, Hurd-Karrer reported that Se absorption capacity of plants belonging to the gramineae family is the lowest compared to leguminoseae and cruciferae families. These differences were later on confirmed by many research workers (Fleming, 1962; Hamilton and Beath, 1963; Dhillon et al., 1977). Plants absorbing the least amount of Se have, thus, been recommended for cultivation in seleniferous areas (Bawa et al., 1992; Dhillon and Dhillon, 1997b). In greenhouse experiments, Se absorption capacity of cereal and leguminous fodder crops commonly grown in the seleniferous region was investigated. Up to a level of 0.25 mg Se/kg soil, the differences in the Se content of different fodders was negligible, but at higher Se levels differences in Se accumulation became apparent. Oat (*Avena sativa*) and sorghum (*Sorghum bicolor*) among cereals and senji (*Melilotus parviflora*) among leguminous crops have been recommended as fodder crops in seleniferous area of northwestern India because these absorb the least amount of Se compared to other fodder crops. In the case of fodders like berseem (*Trifolium alexandrinum*) and lucerne (*Medicago sativa*), the first one/two cuts contain two to three times more Se than the following cuts. The farmers have been advised to avoid feeding of the first cut of berseem and the first two cuts of lucerne to animals.

10.6   Conclusions

Selenium has accumulated to toxic levels in the environment because of natural weathering of Se-containing rocks and additions through anthropogenic activities such as coal combustion, use of fly ash and sewage sludge as a substitute for fertilizer, and changes in cropping pattern and irrigation management. Soils that can supply sufficient Se to produce vegetation containing >5 mg Se/kg, the maximum permissible levels, are referred to as seleniferous soils. In general, cultivated soils containing more than 0.5 mg Se/kg can be potentially toxic. The phytoavailability of Se in soils is related to selenite and selenate, the dominant forms of Se in soils and aquatic environments. In some of the soils around 50% of Se exists in organic forms such as selenomethionine and selenocystine.

Microbial action can change the speciation of Se by mediating oxidation/reduction reaction or through the formation of volatile organic Se compounds. The capacity of microorganisms and plants to change the Se speciation has been advocated as a possible means of restoration of soils containing excessive levels of Se. Bioreduction of selenate, the most labile and highly toxic inorganic form of Se, through assimilatory or dissimilatory reactions results in lowering of Se levels in soils and aquatic environments. A number of micro-organisms specifically involved in the Se reduction pathway have been identified, isolated, and characterized for their potential to reduce soluble Se to elemental form as well as volatile Se compounds. Dissimilatory reduction reaction has been more successfully employed for bioremediation of contaminated waters. Among all the organisms, *Thauera selenatis* has been found to be the most efficient strain for reduction of Se oxyanions. Even in the presence of high nitrate-N levels, Se in the drainage water can be effectively reduced to tolerable levels. Dimethylselenide is the dominant volatile species in seleniferous soils and water. Volatilization of Se under field condition is controlled by the carbon source, aeration, moisture, and temperature.

Two Se accumulating *Astragalus* spp. and two nonaccumulating spp. (tall fescue and Indian mustard) which are quite tolerant to high salinity, B, and Se levels have been found to be the most suitable phytoremediation crops. Cultivation of these species and their removal from the site may result in significant reduction in soil Se inventory even in the presence of high sulphate levels in the soil and groundwater. Higher uptake of Se has been found to be correlated with the rate of Se volatilization of different plant species. Under field conditions, Indian mustard contributed to lowering of soil Se concentration by almost 50% in the 0- to 75-cm layer after 3 years.

For temporary relief, use of soil amendments like gypsum, clay minerals, and iron oxide can reduce entry of Se into the food chain of humans and animals. Permanent flooding, if permissible as per regional soil use, can be the best choice.

10.7 Future Research Needs

Significant advances have been made during the last one and a half decades for decontamination of Se toxic soils, using innovative techniques such as bioremediation, phytoremediation, or immobilization. Discovery of dissimilatory Se reduction can be considered as one major achievement of the last decade. Preliminary research work has been completed regarding identification and isolation of microorganisms and their efficiency in the conversion of oxyanions to elemental Se. However, more needs to be learned regarding organisms that carry out the reaction, and still more efficient strains need to be identified so as to develop cost-effective remediation schemes for on-site management of irrigation waters before their actual disposal on land. When Se is present throughout the soil profile, efforts should be made to increase the effectiveness of microorganisms in the lower layers.

Uptake and volatilization of Se by plants are the two important components determining the effectiveness of phytoremediation technology. Although tall fescue and the Indian mustard have been found to be the most suitable choice as phytoremediation crops, crop rotations that include these crops need to be worked out for seleniferous soils around the world. Using genetic engineering of transgenic plants or by applying chemical modifiers or microbial inoculations, new ways need to be developed for further enhancing Se volatilization by plants in the fields so as to achieve the complete remediation within the shortest possible time. *Brassica* species suffer from one drawback that almost all the leaves are shed at the time of maturity. If these leaves are not removed, the whole purpose of phyto-extraction is forfeited. Next to crucifarae are the leguminoseae plants, which accumulate significant amounts of Se. If high metal uptake and volatilization characteristics are transferred to some multicut leguminous crops like alfalfa through plant breeding and bio-technological approach, annual removal of Se can be substantially increased. With genetic engineering, Se uptake characteristics of plants can be enhanced if the genes for Se accumulation can be identified and manipulated. It would also allow transgressing of genes responsible for Se hyperaccumulation to inedible but very productive and sterile host plants. Protoplast fusion techniques may also be employed to achieve these objectives. Breeding experiments need to be initiated to develop plant species with a deep and most extensive and efficient root system and also with greatest resistance to diseases. It is also important to study the residual fraction of the bioavailable and fixed forms of Se in soils following phytoextraction and the kinetics of their re-equilibration.

The screening of hyperaccumulating natural plants and their conversion into more viable commercial phytoremediation crops using genetic engineering techniques should be emphasized. There is need to develop cultivars which may accumulate Se content more than 10% in the ash, so that a standard commercial Se smelting technology can support a profitable phytoremediation technology.

The main goal of immobilization is to reduce risk of an uncontrolled Se transfer in ground water and the biosphere. To achieve this objective, plant available Se fractions in soil need to be inactivated. This has been achieved by increasing Se binding capacity of soil with addition of clay minerals, iron oxides, barium chloride, organic matter, or change in the water regime. The effect of applied materials on the availability of other nutrient element, however, needs to be investigated.

References

Abrams, M.M., R.G. Burau, and R.J. Zasoski, Organic selenium distribution in selected Californian soils, *Soil Sci. Soc. Am. J.*, 54, 979, 1990.

Abu-Erreish, G.M. and J.N. Lahham, Selenium in soils and plants of the Jordan Valley, *J. Arid Environ.*, 12, 1, 1987.

Abu-Erreish, G.M., W.I. Whitehead, and O.E. Olson, Evolution of volatile selenium from soils, *Soil Sci.*, 106, 415, 1968.

Adriano, D.C., *Trace Elements in Terrestrial Environment*, Springer-Verlag, New York, 1986.

Adriano, D.C., A.L. Page, A.A. Elseewi, A.C. Chang, and I. Straughan, Utilization and disposal of fly ash and other coal residues in terrestrial ecosystems: a review, *J. Environ. Qual.*, 9, 333, 1980.

Ahlrichs, J. S. and L.R. Hossner, Selenate and selenite mobility in overburden by saturated flow, *J. Environ. Qual.*, 16, 95, 1987.

Allaway, W.H., Sulphur-selenium relationship in soils and plants, *Sulphur Inst. J.*, 6, 3, 1970.

Altringer, P.B., D.M. Larsen and K.R. Gardiner, Bench scale process development of selenium removal from wastewater using facultative bacteria, in *Biohydrometallurgy*, Selly, J., McCready, R.J.L., and Wichlacz, W.H., Eds., CANMET, Ottawa, Ontario, Canada, 1989, 643.

Anderson, M.S., H.W. Lakin, K.C. Beeson, F.F. Smith, and E. Thacker, Eds., *Selenium in Agriculture*, USDA Handbook 200, U. S. Government Printing Office, Washington, D.C., 1961.

Arora, S.P., P. Kaur, S.S. Khirwar, R.C. Chopra, and R.C. Ludri, Selenium levels in fodders and its relationship with Degnala disease, *Indian J. Dairy Sci.*, 28, 246, 1975.

Banuelos, G.S., D.W. Meek, and G.J. Hoffman, The influence of selenium, salinity and boron on selenium uptake in wild mustard, *Plant and Soil*, 127, 201, 1990.

Banuelos, G.S., G.E. Cardon, C.J. Phene, L. Wu, S. Akohoue, and S. Zambrzuski, Soil boron and selenium removal by three plant species, *Plant and Soil*, 148, 253, 1993a.

Banuelos, G.S., R. Mead, and G.J. Hoffman, Accumulation of selenium in wild mustard irrigated with agricultural effluent, *Agric. Ecosyst. and Environ.*, 43, 119, 1993b.

Banuelos, G.S., N. Terry, A.M. Zayed, and L. Wu, Managing High Soil Selenium with Phytoremediation, Paper presented at the 1995 National Meeting of the American Society for Surface Mining and Reclamation, Gillette, WY, 1995.

Banuelos, G., H. Ajwa, S. Zambrzuski, and S. Downy, Phytoremediation of selinium-laden field soils, paper presented at the symposium on Phytoremediation of Trace Element Contaminated Soil and Water, University of California, Berkeley, 1997.

Bautista, E.M. and M. Alexander, Reduction of inorganic compounds by soil microorganisms, *Proc. Soil Sci. Soc. Am.*, 36, 918, 1972.

Bawa, S.S., S.K. Dhillon, and K.S. Dhillon, Effect of sulphur application on the absorption of selenium by different fodder crops, *Indian J. Dairy Sci.*, 43, 564, 1990.

Bawa, S.S., K.S. Dhillon, and S.K. Dhillon, Screening of different fodders for selenium absorption capacity, *Indian J. Dairy Sci.*, 45, 457, 1992.

Bell, P.F., D.R. Parker, and A.L. Page, Contrasting selenate-sulphate interaction in selenium accumulating and non-accumulating plant species, *Soil Sci. Soc. Am. J.*, 56, 1818, 1992.

Besser, J.M., J.N. Huckins, E.E. Little, and T.W. La-Point, Distribution and bioaccumulation of selenium in aquatic microcosms, *Environ. Pollut.*, 62, 1, 1989.

Bisbjerg, B. and G. Gissel-Nielsen, The uptake of applied selenium by agricultural plants. I. The influence of soil type and plant species, *Plant and Soil*, 31, 287, 1969.

Boon, D.Y. and P.J. Smith, Carbonaceous materials: problems associated with reclamation, *Am. Soc. Surface Mining and Reclamation*, 2, 248, 1985.

Bramley, R.G.V., Cadmium in New Zealand agriculture, *N.Z. J. Agric. Res.*, 33, 505, 1990.

Brock, T.D. and M.T. Madigan, *Biology of Microorganisms*, Prentice-Hall, Englewood Cliffs, NJ, 1991, 585.

Brooks, R.R., J. Lee, R.D. Reeves, and T. Jaffir, Detection of nickeliferous rocks by analysis of herbarium specimens of indicator plants, *J. Geochem. Exploration*, 7, 49, 1977.

Brown, T.A. and A. Shrift, Identification of selenocysteine in the proteins of selenate-grown *Vigna radiata*, *Plant Physiol.*, 66, 758, 1980.

Brown, T.A. and A. Shrift, Exclusion of selenium from protein of selenium tolerant *Astraglus* species, *Plant Physiol.*, 67, 1051, 1981.

Brown, T.A and A. Shrift, Selenium: toxicity and tolerance in higher plants, *Biol. Rev.*, 57, 59, 1982.

Campbell, J.A., J.C. Laul, K.K. Nielsen, and R.D. Smith, Separation and chemical characterization of finely sized fly ash particles, *Anal. Chem.*, 50, 1032, 1978.

Carter, D.L., C.W. Robbins, and M.J. Brown, Effect of phosphorus fertilization on the selenium concentration in alfalfa, *Proc. Soil Sci. Soc. Am.*, 36, 624, 1972.

Cary, E.E. and W.H. Allaway, The stability of different forms of selenium applied to low-selenium soils, *Proc. Soil Sci. Soc. Am.*, 33, 571, 1969.

Cary, E.E., G.W. Wieczorek, and W.H. Allaway, Reactions of selenite-selenium added to soils that produce low selenium forages, *Proc. Soil Sci. Soc. Am.*, 31, 21, 1967.

Cervinka, V., Agroforestry farming system for the management of selenium and salt on irrigated farmland, in *Selenium in the Environment*, W.T. Frankenberger, Jr. and S. Benson, Eds., Marcel Dekker, New York, 1994, 237.

Chaney, R.L., Plant uptake of inorganic waste constituents, in *Land Treatment of Hazardous Wastes*, Parr, J.F., Marsh, P.B., and Kla, J.M., Eds., Noyes Data Corp., Park Ridge, NJ, 1983, 50.

Chaney, R.L., Final Report of the Workshop on the International Transportation, Utilization or Disposal of Sewage Sludge Including Recommendations, Pan American Health Organization, Washington, D.C., 1985.

Chau, Y.K., P.T.S. Wong, B.A. Silverberg, P.L. Luxon, and G.A. Bengert, Methylation of selenium in the aquatic environment, *Science*, 192, 1130, 1976.

Christensen, B.T., F. Bertelsen, and G. Gissel-Nielsen, Selenite fixation by soil particle-size separates, *J. Soil Sci.*, 40, 641, 1989.

Commins, B., The contribution of drinking water makes to man's overall exposure to inorganic substances, in *Heavy Metals in the Environment*, Amsterdam, 1981, 35.

Davies, E.B. and J.H. Watkinson, Uptake of native and applied selenimum by pasture species. II. Effects of sulphate and of soil type on uptake by clover, *N.Z. J. Agric. Res.*, 9, 641, 1966.

DeMoll-Decker, H. and J.M. Macy, The periplasmic nitrite reductase of *Thauera selenatis* may catalyse the reduction of selenite to elemental selenium, *Arch. Microbiol.*, 160, 241, 1993.

Dhillon, K.S. and S.K. Dhillon, Selenium toxicity in soil-plant-animal system: a case study, *Trans. 14th Int. Cong. Soil Sci.*, 5, 300, 1990.

Dhillon, K.S. and S.K. Dhillon, Selenium toxicity in soils, plants and animals in some parts of Punjab, India, *Int. J. Environ. Stud.*, 37, 15, 1991a.

Dhillon, K.S. and S.K. Dhillon, Accumulation of selenium in sugarcane (*Saccharum officinarum* Linn.) in seleniferous areas of Punjab, India, *Environ. Geochem. Health*, 13, 165, 1991b.

Dhillon, K.S. and S.K. Dhillon, Distribution of seleniferous soils in northwest India and associated toxicity problems in the soil-plant-animal-human continuum, *Land Contamination and Reclamation*, 5, 313, 1997a.

Dhillon, K.S. and S.K. Dhillon, Factors affecting level of selenium in plants grown in seleniferous soils of Punjab, India, in *Ecological Agriculture and Sustainable Development*, Vol. 1, Dhaliwal, G.S., Randhawa, N.S., Arora, Ramesh and Dhawan, A.K., Eds., Centre for Research in Rural and Industrial Development, Chandigarh, India, 1997b, 371.

Dhillon, K.S., S.S. Bawa, and S.K. Dhillon, Selenium toxicity in some plants and soils of Punjab, *J. Indian Soc. Soil Sci.*, 40, 132, 1992.

Dhillon, K.S., N.S. Randhawa, and M.K. Sinha, Selenium status of some common fodders and natural grasses of Punjab, *Indian J. Dairy Sci.*, 30, 218, 1977.

Donovan, J., J.L. Sonderegger, and M.R. Miller, Investigation on Soluble Salt Loads, Controlling Mineralogy and Factors Affecting the Rates and Amounts of Leached Salts, Montana Water Resources Research Center Report 120, Montana State University, Bozeman, 1981.

Doran, J.W., Microorganisms and the biological cycling of selenium, *Adv. Microbial Ecol.*, 6, 1, 1982.

Doran, J.W. and M. Alexander, Microbial formation of volatile Se compounds in soil, *Soil Sci. Soc. Am. J.*, 40, 687, 1977.

Doran, J.W. and D.C. Martens, Molybdenum availability as influenced by application of fly ash to soil, *J. Environ. Qual.*, 1, 186, 1972.

Doyle, P.J. and W.K. Fletcher, Influence of parent material on the selenium content of wheat from west-central Saskatchewan, *Can. J. Plant Sci.*, 57, 859, 1977.

Eisenberg, S.H., M.E. Tittlebaum, H.C. Eaton, and M.M. Soroczak, Chemical characteristics of selected fly ash leachates, *J. Environ. Sci. Health*, Part A, 21, 383, 1986.

El-Bassam, N., H. Keppel, and C. Tietjen, The effect of waste water, sewage sludge and composted refuge on the arsenic and selenium content of cultivated soils, *Landbauforschung-Volkenrode*, 27, 105, 1977.

Ellis, B., On-site and *in situ* treatment of contaminated sites, in *Contaminated Land Treatment Technologies*, J.F. Rees, Ed., Society of Chemical Industry, Elsevier Applied Science, London, 1992.

Elrashidi, M.A., D.C. Adriano, S.M. Workman, and W.L. Lindsay, Chemical equilibria of selenium in soils: a theoretical development, *Soil Sci.*, 144, 141, 1987.

Elseewi, A.A., F.T. Bingham, and A.L. Page, Availability of sulfur in fly ash to plants, *J. Environ. Qual.*, 7, 69, 1978.

Fan, T.W.M. and R.M. Higashi, Biochemical fate of selenium in microphytes: natural bioremediation by volatilization and sedimentation in aquatic environments, in *Environmental Chemistry of Selenium*, W.T. Frankenberger, Jr. and R.A. Engberg, Eds., Marcel Dekker, New York, 1997, 545.

Fleming, G.A., Selenium in Irish soils and plants, *Soil Sci.*, 94, 28, 1962.

Fleming, G.A. and T. Walsh, Selenium occurrence in certain irish soils and its toxic effects on animals, *Proc. Royal Irish Acad.*, 58, 152, 1957.

Flury, M., W.T. Frankenberger, Jr., and W.A. Jury, Long-term depletion of selenium from Kesterson dewatered sediments, *Sci. Total Environ.*, 198, 259, 1997.

Food and Nutrition Board, Recommended Dietary Allowances, NAS-NRC, Washington, D.C., 1980, 162.

Frankenberger, W.T., Jr. and S. Benson, Eds., *Selenium in the Environment*, Marcel Dekker, New York, 1994.

Frankenberger, W.T., Jr. and U. Karlson, Land Treatment to Detoxify Soil of Selenium, U.S. Patent 4,861,482, 1989.

Frankenberger, W.T., Jr. and U. Karlson, Microbial volatilization of selenium from soils and sediments, in *Selenium in the Environment*, W.T. Frankenberger, Jr. and S. Benson, Eds., Marcel Dekker, New York, 1994, 369.

Frankenberger, W.T., Jr. and U. Karlson, Volatilization of selenium from a dewatered seleniferous sediment: a field study, *J. Ind. Microbiol.*, 14, 226, 1995.

Furr, A.K., W.C. Kelly, C.A. Bache, W.H. Gutenmann, and D.J. Lisk, Multielement uptake by vegetable and millet grown in pots on fly ash amended soil, *J. Agric. Food Chem.*, 24, 885, 1976.

Furr, A.K., G.S. Stoewsand, C.A. Bache, W.A. Gutenmann, and D.J. Lisk, Multielement residues in tissues of guinea pigs fed sweet clover grown on fly ash, *Arch. Environ. Health*, 30, 244, 1975.

Furr, A.K., T.F. Parkinson, C.L. Heffron, J.T. Reid, W.M. Haschek, W.H. Gutenmann, C.A. Bache, L.E. St. John, and D.J. Lisk, Elemental content of tissues and excreta of lambs, goats, and kids fed white sweet clover grown on fly ash, *J. Agric. Food Chem.*, 26, 847, 1978a.

Furr, A.K., T.F. Parkinson, W.H. Gutenmann, I.S. Pakkala, and D.J. Lisk, Elemental content of vegetables, grains and forages field grown on fly ash amended soils, *J. Agric. Food Chem.*, 26, 357, 1978b.

Ganje, T.J. and E.I. Whitehead, Evolution of volatile selenium from Pierre shale supplied with ^{75}Se as selenite or selenate, *Proc. South Dakota Acad. Sci.*, 37, 81, 1958.

Gerhardt, M.B., F.B. Green, R.D. Newmen, T.J. Lundquist, R.B. Tresan, and W.J. Oswald, Removal of selenium using a noval algal bacterial process, *Res. J. Water Pollut. Control Fed.*, 63, 799, 1991.

Ghosh, A., S. Sarkar, A.K. Pramanik, S.P. Chowdhury, and S. Ghosh, Selenium toxicosis in grazing buffaloes and its relationship with soil and plant of West Bengal, *Indian J. Anim. Sci.*, 63, 557, 1993.

Giedrojc, B., J. Fatyga, and Z. Hryncewiez, The effect of fertilization with ashes from black coal burned in electric power stations on the properties of sandy soil and crops, *Polish J. Soil Sci.*, 13, 163, 1980.

Gissel-Nielsen, G., Selenium content of some fertilizers and their influence on uptake of selenium in plants, *J. Agric. Food Chem.*, 19, 564, 1971.

Gustafsson, J.P. and L. Johnsson, Selenium retention in the organic matter of Swedish forest soils, *J. Soil Sci.*, 43, 461, 1992.

Gutenmann, W.H., C.A. Bache, W.D. Youngs, and D.J. Lisk, Selenium in fly ash, *Science*, 191, 966, 1976.

Hamdy, A.A. and G. Gissel-Nielsen, Fixation of selenium by clay minerals and iron oxides, *Z. Pflanzenernaehr. Bodenk.*, 140, 63, 1977.

Hamilton, J.W. and O.A. Beath, Selenium uptake and conversion by certain crop plants, *Agron. J.*, 55, 528, 1963.

Haygarth, P.M., A.I. Cooke, K.C. Jones, A.F. Harrison, and A.E. Johnston, Long-term change in the biochemical cycling of atmospheric selenium: deposition to plants and soils, *J. Geophys. Res.*, 98, 16769, 1993a.

Haygarth, P.M., A.F. Harrison, and K.C. Jones, Geographical and seasonal variation in deposition of selenium to vegetation, *Environ. Sci. Technol.*, 27, 2879, 1993b.

Haygarth, P.M., K.C. Jones, and A.F. Harrison, Selenium cycling through agricultural grasslands in the U.K.: budgeting the role of the atmosphere, *Sci. Total Environ.*, 103, 89, 1991.

Heider, J. and A. Bock, Selenium metabolism in microorganisms, *Adv. Microbial Physiol.*, 35, 71, 1993.

Horne, A.J., Selenium detoxification in wetlands by permanent flooding. I. Effects on a microalga, an epiphytic herbivore and an invertebrate predator in the long-term mesocosm experiment at Kesterson Reservoir, California, *Water, Air and Soil Pollut.*, 57/58, 43, 1991.

Hurd-Karr, A.M., Relation of sulphate to selenium absorption by plants, *Am. J. Bot.*, 25, 666, 1938.

Jacobs, L.W., Ed., *Selenium in Agriculture and the Environment*, SSSA Special Publication no. 23, Madison, WI, 1989.

Jefferis, S.A., Remedial barriers and containment, in *Contaminated Land Treatment Technologies*, J.F. Rees, Ed., Society of Chemical Industry, Elsevier Applied Science, London, 1992.

Kabata-Pendias, A. and H. Pendias, *Trace Elements in Soils and Plants*, CRC Press, Boca Raton, FL, 1984.

Kansal, B.D., H. Singh, and H.S. Randhawa, Land application of fly ash to sustain productivity, in *Proc. Int. Conf. Sustainable Agricult. and Environ.*, Haryana Agricultural University, Hissar, India, 1995, 75.

Karlson, U. and W.T. Frankenberger, Jr., Effects of carbon and trace element addition on alkylselinide production by soil, *Soil Sci. Soc. Am. J.*, 52, 1640, 1988.

Korkman, J., *Selenium in Fertilizers*, Bulletin 28, Selenium-Tellurium Development Association, Darien, CT, 1985.

Kovalski, V.V., V.V. Ermakor, and S.V. Letunova, Geochemical ecology of microorganisms in soils with different selenium content, *Microbiology*, 37, 103, 1968.

Kumar, V., G. Goswami, and K.A. Zacharia, Fly ash use in agriculture: issues and concerns, in *Proc. Int. Conf. on Fly Ash Disposal and Utilization*, CBIP, New Delhi, 1, Session 6, 1, 1998.

Kumar, V. and P. Sharma, Mission mode management of fly ash: Indian experiences, in *Proc. Int. Conf. on Fly Ash Disposal and Utilization*, CBIP, New Delhi, 1, Session 1, 1, 1998.

Lakin, H.W. and D.F. Davidson, Selenium, U. S. Geological Survey Professional Paper, 820, 573, 1973.

Levesque, M., Some aspects of selenium relationships in eastern Canadian soils and plants, *Can. J. Soil Sci.*, 54, 205, 1974.

Lewis, B.G., C.M. Johnson, and C.C. Delwiche, Release of volatile selenium compounds by plants, *J. Agric. Food Chem.*, 14, 638, 1966.

Logan, T.J., A.C. Chang, A.L. Page, and T.J Ganje, Accumulation of selenium in crops grown on sludge treated soil, *J. Environ. Qual.*, 16, 349, 1987.

Lortie, L., W.D. Gould, S. Rajan, R.G.L. McCready, and K.J. Chang, Reduction of selenate and selenite to elemental selenium by a *Pseudomonas stutzeri* isolate, *Appl. Environ. Microbiol.*, 58, 4042, 1992.

Losi, M.E. and W.T. Frankenberger, Jr., Bioremediation of selenium in soil and water, *Soil Sci.*, 162, 692, 1997a.

Losi, M.E. and W.T. Frankenberger, Jr., Reduction of selenium oxyanious by *Enterobactor cloacae* strain SLD 1a-1, in *Environmental Chemistry of Selenium*, W.T. Frankenberger, Jr. and R.A. Engberg, Eds., Marcel Dekker, New York., 1997b, 515.

Lundquist, T.J., F.B. Green, R.B. Tresan, R.D. Newman, and W.J. Oswald, The algal-bacterial selenium removal system: mechanism and field study, in *Selenium in the Environment*, W.T. Frankenberger, Jr. and S. Benson, Eds., Marcel Dekker, New York, 1994, 251.

Macy, J.M., Biochemistry of selenium metabolism by *Thauera selenatis* gen. nov. sp. nov. and use of the organism for bioremediation of selenium oxyanions in San Joaquin drainage waters, in *Selenium in the Environment*, W.T. Frankenberger, Jr. and S. Benson, Eds., Marcel Dekker, New York, 1994, 421.

Macy, J.M., S. Rach, G. Auling, M. Dorsch, E. Stackebtandt, and L. Sly, *Thauera selenatis* gen. nov. sp. nov., a member of the beta sub-class of proteobacteria with a noval type of respiration, *Int. J. Systematic Bacteriol.*, 43, 135, 1993.

Masscheleyn, P.H., R.D. Delaune, and W.H. Patrick, Transformation of selenium as affected by sediment oxidation–reduction potential and pH, *Environ. Sci. Technol.*, 24, 91, 1990.

Mayland, H.F., Selenium in plant and animal nutrition, in *Selenium in the Environment*, W.T. Frankenberger, Jr. and S. Benson, Eds., Marcel Dekker, New York, 1989, 29.

McDowell, L.R., *Minerals in Animal and Human Nutrition*, Academic Press, New York, 1992.

McCready, R.G.L., J.N. Campbell, and J. Payne, Selenite reduction by *Salmonella heidelberg*, *Can. J. Microbiol.*, 12, 703, 1966.

McLaughlin, M. J., K.G. Tiller, R. Naidu, and D.P. Stevens, The behaviour and environmental impact of contaminants and fertilizer, *Aust. J. Soil Res.*, 34, 1, 1966.

McNeil, K.R. and S. Waring, Vitrification of contaminated soils, in *Contaminated Land Treatment Technologies*, J.F. Rees, Ed., Society of Chemical Industry, Elsevier Applied Science, London, 1992, 143.

Mikkelsen, R.L., A.L. Page, and G.H. Haghnia, Effect of salinity and its composition on the accumulation of selenium by alfalfa, *Plant and Soil*, 107, 63, 1988a.

Mikkelsen, R.L., G.H. Haghnia, A.L. Page, and F.J. Bingham, The influence of selenium, salinity and boron on alfalfa tissue composition and yield, *J. Environ. Qual.*, 17, 85, 1988b.

Mikkelsen, R.L. and H.F. Wan, The effect of selenium on sulphur uptake by barley and rice, *Plant and Soil*, 121, 151, 1990.

Moxon, A.L. and M. Rhian, Selenium poisoning, *Physiol. Rev.*, 23, 305, 1943.

Neal, R.H., Selenium, in *Heavy Metals in Soils*, B.J. Alloway, Ed., Blackie, Glasgow, 1990.

Nriagu, J.O., A global assessment of natural sources of atmospheric trace metals, *Nature*, 338, 47, 1989.

Nriagu, J.O. and J.M. Pacyna, Quantitative assessment of worldwide contamination of air, water and soils by trace metals, *Nature*, 333, 134, 1988.

Nriagu, J.O. and H.K. Wong, Selenium pollution of lakes near the smelters of Sudbury, Ontario, *Nature*, 301, 55, 1983.

Ohlendorf, H.M., D.J. Hoffman, M.K. Saiki, and T.W. Aldrich, Embryonic mortality and abnormalities of aquatic birds: apparent impacts of selenium from irrigation drain water, *Sci. Total Environ.*, 52, 49, 1986.

Oremland, R.S., Selenate Removal from Wastewater, U.S. Patent 5,009,786, 1991.

Oremland, R.S., Biogeochemical transformations of Se in anoxic environments, in *Selenium in the Environment*, W.T. Frankenberger, Jr. and S. Benson, Eds., Marcel Dekker, New York, 1994, 369.

Oremland, R.S., N.A. Steinberg, T.S. Presser, and L.G. Miller, *In situ* bacterial selenate reduction in the agricultural drainage systems of western Nevada, *Appl. Environ. Microbiol.*, 57, 615, 1991.

Owens, L.P., Bioreactors in removing selenium from agricultural drainage water, in *Environmental Chemistry of Selenium*, W.T. Frankenberger, Jr. and R.A. Engberg, Eds., Marcel Dekker, New York., 1997, 501.

Parker, D.R. and A.L. Page, Vegetation management strategies for remediation of selenium-contaminated soils, in *Selenium in the Environment*, W.T. Frankenberger, Jr. and S. Benson, Eds., Marcel Dekker, New York, 1994, 327.

Parker, D.R., A.L. Page, and D.N. Thomas, Salinity and boron tolerances of candidate plants for the removal of selenium from soils, *J. Environ. Qual.*, 20, 157, 1991.

Pattishall, J.C., Building the future with high volume low tech ash products, in *Proc. Int. Conf. on Fly Ash Disposal and Utilization*, CBIP, New Delhi, Vol. 1, Session I, 62, 1998.

Peters, R.W. and L. Shem, Use of chelating agents for remediation of heavy metal contaminated soil, in *Environmental Remediation Removing Organic and Metal Ion Pollutants*, G.F. Vandegrift, D.T. Reed, and I.R. Tasker, Eds., American Chemical Society, Washington, D.C., 1992, 70.

Pierzynski, G.M., J.T. Sims, and G.F. Vance, *Soils and Environmental Quality*, CRC Press, Boca Raton, FL, 1994.

Pilley, K.K.S., C.C. Thomas, Jr., and J.W. Kaminiski, Neutron activation analysis of the Se content of fossil fuels, *Nuclear and Appl. Technol.*, 7, 478, 1969.

Pratley, J.E. and J.D. McFarlane, The effect of sulphate on the selenium content of pasture plants, *Aust. J. Exp. Agric. Anim. Husb.*, 14, 533, 1974.

Presser, T.S. and I. Barnes, Dissolved constituents including selenium in waters in the vicinity of Kesterson National Wildlife Refuge and the West Grassland, Fresno and Merced Counties, California, Water Resources Investigation Report 85-4220, U.S. Geological Survey, Menlo Park, CA, 1985.

Rajan, S.S.S. and J.H. Watkinson, Adsorption of selenite and phosphate on an alloplane clay, *Soil Sci. Soc. Am. J.*, 40, 51, 1976.

Ravikovitch, S. and M. Margolin, The effect of barium chloride and calcium sulphate in hindering selenium absorption by lucerne, *Emp. J. Exp. Agric.*, 27, 235, 1959.

Retana, J., D.R. Parker, C. Amerhein, and A.L. Page, Growth and trace element concentrations of five plant species grown in highly saline soil, *J. Environ. Qual.*, 22, 805, 1993.

Robberecht, H. and R. van Grieken, Selenium in environmental waters: determination, speciation and concentration levels, *Talanta*, 29, 823, 1982.

Robberecht, H., R. van Grieken, M.V. Sprundel, D.V. Berghe, and H. Deelstra, Selenium in environmental and drinking waters of Belgium, *Sci. Total Environ.*, 26, 163, 1983.

Robbins, C.W. and D.L. Carter, Selenium concentration in phosphorus fertilizer materials and associated uptake by plants, *Soil Sci. Soc. Am. Proc.*, 34, 506, 1970.

Rosenfeld, I. and O.A. Beath, Selenium: geobotany, biochemistry, in *Toxicity and Nutrition*, Academic Press, New York, 1964.

Ross, H.B., Speciation and Possible Transformation of Atmospheric Selenium, *Atmospheric Selenium Report* CM-66, Department of Meteorology, University of Stockholm, International Meteorological Institute for Stockholm, 1984, 1.

Sarathchandra, S.U. and J.H. Watkinson, Oxidation of elemental selenium to selenite by *Bacillus megaterium*, *Science*, 211, 600, 1981.

Sauerbeck, D., *Scientific Basis for Soil Protection in the European Community*, Elsevier, London, 1987, 181.

Senesi, N., M. Polemio, and L. Lorusso, Content and distribution of arsenic, bismuth, lithium and selenium in mineral and synthetic fertilizers and their contribution to soil, *Commun. Soil Sci. and Plant Anal.*, 10, 1109, 1979.

Severson, R.C. and L.P. Gough, Selenium and sulphur relationship in alfalfa and soil under field conditions, San Joaquin Valley, California, *J. Environ. Qual.*, 21, 353, 1992.

Shubik, P., D.B. Clayson, and B. Terracini, *The Quantification of Environmental Carcinogens*, International Union Against Cancer, Technical Report Series 4, 1970.

Singh, M. and P. Kumar, Selenium distribution in soils of bio-climatic zones of Haryana, *J. Indian Soc. Soil Sci.*, 24, 62, 1976.

Singh, M. and P.K. Malhotra, Selenium availability in berseem (*Trifolium alexandrinum*) as affected by selenium and phosphorus application, *Plant and Soil*, 44, 261, 1976.

Singh, M., N. Singh, and P.S. Relan, Adsorption and desorption of selenite and selenate selenium on different soils, *Soil Sci.*, 132, 134, 1981.

Smith, B., Remediation update finding the remedy, *Waste Manage. and Environ.*, 4, 24, 1993.

Soltanpour, P.N. and S.M. Workman, Use of NH_4HCO_3 DTPA soil test to assess availability and toxicity of selenium to alfalfa plants, *Commun. Soil Sci. and Plant Anal.*, 11, 1147, 1980.

Squires, R.C., G.R. Groves, and W.R. Johnston, Economics of selenium removal from drainage water, *J. Irrigation and Drainage Eng.*, 115, 48, 1989.

Srikanth, R., C.S. Kumar, and A. Khanum, Heavy metal content in forage grass grown in urban sewage sludge, *Indian J. Environ. Health*, 34, 103, 1992.

Steudal, R., E.M. Strauss, M. Pappavassiliou, P. Bratter, and W. Gatschke, Selenium content of naturally occurring elemental sulfur, of industrially produced sulfur and of the mineral selen-sulfur, *Phosphorus and Sulfur*, 29, 17, 1987.

Stoewsand, G.S., W.H. Gutenmann, and D.J. Lisk, Wheat grown on fly ash: high selenium uptake and response when fed to Japanese quail, *J. Agric. Food Chem.*, 26, 757, 1978.

Tan, J.A., W.Y. Wang, D.C. Wang, and S.F. Hou, Adsorption, volatilization and speciation in different types of soils in China, in *Selenium in the Environment*, W.T. Frankenberger, Jr. and S. Benson, Eds., Marcel Dekker, New York, 1994, 47.

Terry, N., C. Carlson, T.K. Raab, and A.M. Zayed, Rates of selenium volatilization among crop species, *J. Environ. Qual.*, 21, 341, 1992.

Terry, N. and A.M. Zayed, Selenium volatilization by plants, in *Selenium in the Environment*, W.T. Frankenberger, Jr. and S. Benson, Eds., Marcel Dekker, New York, 1994, 343.

Terry, N. and A. Zayed, Remediation of Selenium Contaminated Soils and Waters by Phytovolatilization, Paper presented at the Symposium on Phytoremediation of Trace Element Contaminated Soil and Water, University of California, Berkeley, 1997.

Thomson, B.M. and R. J. Heggen, Water quality and hydrologic impacts of disposal of uranium mill tailings by back filling, in *Management of Wastes from Uranium Mining and Milling*, IAEA, Vienna, 1982, 373.

Thompson-Eagle, E.T. and W.T. Frankenberger, Jr., Selenium biomethylation in an alkaline, saline environment, *Water Res.*, 25, 231, 1991.

Tokunaga, T.K., P.T. Zawislanski, P.W. Johannis, D.S. Lipton, and S. Benson, Field investigations of selenium speciation, transformation and transport in soils from Kesterson Reservoir and Lahontan Valley, in *Selenium in the Environment*, W.T. Frankenberger, Jr. and S. Benson, Eds., Marcel Dekker, New York, 1994, 119.

U.S. Environmental Protection Agency, Summary of Environmental Profiles and Hazard Indices for Constituents of Municipal Sludge, U.S. EPA, Office of Water Regulations and Standards, Washington, D.C., 1985.

Vokal-Borek, H., *Selenium*, University of Stockholm, Institute of Physics, Rep. 79-16, Stockholm, Sweden, 1979.

vom Berg, W., Utilization of fly ash in Europe, in Proc. Int. Conf. on Fly Ash Disposal and Utilization, CBIP, New Delhi, Vol. 1, Session 1, 8, 1998.

Wagde, A., M. Hutton, and P.J. Peterson, The concentrations and particle size relationships of selected trace elements in fly ashes from U.K. coal-fired power plants and a refuge incinerator, *Sci. Total Environ.*, 54, 13, 1986.

Wang, D., G. Alfthan, A. Arro, L. Kauppi, and J. Soveri, Selenium in tap water and natural water ecosystems in Finland, in *Trace Elements in Health and Disease*, A. Aitio, A. Arro, J. Jarvisalo, and H. Vainio, Eds., Royal Society of Chemistry, Cambridge, 1991, 49.

Webb, J.S., I. Thornton, and K. Fletcher, Seleniferous soils in parts of England and Wales, *Nature*, 5046, 377, 1966.

Williams, C. and I. Thornton, The effect of soil additives on the uptake of molybdenum and selenium from soils from different environments, *Plant and Soil*, 36, 395, 1972.

Wu, L., Selenium accumulation and colonization of plants in soils with elevated selenium and salinity, in *Selenium in the Environment*, W.T. Frankenberger, Jr. and S. Benson, Eds., Marcel Dekker, New York, 1994, 279.

Wu, L., Z.Z. Huang, and R.G. Burau, Selenium accumulation and selenium salt tolerance in five grass species, *Crop Sci.*, 28, 517, 1988.

Yang, G., S. Wang, R. Zhar, and S. Sun, Endemic selenium intoxication of humans in China, *Am. J. Clin. Nutr.*, 37, 872, 1983.

Ylaranta, T., Volatilization and leaching of selenium added to soils, *Ann. Agric. Fenn*, 21, 103, 1982.

Ylaranta, T., Effect of added selenite and selenate on the selenium content of Italian ryegrass (*Lolium multiflorum*) in different soils, *Ann. Agric. Fenn*, 22, 139, 1983a.

Ylaranta, T., Sorption of selenite and selenate in the soil, *Ann. Agric. Fenn*, 22, 29, 1983b.

Zehr, J.P. and R.S. Oremland, Reduction of selenate to selenite by sulphate-respiring bacteria: experiments with cell suspensions and estuarine sediments, *Appl. Environ. Microbiol.*, 53, 1365, 1987.

Zingaro, R.A. and W.C. Cooper, *Selenium*, Van Nostrand Reinhold, New York, 1974.

11

Trace Metals in Soil-Plant Systems under Tropical Environment

Sultana Ahmed and S. M. Rahman

CONTENTS

1-56670-457-X/00/$0.00+$.50

11.1 Introduction

Soils, as a part of the environment, need protection against metal contamination. Trace metals are widely distributed in nature, soils, plants, and also within living beings. The concept of trace metals in relation to our environment commonly implies some negative effect of the metal on the biological/living system. With reference to the soil–plant–environment, according to Leeper (1972), trace metals include zinc (Zn), copper (Cu), manganese (Mn), molybdenum (Mo), iron (Fe), chromium (Cr), cobalt (Co), nickel (Ni), cadmium (Cd), lead (Pb), and mercury (Hg). The first five (Zn, Cu, Mn, Mo, and Fe) are essential trace metals called the micronutrients. However, there is no conclusive evidence on the essentiality of Cr, Co, and Ni in any living system, while Cd, Pb, and Hg are absolutely nonbiogenic and rather hazardous to any form of life. To sustain a healthy environment, it is imperative to adopt measures that help in maintaining soil health and, in turn, human health. Trace metals, either essential or nonessential, are generally toxic in nature when present in available forms because of their usual long biological half-lives. Hence the concentrations of such metals in any biologically living system needs to be maintained within a critical level to achieve optimum biological functions of plants, animals, and human beings.

Global climatic change has led to an increasing concern in recent years regarding the abundant entry of some trace metals into the soil and their probable adverse effects that might be reflected on plants, animals, and, in turn, on human health through the food chain. Thus preservation of the environment and at the same time restoration of metal-contaminated soil are very essential for sustainable agriculture in the context of the tropical environment, as tropical climate is quite receptive/susceptible to any contamination/pollution.

11.2 Trace Metals in Soils and Crops

The overall content of trace metals in any soil depends initially on the nature of parent materials because a soil inherits from its parent material a certain stock of elements that is redistributed by pedological processes. The size and quality are determined by the geochemical history of the parent rock (Davies, 1980). Soil erosion is probably the major pathway through which trace metals may be lost from surface soil. In general, trace metals are less mobile and thus less bioavailable in soils with higher organic matter and clay content (Cottenie, 1983). Table 11.1 includes the specific important trace metals of our interest which may be useful in predicting the probability of significant alterations of trace metal concentrations in food and feeds. However, the transport of trace metals in soil and their uptake by plants are governed by their mobility (Cottenie, 1980).

The present section deals in biological functions, essentiality, and toxicity of trace metals. Trace metals included herein are zinc, copper, manganese, iron, chromium, lead, cadmium, nickel, and mercury.

11.2.1   Trace Metals

11.2.1.1   Zinc

All living organisms require Zn insofar as is known today. It is an important constituent of all cells. Its deficiency is dramatically demonstrated through combination of chlorosis,

TABLE 11.1

Some Heavy Metal Concentrations and Their Variations in Soils, Lithosphere, and Rocks ($\mu g\ g^{-1}$ Dry Material)

Trace Metals	Soils Usual Range	Soils Average	Lithosphere Average	Rocks Igneous	Rocks Limestone	Rocks Sandstone	Rocks Shale
Zinc	10–300	50	80	80	4–20	5–20	50–3000
Copper	2–100	20	70	70	5–20	10–40	30–150
Manganese	200–300	850	1000	1000	1300	385	—
Iron	14000–40000	—	—	40600	13000	31000	43000
Molybdenum	0.2–0.5	2850	2.3	1.7	0.1–0.5	0.1–1.0	1.0
Chromium	5–1000	200	200	117	5	10–100	100–400
Cobalt	1–40	8	40	18	0.2–2.0	1–10	10–50
Nickel	5–50	40	100	100	3–10	2–10	20–100
Cadmium	0.01–0.70	0.50	0.18	0.13	—	—	0.3
Lead	2–200	10	16	16	5–10	10–40	20
Mercury	0.03–3.0	0.03	0.5	0.06	0.03	0.03–0.1	0.4

From Swaine, D.J., The trace element content of fertilizers, *Tech. Communication*, no. 48 Commonwealth Agr. Bur. Soils, Harpenden, Farnham Royal, Bucks, England, 1955, 157.

rosette leaves, abnormal vegetative growth, lower yield, and lack of chlorophyll in plant. Zn application to some extent increases N uptake by crop (Mishra and Singh, 1996). Zn is present in a prosthetic group of several enzymes. The role of Zn as essential components of a variety of dehydrogenase, proteinases, and peptidases was identified (Vallee and Wacker, 1970). A number of these dehydrogenases show sensitivity to Zn deficiency so that metabolism can be strongly and specifically affected (Price, 1970). There are several reports on trace metal in soil–plant systems that show that the earliest possible causal effect of Zn deficiency is a sharp decrease in the levels of RNA and the ribosome contents of cells. It was found that the cytoplasmic ribosomes of *Euglena gracilis* normally contain substantial amounts of Zn and that these organelles become extremely unstable due to Zn deficiency (Prask and Plocke, 1971).

Soils deficient in available Zn have been reported in a number of areas where food and feed crops are grown. In the tropical climate in Asia, available Zn status was reported to be 66 mg kg^{-1} (Domingo and Kyuma, 1983), while 40 mg kg^{-1} was recorded for world soils (Berrow and Reaves, 1984). A voluminous literature on zinc in relation to the soil–plant system is available today involving the visual symptoms of its deficiency and toxicity, the concentration in different plant parts under extreme conditions, total and available Zn content of soils, and methods for Zn determination, as well as physiological function and chemistry of Zn.

11.2.1.2 Copper

The conclusive evidence of the essentiality of Cu for green plants was reported (Lipman and Mackinney, 1931). There are many indications that secondary influences of Cu on biological processes are important. The essentiality of Cu has been evidenced in the growth of plants and reported in many scientific world literatures. Reviews on the role and contamination aspects of Cu have been made by Reuther and Labanauskas (1966) and for soils by Lagerwerff (1967). It is now clearer that Cu is an essential trace metal for plants but the amount needed is very small for optimum plant growth.

Cu exists as a series of Cu proteins in vertebrate blood. It is also known to be an essential constituent of many enzymes. It is important in the synthesis of hemoglobin. It is not, however, certain whether cytochrome oxides or any other identified Cu proteins ever become limiting to plants because of Cu deficiency (Price et al., 1971).

11.2.1.3 Manganese

Although Mn activates a number of enzymes nonspecifically, only one mangano-enzyme has been clearly identified as pyruvate carboxylase (Scrutton et al., 1966). There are well-recognized diseases and normal growth disorders which may develop in plants and agricultural crops on account of Mn deficiencies. On the other hand, several cases of Mn toxicity in plants have also been reported because of excess of available Mn in soils. Mn is known to be frequently concentrated in soil horizons rich in organic matter where it is presumably immobilized by complex formation. The usual range of the element in soil was recorded from 100 to 4000 mg kg^{-1} (Adriano, 1986).

11.2.1.4 Iron

Iron is one of the essential microelements for human beings, plants, and animals. The literature reviews on the role and functions of Fe in plants (Wallihan, 1966) and in animal and human nutrition (Underwood, 1971) have been reported. The availability of Fe in soils may vary from deficient to excessive range for the growth of plants. However, for animals and man its availability is normally below optimum. Fe is not absorbed by plants in amounts toxic to animals (Underwood, 1971). Factors affecting the availability of micronutrient cations including Fe in relation to soil composition were reviewed by Viets and Lindsay (1973).

Although Fe toxicity is rather rare under natural conditions, it has, however, been reported to occur in plants that have received soluble Fe-salts either as sprays or as soil amendments in excess quantities. Initial toxicity symptoms in plants appear as easily recognizable necrotic spots. On the other hand, during the moderate to acute stage of Fe deficiency, a characteristic type of leaf chlorosis occurs in most plants.

Fe is an essential component of the many heme and nonheme Fe enzymes and carriers, but it is now generally accepted that Fe does not play a role in the enzymatic synthesis of porphyries either in plants (Carell and Price, 1965) or in porphyrin-secreting bacteria (Kortstee, 1970).

11.2.1.5 Molybdenum

Molybdenum as a trace element for plants has long been known to be required for the normal assimilation of nitrogen in plants. Among the four enzymes found to contain Mo — aldehyde oxidase, xanthine oxidase, and nitrogenous and nitrate reductase, only the latter two are found in plants. Nitrate reductase is found in most plant species as well as fungi and bacteria, and is probably a key factor in plant dispersion under varying nitrogen environments. The increased Mo requirement of most plants grown on nitrate nitrogen can almost completely be accounted for by the Mo in nitrate reductase (Evans, 1956). This enzyme is essential in the assimilation of nitrates because it catalyzes the first step of reduction of nitrate to ammonia.

Although Mo is listed as one of the essential trace metals for animals, the required levels for plants have been considered to be less than 0.2 ppm (Reisenauer, 1965).

11.2.1.6 Chromium

There is still no conclusive evidence that Cr is essential for the growth of plants. However, Cr is widely distributed in soils, water, and biological materials. In a comprehensive review it was reported that Cr in soils is usually in the range of 5 to 1000 ppm (Swaine, 1955). The most interesting work resulting in the identification of Cr as an essential component part of a "glucose tolerant factor" was recorded (Mertz, 1969). Mertz also indicated that human patients suffering from diabetes in some cases responded to chromium treatment. Limited information as to the absorption of Cr and its retention in plants is reported (Allaway, 1968).

11.2.1.7 Cobalt

Cobalt is apparently not required by plants. No precise evidence is yet available on its essentiality for plants. Ahmed et al. (1981), however, reported some irregular uptake of Co in rape and pea. The Co uptake by both rape and pea was retarded by the application of phosphate and/or lime. The essential biological value of Co has been enlarged by the discovery of the fact that Co is an important component of vitamin B_{12}, which, in turn, is believed to be an essential diet ingredient of all animals (Smith, 1948). It seems, therefore, that available Co in agricultural soils as well as its uptake by plants to a certain optimum level are quite desirable.

11.2.1.8 Nickel

The function of nickel in plants is not well established, although there are a few sources that indicate that Ni may be essential for growth and reproduction of plants. Nickel has not yet been proven to be directly essential for plants. It was also stated that increased yield of grapes was recorded because of application of Ni, but mustard is a poor absorber of Ni and also tolerant to Ni toxicity (Gupta et al., 1996). This is in agreement with the finding of Ahmed et al. (1981) who reported that application lime and phosphate reduced the uptake of Ni by rape and pea.

Nickel content of soil usually varies between 5 and 500 ppm, with an average of about 100 ppm (Swaine, 1955). It was noted that soil levels of 20 to 34 ppm of 0.5 N acetic acid extractable Ni were toxic to oats (Page, 1974). The maximum tolerance level of Ni in plant was recommended as 3 ppm (Melstead, 1973). Ni adsorption through diet by animals has been reported to be low (Underwood, 1971).

11.2.2 Biogenic Trace Metals

11.2.2.1 Cadmium

Concern about the effects of Cd stems from the metal's tendency to be accumulated by mammals. It is potentially harmful and relatively mobile in the environment. The environmental presence of Cd is normally linked to that of Zn because of their geochemical kinship and incomplete technical separation. Cd thereby may replace certain enzymes causing disease (Lagerwerff, 1971). As an aerosol constituent, Cd like other metals reaches plant and soil through precipitation and by direct deposition. Soluble Cd is easily absorbed through the roots of important food crops, especially the major grains like wheat, corn, rice, oats, and millets. It is also present in vegetable crops like peas, beets, and lettuce (Schroeder and Balassa, 1963). As with some of the heavy metals, an increase in soil pH by liming somewhat suppressed uptake of Cd (Lagerwerff, 1971). The inhibitory effect of Zn on translocation of Cd in rice grain, and a higher level of Zn needed to check the absorption of Cd (Sarkunan et al., 1996) were also reported.

11.2.2.2 Lead

Among the heavy metals impairing the quality of our environment, Pb is physiologically unessential and potentially hazardous. Its distribution in the atmosphere, soil, sea, and groundwater is known to be wide. In the last two or three decades man's continuous use of Pb in industry and everyday life has emerged to be much higher in the environment than it should otherwise exist naturally.

Lead is a natural consitituent of soil, water, vegetation, animals, and air. The natural sources of Pb include dust from soils and particles from volcanoes. Patterson (1965) postulated that natural Pb concentration of air is caused by erosion of Pb-containing soils. Exhausted gases from earth's crust have been reported to be the other important natural source of Pb (Blanchard, 1966). Lead absorption and accumulation have been identified in most plant and animal tissues and to a far greater extent particularly in bones, liver, and kidneys (Thomas et al., 1967). In recent years, the distinct rise of Pb concentration in water, air, and soil has been attributed because of intensive use of gasolines (Chow and Earl, 1970).

Concentrations of Pb in plant material are known to vary widely. Pb uptake by plant is known to be a function of soil properties (Maclean et al., 1969). The functional role of Pb on plant enzymes is not completely understood. It is, however, certain today that due to air and soil contamination plants can accumulate significant amounts of Pb, which is likely to cause serious problems in our food chain (McGrath, 1994).

11.2.2.3 Mercury

Similar to Cd and Pb, Hg is considered to be a nonbiogenic heavy metal which has no known essential functions to man, animals, and plants. As compared to Pb, Hg occupies almost a similar position in its extent of pollution threat to our ambient environment. Mercury occurs naturally in the environment and is well contributed to by numerous industrial and agricultural activities.

The toxicity of Hg to plants apparently depends on the chemical state of the element. Both organic and inorganic Hg compounds have been used for many years in seed disinfection as herbicides in order to control plant diseases. Translocation of Hg to plant tissues including those of the leaves and fruits of apples has been reported (Ross and Stewart, 1969).

11.2.2.4 Factors Affecting Trace Metals Accumulation in Soils and Plants

11.2.2.5 Soil and Fertilization Effects

Soil composition is undoubtedly the basic factor determining the concentrations of the trace metal uptake in plants, and so in the ultimate food chain. Soil composition, in turn, depends on a number of factors such as the parent rocks, the geographical and weathering conditions, the history of soil amendments and fertilizer treatments, the drainage and aeration status of soil, the history of crop cultivation, etc. The marked influence of soil factors on the uptake of Co and Ni by pasture plants was first demonstrated by Mitchell (1957). Alsike clover favored the uptake of Mo and Co but not Cu under a wetland condition as observed (Kubota et al., 1963). However, trace metal accumulations in soil are generally more around the industrialized area where their concentrations exceed ecological concentration. Furthermore, the uptake of mineral elements including trace metals by plants is governed by soil reactions. Generally, the transformation, mobilization, and immobilization of trace metals in soils depend on the type of metals, soil, climatology, geomorphology of the area, and flora and fauna of the ecosystems, as well as the dissolved constituents (Cruañas, 1992). Increase in soil pH increases fixation capacity of soil for most trace elements (Adriano, 1986). Climatic change also plays a dominant role in changing physicochemical properties of the soil, thus changing the metal dynamics as a whole.

Cobalt and Ni, and to certain extent Cu and Mn, are poorly taken up by plants from calcareous soils, whereas Mo uptake is reported to be higher from calcareous than neutral or acid soil (Underwood, 1971). The incremental dressings of calcium carbonate depressed the Co, Ni, and Mn level, while the Mo level was enhanced in the test plant. Liming reduces mobility of many trace metals, in general (Mitchell, 1957). Although Cr and Pb are rated as more phytotoxic in soil under wetland conditions, they are quite insoluble in most soils and toxicity from these metals is rarely observed. The risk of Pb movement from soils to edible plant parts is believed to be largely from 10 to 84 ppm soil Pb (range of soil means worldwide) according to McBride (1994). However, controlled application of essential trace metals containing fertilizers to soil is known to have beneficial effects with respect to increased yield and trace metal composition of plants. Some sources of soil contaminants are as follows:

Lead — Combination of coal, gasoline additives, iron and steel production, lead base paints, pesticides, batteries

Mercury — Metallurgy, pesticides, thermometers

Nickel — Batteries, electroplating, gasoline

Cadmium — Pigments in paints, batteries, and plastic stabilizers

Copper — Fertilizers, fly ash, dust

Zinc — Galvanized iron and steel, brass, alloys of metals

11.2.2.6 Accumulation of Trace Metal in Plants

Plant materials are the major source of all mineral elements, including hazardous trace metals to human and animals. The factors influencing the mineral concentrations in plants may, therefore, be taken as the major determinants of dietary intakes of these mineral elements. Such basic interrelated factors as stated by Underwood (1971) are (1) genetic difference and (2) stages of maturity of plant and seasonal influences.

Genetic differences: Specific plant species are able to accumulate high concentrations of a particular metal element; as, for example, the leguminous plants under similar growing conditions and maturity are reported to contain significantly higher concentrations of Cu, Zn, Fe, Co, and Ni than cereals or grasses (Gladstones and Loneragan, 1967).

Stages of plant maturity and seasonal influence: There is perhaps no consistent relationship between the total trace metal concentration in plants and stages of plant maturity. A number of investigators have reported an increase in metal concentrations with advancing plant maturity. Seasonal changes also influence the concentration and uptake of trace metals in plants (Shuman, 1980).

Cadmium, Pb, and Hg are accepted to be nonbiogenic in nature and are absolutely dangerous pollutants of our environment. They stand for no known biological value. Furthermore, the pollution problem with these heavy metals in our total environment is of special interest because very small amounts of them can affect human health seriously. Natural soil and plant concentrations of several trace metals considered toxic to living beings are presented in Table 11.2.

11.3 Trace Metals and Environmental Problems

Apart from natural sources, the most common distinguished sources of trace metal contamination of environment are man's industrial or urban activities, which may be grouped as follows:

TABLE 11.2

Natural Soil and Plant Concentrations of Trace Metals Which Are Considered as Being Toxic to Living Beings

Element	Total Soil (range) (mg kg⁻¹)	Plant (range) (mg kg⁻¹)
Zn	10–300	8–15
Cu	2–200	0.1–10
Mn	100–400	15–1000
Co	1–40	0.05–0.5
Cr	5–1000	—
Cd	0.01–7	0.2–0.8
Ni	10–1000	—
Hg	0.02–0.02	—

From Bohn, H.L., B. McNeal, and G.A. O'Connor, *Soil Chemistry*, Wiley Interscience, New York, 1979.

1. Urban and industrial aerosol — a system consisting of colloidal particles dispersed in gas, smoke, or fog
2. Industrial and agricultural chemicals
3. Mining wastes
4. Sewage sludges

In certain circumstances, there are direct atmospheric inputs to plants, accumulation in soil, and subsequent transfer from soil to plants.

A brief description of environmental pollution by trace metals as affected by different sources are presented below.

11.3.1 Aerosols

Aerosols mainly collect their heavy metals from oxidation processes such as gasoline combustion, coal burning, and metal smelting. Pb and Cd, for instance, are easily volatilized at temperatures prevailing in the Pb smelting process, and substantial amounts of these heavy metal ions may thereby be released into the atmosphere if there is no control action. It can also be noted that various heavy metal ions such Zn, Cu, Ni, Cd, Pb, and Hg are emitted into the environment through various industrial processes. These heavy metal ions are accumulated in the atmosphere, or after deposition in the soil may enter into plants either by their root system or through foliar uptake (Lagerwerff and Specht, 1970). It was reported that Pb and Cd contamination were prevalent in the surroundings of a lead smelter. The smelter was the source of Pb and Cd in the ambient air, which in turn polluted the local soil and vegetation around the smelter plant, thus also becoming a contamination threat to grazing animals. Reports on environmental buildup of Zn, Cu, Pb, and Cd around Pb-producing areas within a period of 1 year were recorded. The influence of airborne contaminants charged with heavy metals on the heavy metal content of a number of slow growing plants was reported by Chang et al. (1992).

11.3.2 Industrial and Agricultural Chemicals

The nonbiogenic trace metals, mainly Cd, Pb, and Hg, are contributed by different industrial and agricultural activities. The application of limestone and phosphatic fertilizers to agricultural land implies an inevitable corporation of heavy metals (Caro, 1964). It was reported that cadmium content of several vegetable species was increased because of heavy application of super-phosphate (Schroeder and Balassa, 1963). The role of commercial fertilizers as one of the sources of trace metal contamination has been emphasized in the world scientific literature.

Pest control is an integral part of modern agriculture for protecting crops, but it has a considerable impact on the environment. Pesticides including insecticide and fungicide are often reported to be dangerous sources of biotransfer of trace metals to environment, where they are indiscriminately used for pest and disease management of crop plants.

Copper-containing fungicide applied as sprays has been recorded as damaging to the citrus belt in Florida (Reuther and Smith, 1953). Mercury-containing fungicide has been known to be a source of soil contamination with mercury. The accumulation of mercury in soil after annual application of organo-mercurial fungicide was identified.

A list of chemical compounds containing significant amounts of injurious nonessential trace metals in a group of chemicals for plant disease control was recorded. For example, past use of lead arsenate pesticide in agriculture is one of the prime sources of soil contamination with lead. The occurrence of toxic trace metal buildup in soil from the use of pesticides has been widely reported (McGrath et al., 1995).

11.3.3 Mining Wastes

Generally, geologists explore for metal ore within the areas containing high concentrations of trace elements in soils and plants. Metal ore bodies in the rocks below the soil profile contribute to its parent materials and thereby increase the soil trace element contents in their immediate vicinity. The major impact arises when the ores are mined and processed. Losses of elements to the environment are possible at all stages of processing and the agencies of dispersal are air, water, and gravitation. Soil contamination around metal mining areas can be widespread. Pb and Zn concentrations as high as 20,000 mg kg^{-1} and Cd contents up to 1000 mg kg^{-1} have been recorded in some mining areas (Davies and Jones, 1988).

11.3.4 Sewage Sludges

Protection of environment requires mostly recycling of organic wastes. The problem of sewage sludge disposal in industrial and urban areas as landfill or in agricultural fields is likely to be a source of heavy metal contamination because sludges often contain large amounts of trace metals such as Zn, Cu, Pb, Ni, Cd, etc. Thus maximum heavy metal concentration recommended for municipal sewage sludge for application in agricultural land is presented in Tables 11.3 and 11.4.

11.4 Management of Trace Metals in Soils, Crops, and Environment

11.4.1 Soils

Proper management of trace metals in soils and environment can easily reduce their uptake by crops. Toxic trace elements are partly ingested through the edible parts of the crops. In order to save the environment as well as the human population, approaches to limit the trace metal loading on sludge-amended soils were considered. Regulations for sewage sludge application to land have been put forth by the U.S. Environmental Protection Agency (1993), which is shown in Table 11.4. Maximum allowable levels of specific heavy

TABLE 11.3

Maximum Values for Metal Concentration in Sewage Sludge Used in Agriculture, Their Rate of Application in Sludge Treated Soil in the Comm. of the European Communities

Trace Metals	Maximum Permitted Conc in Sewage Sludge (mg kg^{-1})	Maximum Permitted Conc in Agricultural Soil, pH Range 6–7 (mg kg^{-1})	Maximum Average Rate of Addition over 10 years (kg ha^{-1} year^{-1})
Cd	20–40	1–3	0.15
Cr	—	100–150	—
Cu	1000–1750	150–140	12
Hg	16–25	1–1.5	0.1
Pb	750–1200	50–300	15
Ni	300–400	30–75	3
Zn	2500–4000	150–300	30

From McGrath, S.P., A.C. Chang, A.L. Page, and E.W. Witter, Land application of sewage sludge: scientific perspectives of heavy metal loading limits in Europe and United States, *Environ. Rev.*, 2, 108, 1994. With permission.

TABLE 11.4

Summary of the U.S. Regulations for Sewage Sludge Applied to Land

	Maximum Permitted Conc in Sewage Sludge (mg/kg)	Maximum Conc in Clean Sludge (mg/kg)	Maximum Annual Loading (mg/kg/year)	Maximum Cumulative Pollutant Loading (kg/ha)
As	75	41	2.0	41
Cd	85	39	1.9	39
Cr	3000	1200	150	3000
Cu	4300	1500	75	1500
Pb	840	300	15	300
Hkg	57	17	0.85	17
Mo	75	18	0.90	18
Ni	420	420	21	420
Se	100	36	5.0	100
Zn	7500	2800	140	2800

From McGrath, S.P., A.C. Chang, A.L. Page, and E.W. Witter, Land application of sewage sludge: scientific perspectives of heavy metal loading limits in Europe and United States, *Environ. Rev.*, 2, 108, 1994. With permission.

TABLE 11.5

Maximum Concentrations of Metals Allowed in Sludge-Amended Agricultural Soils (mg kg^{-1})

Country	Year	Cd	Cu	Cr	Ni	Pb	Zn	Hg
U.K.	1989	3	135	—	75	300	300	1
U.S.	1993	20	750	1500	210	150	1400	8

From McGrath, S.P., A.C. Chang, A.L. Page, and E.W. Witter, Land application of sewage sludge: scientific perspectives of heavy metal loading limits in Europe and United States, *Environ. Rev.*, 2, 108, 1994. With permission.

metals in sewage sludge destined for use on agricultural soils in the U.K. and U.S. are presented in Table 11.5.

Soil is the most important resource which supports plant growth. The soil offers mechanical anchorage along with plant nutrients for survival of the plant. These plant nutrients include both major and trace elements. All trace elements, however, are not essential to plants. There are certain nonbiogenic trace metals such as Pb, Hg, and Cd that are known to be toxic to crops and as well to impair the quality of our environment if present in excess amounts.

11.4.2 Environment

With a view to restoring the environment from contamination, it is very essential to maintain the natural ecosystem. The trace metals are emitted into the environment through various industrial processes. These metals accumulate in the atmosphere, and to some extent are deposited in the soil; from there the potential hazardous metals enter into the plant root system. Also, organic wastes containing heavy metals impair the quality of the environment. Thus disposal of industrial and organic waste is necessary, and emission of black smoke from vehicles should be prevented. Monitoring steps should be taken to control excess concentrations of trace metals on crop land. Our understanding on the mechanisms of soil retention and release involving trace elements and their uptake by plants is incomplete. Also, a more complete understanding of the biological and chemical transformation of trace metal pollutants in soil–plant systems is needed. Further, the maximum allowable level of trace metals in agricultural soils of the European Community and the United States as shown in Table 11.5 must also be followed for countries in a tropical climate where chances of metal contaminations in soil–plant systems are more favorable. Such a program will certainly minimize the trace metal contamination of soil and at the same time accord possibilities to preserve/restore the environment.

11.5 Conclusion

Since restoration of metal-contaminated soil and environment is a matter of global concern, measures must be taken to reduce contamination from potential hazardous trace metal in the soil–plant and environment which, in turn, can adversely affect animal and human health.

- The chemistry of soil is very complex and appropriate management of trace metals in soils is very much dependent on soil reactions. Alternate wetting and drying of soil help reduce trace metal concentration in soil to a certain extent.
- There must be metal limits which regulate land application of sewage sludge such as the EC directive (1986) and/or U.S. Environmental Protection Agency (1993) (McGrath et al., 1994).
- Balance the use of chemical fertilizers and agrochemicals (pesticides and insecticides).
- Develop of analytical techniques to measure bioavailability of trace metals.

In order to avoid trace metal toxicities in plants, the following points need consideration:

- Efficient soil management followed by precise assessment for plant requirement of essential trace metals
- Use of balanced fertilizer and agrochemicals
- Necessary measures to be taken to place apart the industrial source of non-biogenic trace metals (Pb, Cd, Hg, and Ni) which are present in commercial fertilizers and agrochemicals
- Finally, cultivars resistant to certain trace metals to be developed which may help to restore metal contamination from plant to soil and vice versa

References

Adriano, D.C., Ed., *Trace Elements in Terrestrial Environment*, Springer-Verlag, New York, 1986.

Ahmed, S., S. Ghosal, and S.L. Jansson, Isotope aided studies on the relative uptake of Zn, Mn, Co and Ni by rape and pea, *Bangladesh J. Agric. Sci.*, 1, 45, 1981.

Allaway, W.H., Agronomic controls over the environmental cycling of trace metals, *Adv. Agron.*, 20, 235, 1968.

Berrow, M.L. and G.A. Reaves, *Proc. Int. Conf. Environ. Contamination*, CEP Consultants Ltd., Edinburgh, U.K., 1984, 333.

Blanchard, R.L., Radio chemical concentration process, in Proc. Int. Symp., New York, 1966, 281.

Bohn, H.L., B. McNeal, and G.A. O'Connor, *Soil Chemistry*, Wiley Interscience, New York, 1979.

Carell, E.F. and C.A. Price, Porphyrins and the iron requirement for chlorophyll formation in Euglena, *Plant Physiol.*, 40, 1, 1965.

Caro, J.H., Characterization of Superphosphate, Superphosphate: Its History, Chemistry and Manufacture, U.S. Department of Agriculture and TVA, 1964, 273.

Chang, A.C., T.C. Granato, and A.L. Page, A methodology for establishing phytotoxicity criteria for chromium, copper, nickel and zinc in agricultural land application of municipal sewage sludge, *J. Environ. Qual.*, 13, 521, 1992.

Chow, T.J. and J.L. Earl, Lead aerosols in the atmosphere: increasing concentrations, *Science*, 169, 577, 1970.

Cottenie, A.L., Behaviour and biological importance of heavy metals in the soil, *Chimia*, 34, 344, 1980.

Cottenie, A.L., Kiekens, and G. Van Landschoot, Problems of the mobility and predictability of heavy metal uptake by plants, in *Processing and Use of Sewage of Sludge*, L' Hermite, P. and H. Ott, Eds., Reidel, Bordrecht, 1983, 124.

Cruanas, R., *Aspects fisco — quimicos implicates en la contamination yen el Saneamiento de los suelos*, K.I.M.P, Valencia, 1992, 13.

Davies, B.E., *Applied Soil Trace Element*, Vol. 5, 2nd ed., Vallee, B.L. and W.E.C. Wacker, Eds., Wiley Interscience, Academic Press, New York, 1980, 192.

Davies, B.E. and L.H.P. Jones, Micronutrients and toxic elements, in *Russells Soil Condition and Plant Growth*, 11th ed., Wild, A., Ed., Longman Scientific and Tech., Essex CM 20 2JE, England, 1988, 780.

Domingo, L.E. and K. Kyuma, Mean and range of concentrations of selected trace elements in tropical Asian paddy soils, *Soil Sci. Plant Nutr.*, 29, 439, 1983.

Dowdy, R.H., W.E. Larson, J.M. Titrud, and J.J. Latterell, Growth and metal uptake of snap beans grown on sewage sludge amended soil, *J. Environ. Qual.*, 7, 252, 1978.

Evans, H.J., Role of Mo in plant nutrition, *Soil Sci.*, 81, 199, 1956.

Gladstones, J.S and J.F. Loneragan, Mineral elements in temperate crops and pasture plants. I. Zinc, *Aust. J. Agric. Res.*, 18, 427, 1967.

Gupta, V.K., Ramkala, and S.P. Gupta, Effect of nickel on yield and its concentration in some Rabi crops grown on Typic Ustipsamment, *J. Indian Soc. Soil Sci.*, 44(2), 348, 1996.

Kortstee, G.J.J., Iron deficiency and porphyrin formation by Arthrobacter globinformis, *Antonie van Leeuwenhoek*, 36, 579, 1970.

Kubota, J., E.R. Lemon, and W.H. Allaway, The effect of soil moisture content upon the uptake of molybdenum, copper and cobalt by Alsike clover, *Soil Sci. Soc. Am. Proc.*, 27, 679, 1963.

Lagerwerff, J.V., Heavy metal contamination in soils, in agriculture and the quality of our environment, *Am. Assoc. Adv. Sci. Publ.*, 85, 343, 1967.

Lagerwerff, J.V. and A.W. Specht, Contamination of roadside soil and vegetation with cadmium, nickel, lead and zinc, *Environ. Sci. Technol.*, 4, 538, 1970.

Lagerwerff, J.V., Uptake of cadmium, lead and zinc by radish from soil and air, *Soil Sci.*, 111, 129, 1971.

Leeper, G.W., Jr., Department of Army Corps of Engineers, DACW 73-73-C-0026 (Report), 70, 1972.

Lipman, C.B. and G. Mackinney, Proof of essential nature of copper for higher green plants, *Plant Physiol.*, 6, 593, 1931.

Maclean, A.J., R.L. Halstead, and B.J. Finn, Extractability of added lead in soils and its concentration in plants, *Can. J. Soil Sci.*, 49, 327, 1969.

McBride, M.B., *Environmental Chemistry of Soils*, Oxford University Press, New York, 1994, 325.

McGrath, S.P, A.C. Chang, A.L. Page, and E.W. Witter, Land application of sewage sludge: scientific perspectives of heavy metal loading limits in Europe and the United States, *Environ. Rev.*, 2, 108, 1994.

McGrath, S.P., A.M. Caudri, and K.E. Giller, Long-term effects of metals in sewage sludge on soils, microorganisms and plants, *J. Ind. Microbiol.*, 14, 94, 1995.

Melstead, S.W., Recycling municipal sludges effluent land, *Proc. Joint Conf.* (Illinois), 121, 1973.

Mertz, W., Chromium occurrence and functions in biological systems, *Plant Physiol. Rev.*, 49, 163, 1969.

Mishra, J. and R.S. Singh, Effect of nitrogen and zinc on the growth and uptake of N and Zn by linseed, *J. Indian Soc. Soil Sci.*, 44(2), 328, 1996.

Mitchell, R.L., Spectrochemical methods in soil investigation, *Soil Sci.*, 83, 13, 1957.

Patterson, C.C., Contaminated and natural lead environments of man, *Arch. Environ. Health*, 344, 1965.

Page, A.L., Fate and effects of trace elements in sewage sludge when applied to agricultural land, U.S. Environmental Protection Agency, Environ. Res. Center, Ohio, *Nat.* 97, 1974.

Prask, J.A. and D.J. Plocke, A role for zinc in the structural integrity of cytoplasmic ribosomes of *Euglena gracilis*, *Plant Physiol.*, 48, 150, 1971.

Price, C.A., *Molecular Approaches to Plant Physiology*, McGraw-Hill, New York, 1970, 398.

Price, C.A., H.E. Clark, and E.A. Fundkhouser, Functions of micronutrients in plants, in *Micronutrients in Agriculture*, Mortvedt, J.J., P.M. Giordano, and W.L. Lindsay, Eds., Soil Sci. Soc. Am. Inc., Madison, WI, 1971, 231.

Reisenauer, H.M., Molybdenum, in *Methods of Soil Analysis*, Black, C.A., Ed., II Agronomy 9, 1965, 1050.

Reuther, W. and C.K. Labanauskas, Copper, in *Diagnostic Criteria for Plants and Soils*, Chapman, H.D., Ed., 1966, 157.

Reuther, W. and P.F. Smith, Effects of high copper content of sandy soil on growth of citrus seedlings, *Soil Sci.*, 75, 219, 1953.

Ross, R.G. and D.K.R. Stewart, Cadmium residues in apple fruit and foliage following a cover spray of cadmium chloride, *Can. J. Plant Sci.*, 49, 49, 1969.

Sarkunan, V., A.K. Misra, and A.R. Mohapatra, Effect of Cd and Zn on yield and Cd and Zn content in rice, *J. Indian Soc. Soil Sci.*, 44(2), 346, 1996.

Schroeder, H.A. and J.J. Balassa, Cadmium: uptake of vegetables from superphosphate by soils, *Science*, 140, 819, 1963.

Scrutton, M.C., M.F. Utter, and A.S. Mildvan, Pyruvate carboxylase. VI. The presence of tightly bound manganese, *J. Biol. Chem.*, 241, 3480, 1966.

Shuman, L.M., Effect of soil temperature, moisture, and air drying on extractable manganese, iron, copper and zinc, *Soil Sci.*, 130, 336, 1980.

Smith, E.L., Presence of cobalt in the anti-pernicious anemia factor, *Nature*, 162, 144, 1948.

Swaine, D.J., The trace element content of fertilizers, *Tech. Communication*, no. 48 Commonwealth Agr. Bur. Soils, Harpenden, Farnham Royal, Bucks, England, 1955, 157.

Thomas, H.V., B.K. Milmore, G.A. Heidbreder, and B.A. Kogan, Blood lead of persons living near freeways, *Arch. Environ. Health*, 15, 695, 1967.

Underwood, E.J., *Trace Element in Human and Animal Nutrition*, 3rd ed., Academic Press, New York, 1971, 429.

U.S. Environmental Protection Agency, Standards for the use of disposal of sewage sludge final rules, *Fed. Regist.*, 58, 9248, 1993.

Vallee, B.L. and W.E.C. Wacker, Metalloproteins, in *The Proteins*, Vol. 5, 2nd ed., Neurath, H., Ed., Academic Press, New York, 1970, 192.

Viets, F.G., Jr. and W.L. Lindsay, Testing soils for zinc, copper, manganese and iron, in *Soil Testing and Plant Analysis*, Walsh, L.M. and J.D. Beaton, Eds., Soil Sci. Soc. Am., Madison, WI, 1973.

Wallihan, E.F., Iron, in *Diagnostic Criteria for Soils and Plants*, Chapman, H.D., Ed., University of California Press, Div. Agric. Sci., 1966, 203.

12

Polyamino Acid Chelation for Metal Remediation

Maury Howard and James A. Holcombe

CONTENTS

12.1 Introduction

12.1.1 Role of Chelators in Homogeneous Solution Processes vs. Column Elution

One common mode of soil remediation involves mobilization of the target metal via leaching or bulk processing of the soils. If the metal is in the correct oxidation state, metal chelation (e.g., with EDTA) can be used in mobilizing metals from the condensed state (e.g., metal oxide) in the soil. When not in the correct oxidation state for chelation, oxidation of the material often precedes chelation. "Remediation" requires the transfer of the target material from the contaminated material into another medium, often with the implication that the contaminant concentration in the medium be greater than in the original site. For example, a metal chelate can be concentrated via liquid-liquid extraction for purposes of metal recovery. Alternatively, the solution containing the mobilized metal can be passed through an appropriate column containing, for example, an ion exchanger to again effectively preconcentrate the target metal. Recovery, in the case of the latter, then involves elution of the column with the appropriate reagent to strip the metal into a solution of considerably lower volume than that represented by the original solution.

Both complexation with liquid-liquid extraction and preconcentration using column methodologies have several prerequisites for optimal employment as a remediation tool:

1. The chelator should be highly selective for the target metal.
2. The complex should have a large formation constant (small K_d) when extraction is taking place.
3. The K_d should be "adjustable" to permit easy release of metal if the complexing reagent or column material is to be reused.
4. Recovery should produce the largest possible ratio of contaminant mass to recovered solution (or solid) mass.

To be cost-effective, the chelating material should be as inexpensive as possible or reusable with a lifetime proportional with the initial material cost.

Most ion exchange resins are represented by functionalized polymers. While these materials often have capacities in the low multiequivalent/gram range, they are often subject to shrinking and swelling, especially as the pH is changed. This can impact the flow characteristics leading to blockage or flow channeling within the column. An alternative is the use of a mechanically stable support with a bonded phase. An increasingly common support is controlled pore glass which has the basic stability and chemical reactivity/inertness of glass at pH<10, but possesses a very high surface area (ca. 100 m^2/g in many cases) as a result of its high porosity. While the material is not prone to shrinking, it is somewhat fragile and the more porous of these materials may not likely stand up to the pressures associated with, for example, HPLC. These support materials are often available with active functionalities previously attached or linker arms can be added to the silica surface to accomplish immobilization (i.e., bonding) of an active complexing agent to the surface.

12.1.2 Requirements For Successful Use of Immobilized Chelators

As is the case with complexation in homogeneous solutions, polydentate ligands offer advantages over monodentate ligands via the "chelating effect." With properly spaced binding functionalities on a single chelating ligand, once an initial alignment of the target metal and one of the binding functionalities has been established, the rate of attachment of the second functionality of the same chelator to the target metal is considerably faster than would be the attachment of a second monodentate ligand. However, it is imperative that the steriochemistry of immobilized polydentate ligand that provided the chelate effect not be perturbed as a result of the immobilization. Additionally, the immobilization should not significantly impact the charge density, stereochemistry, etc. of the complexing functionality if the complexation strength that was exhibited in solution is not to be significantly altered once the chelator has been immobilized. It is obvious that immobilized monodentate ligands might not yield the same overall formation constant as is observed in homogeneous solutions when large coordination numbers (n > 1) were responsible for the complexation strength. This is a simple result of the geometric constraints imposed by surface immobilization.

12.2 Nature's Metal Binding System: Proteins

12.2.1 Using Amino Acids as "Building Blocks" for Chelator Design

It is well known that proteins have a propensity for metal binding and, in fact, metal incorporation is required for many enzymatic systems to be active. While proteins have not been used directly for metal preconcentration or soil remediation, they are obviously an inherent part of biological systems (e.g., plants, microbes, etc.) that are actively pursued for remediation activities. In the areas of metal binding, Crist et al. (1981) were the first to suggest possible binding moieties based simply on the amino acid functionalities. Two NMR studies further elaborated on Cd binding to algae and conflictingly report that Cd-S (Majidi et al., 1990) and Cd-carboxylate (Rayson and Drake, 1996) binding was active.

While a number of proteins incorporate a metal as part of the active center, metallothioneins are among the better known class of metal binding proteins (e.g., Stillman et al., 1992; Sigel and Sigel, 1989; Harrison, 1985). A majority of the studies of metallothioneins have been directed at the *in situ* metal binding characteristics. However, Anderson (1994) isolated copper metallothionein from yeast cells and immobilized the material on silica. He reported that *ex situ* behavior of this protein showed minimal preferential selectivity for copper and comparably strong binding to other metals. Protein binding affinity for several metals was also reported for intracellular binding (Kagi and Kojima, 1981; Kagi and Vallee, 1961; Li et al., 1980; Hamer, 1986; Winge et al., 1985). Additionally, Anderson's studies suggest only modest binding strengths once the material was immobilized on silica outside the cell. This lack of binding strength and selectivity may be due, in part, to the uniqueness of intracellular chemical environment which has a moderately high ionic strength and is chemically reducing.

12.2.2 Metallothioneins

Metallothioneins, like other metal-binding proteins, rely on the tertiary structure of the polypeptide to provide a binding pocket to enhance selectivity based on the size of the metal ion. The principle is not dissimilar from the design philosophy for "cage molecules" such as the crown ethers. In these proteins, the tertiary shape is governed by covalent cross-linking (e.g., disulfide bonds) as well as electrostatics, which includes hydrogen bonding. As one might expect, the hydrogen bonding of ion-ion and ion-dipole interactions are very dependent on the local solution environment. While the covalent disulfide bonds are significantly stronger, they too can be impacted by the chemical potential of the local solution environment. It is likely that the combination of these factors contributed to the altered binding behavior observed by Anderson (1994) once the metallothionein was immobilized and used outside the cell. It is this altered "activity" that often brands proteins as "fragile" and has made questionable their utility for tasks (e.g., metal binding) once removed from the cellular environment. In reality, the fragility might be more correctly assigned to the tertiary structure of the protein rather than to the intramolecular bonds within the polypeptide chain, which are quite robust. It remains, then, to explore the possibility of employing the varied functionalities associated with amino acids that make up polypeptides as potential metal-binding chelators while devising a polypeptide design that is less reliant on the prearranged tertiary structure dictated by the weaker intramolecular forces of H-bonding and electrostatic interactions that is typical of natural proteins.

METALLOTHONEIN

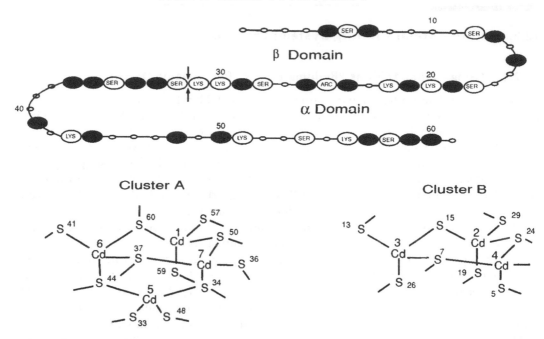

FIGURE 12.1

Metallothionein; cysteines represented by "–●–." (Stillman, M.J., C.F.S. Shaw, and K.T. Suzuki, VCH Publishers, New York, 1992, 1.)

12.3 Model System: Poly-L-Cysteine

12.3.1 Characteristics of the Cysteine Homopolymer

The chemical disadvantages inherent to natural proteins are only one practical barrier to their employment for metal chelation. Additionally, naturally occurring proteins are costly to purchase or time-consuming in their production or isolation. Figure 12.1 shows a schematic diagram of a Cd-binding metallothionein. As can be seen, cysteine is a primary metal-coordinating constituent of this and many metallothioneins and is the logical starting place for the investigation of metal complexation by a simple polyamino acid.

There have been studies focusing on the metal binding behavior of cysteine (Elmahadi and Greenway, 1993; Cherifi et al., 1990; Li and Manning, 1955), but few have pursued compounds containing multiple cysteine or sulfhydryl residues. Poly-L-cysteine (PLC) is such a system and was first synthesized by Berger et al. (1956) to study the reactivity of the sulfhydryl group. PLC is now available commercially (Sigma) as Poly-S-CBZ-L-cysteine, a homopolymer of cysteines blocked with a carbobenzoxy ring at each sulfhydryl group that must be removed prior to use. The polymer is prepared by base-initiated polymerization of its N-carboxyanhydride resulting in a blocked cysteine polymer of M.W. 5,000 to 15,000 Da (approximately 50 cysteine residues long). Berger also noted in his early studies that PLC is only soluble in alkaline solutions and can be oxidized by hydrogen peroxide, aerated solutions, and copper (II). Recent studies of PLC in homogeneous solution have confirmed the importance of pH on the polymer dissolution (Jurbergs and Holcombe, 1997) and its sensitivity to oxidizing and reducing agents (Howard et al., 1998).

12.3.2 Characterization of Homogeneous PLC

The pK_a of PLC is ca. 6 (Howard et al., 1999), which is about two pH units lower than that of the sulfhydryl functionality of cysteine ($pK_a = 8.36$) (Edsall and Wyman, 1958). Peptide bonding can have a large impact on the pKa of the amino acid functional group, causing it to become more acidic (Edsall and Wyman, 1958), and the proximity of charged groups can also cause the pH to shift due to electrostatic interferences (Sela et al., 1962; Voet and Voet, 1990). For the cysteine homopolymer where the thiol is the only potential charged group within the chain (excluding the amine and carboxylic terminals), the material has only limited solubility in solutions when the pH< pK_a. Unfortunately, this limits the homogeneous solution use of PLC to applications where the pH is greater than ca. 5 to 6. At lower pH values, the material precipitates and is slow to solubilize even after the pH is elevated. It is likely that when the pH is reduced below the pKa, the previous ion-dipole interactions which maintained PLC's solubility disappear as the sulfides become protonated. Denaturation and likely formation of a tight random coil result in the precipitation of the PLC. Circumvention of this problem through immobilization of the PLC on a solid support will be discussed in a later section.

The "soft base" character of the thiol group predicts a preference for "soft-acid" metal binding, which suggests that PLC will be inherently selective in binding/complexing metals such as Hg, Cd, and Pb. Torchinskii and Moiseevich (1974) state that the sulfide group binds univalent cations: Hg^+, Ag^+, Cu^+, Au^+; divalent cations: Hg^{2+}, Pb^{2+}, Cu^{2+}, Cd^{2+}, Zn^{2+}; and even tervalent ions: As^{3+} and Sb^{3+}. As seen by Li and Manning (1955), the binding affinity of thiol compounds glutathione and 2-mercaptoethylamine seems to follow the order $Pb^{2+}>Cd^{2+}>Zn^{2+}$. Homogeneous solution titrations of the PLC with these metals appear to support this natural selectivity (Autry and Holcombe, 1995). In these titrations 4(2-pyridylazo) resorcinol (PAR) was chosen as the metal-binding indicator (Shibata, 1972). The absorbance maxima for the PAR complex with Zn, Cd, and Cu appear at 490, 494, and 498 nm, respectively. Figure 12.2 shows the spectrophotometric titration results where the absorbance has been converted to metal-PAR concentrations for the ordinate.

Since PAR represents a competing ligand for PLC complexation of the metal, these titrations confirmed the relative strength of some of the sites for the metal-PLC complex and showed that multiple binding sites exist on each of the PLC polymer chains as shown in Table 12.1. Not unexpectedly, the presence of excess Na^+ (0.016 M) or Ca^{2+} (5.0×10^{-4} M) resulted in no displacement, for example, of Zn^{2+} from the Zn-PLC complex, indicating a strong preference for Zn^{2+}. In fact, this is consistent with the results from cysteine immobilized on silica, which also showed little affinity for alkali and alkaline earth metals (Elmahadi and Greenway, 1993). These two studies clearly indicate the potential selectivity inherent to the thiol functionality associated with cysteine and the polymer PLC. This binding preference may be exploited for metal-specific extraction, preconcentration and remediation, even in the presence of high salt matrices.

12.3.3 Immobilized Poly-L-Cysteine

As noted previously, the limited solubility (especially at pH < 6) of PLC gives the material limited utility for metal chelation in homogeneous solution applications. However, by immobilizing PLC on a support (e.g., porous glass), many of these problems disappear and, in fact, may serve to enhance some performance capabilities as will be shown in subsequent sections.

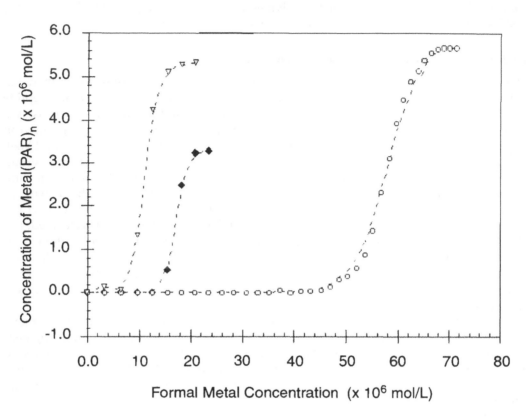

FIGURE 12.2
Spectrophotometric titration of PLC with Zn(♦), Cd (■), Cu (∇) using the metal-PAR colored complexes to calculate the concentration of PLC's strong binding sites for Zn, Cd, and Cu.

TABLE 12.1

Metal Capacity for PLC

Metal	PLC (mmol/g)	PLC (sites/mol)
Zn^{2+}	0.40	2.1
Cd^{2+}	1.4	7.4
Cu^{2+}	0.14	0.76

12.3.4   Immobilization Procedure

Showing a potential for metal extraction and separation, PLC was immobilized to study its effectiveness in "on-line" column applications. Proteins are immobilized routinely on various types of substrates including agarose and polyacrylamide gels, cross-linked polymer beads, and silica. Controlled pore glass (CPG) was selected as the support for PLC due to its stability and high surface area. CPG is rugged enough to stand up to high flow rates and back pressures that occur in flow injection systems, viz., low pressure chromatography. The glass is also resistant to pH effects and does not shrink or swell when the pH is drastically altered. Finally, CPG boasts a surface area of ca. 94 m^2/g due to its high porosity. The immobilization of PLC on CPG closely follows a procedure described by Masoom and Townshend (1984) for the immobilization of enzymes on CPG. However, prior to its immobilization, the PLC must be deblocked by removing the carbobenzoxy ring using a sodium ammonia

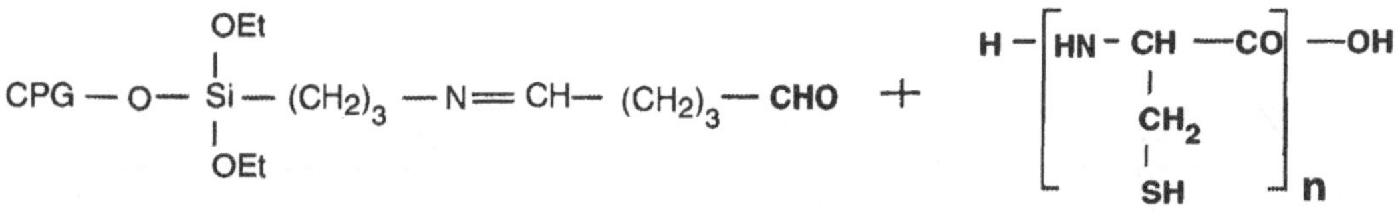

$$CPG—O—\underset{\underset{OEt}{|}}{\overset{\overset{OEt}{|}}{Si}}—(CH_2)_3—N{=}CH—(CH_2)_3—CHO \quad + \quad H—\left[\underset{\underset{SH}{\underset{|}{CH_2}}}{\overset{\overset{HN-CH—CO}{}}{}}\right]_n—OH$$

FIGURE 12.3
Attachment of PLC to controlled pore glass (CPG) through gluteraldehyde bifunctional linker.

reduction reaction (Berger et al., 1956), which was executed with dithiothreitol (DTT) (Cleland, 1964) present during reaction as a protecting agent to prevent oxidation of cysteine's sulfhydryl residues.

In brief, CPG is prepared for immobilization by activating the glass using dilute HNO_3 and heating. The activated glass is silanized via reaction with an aminoalkylating agent, 3-aminopro-pyltriethoxysilane. Gluteraldehyde, a bifunctional reagent, links the amine terminus of the PLC to the amine terminus of the surface silane group completing the immobilization of PLC on CPG as shown in Figure 12.3.

The efficiency of this reaction has been measured as a function of the surface coverage of PLC on the glass surface. An acid titration of the sulfides of the immobilized PLC was run to estimate the surface coverage, assuming 50 residues/polypeptide, and yielded an estimate of ca. 1 to 5% surface coverage with ca. 6.8×10^{-6} moles PLC/g PLC-CPG (Jurbergs and Holcombe, 1997). This number was supported by a sulfur analysis using ICP which showed a concentration of 5.3×10^{-6} moles PLC/g PLC-CPG.

12.3.5 Description of Flow Injection System

Some of the reasons for use of a flow injection analysis (FIA) system are its low sample/reagent consumption and high sample throughput (Fang, 1993) These advantages are further augmented when the system is used with a preconcentration column for analysis. In this case, the detectibility can be significantly improved through analyte preconcentration on the column and interferences reduced via analyte separation from the matrix (Burguera, 1989). FIA can be used to evaluate separation processes that might be used in a larger scale operation and is easily automated to enhance sample throughput. FIA is also easily interfaced with detection systems like a flame atomic absorption spectrometer (FAAS), which makes it very useful for metal separation and preconcentration studies.

Figure 12.4 is a diagram of a typical FIA system in which samples and reagents are selected through a valve system and delivered to a microcolumn by a peristaltic pump. Finally, the effluent is introduced to the FAAS where air compensation is provided by a T-junction prior to the nebulizer to make up for the difference between the nebulizer uptake rate and the peristaltic pump rate. Table 12.2 outlines a typical protocol for the use of the system shown in Figure 12.4.

12.3.6 Characterization of Immobilized PLC

Once PLC was immobilized on CPG, it became a more versatile material with which to work. For instance, under acidic conditions PLC undoubtedly still undergoes drastic conformation change (e.g., formation of a tight random coil); however, it can no longer agglomerate since it is bonded to the glass. Alteration to its more open form with an increase in solution pH occurs reversibly and rapidly (Jurbergs and Holcombe, 1997). Additionally, it can be oxidized and reduced reversibly by passing suitable agents through the column

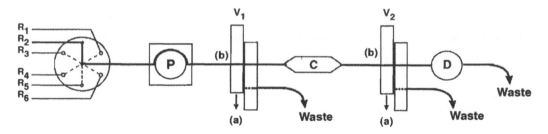

FIGURE 12.4

Typical flow injection system. R_{1-6}, reagents or sample as selected by 6-position Teflon rotary valve; P, peristaltic pump; $V_{1,2}$, 3-way slider valves; C, microcolumn (e.g., Omnifit 25 mm × 3 mm i.d.); D, detector (e.g., flame AA).

TABLE 12.2

Typical Protocol for FIA System

Operation	Solution Input	Valve 1 Position	Valve 2 Position	Notes
Column conditioning	Buffer	b	b	
Breakthrough curve	Sample	b	b	FAA data collection
Strip (line fill)	Acid	a	a	
Strip	Acid	b	**	**Sample sent to D (e.g., FAA) if preconcentration mode used; otherwise, sent to waste or fixed volume container for analysis
Calibration curve	Acidified stds	b	b	FAA readings taken
Column prep	Acid	b	b	

(Howard et al., 1998). PLC is also very strongly attached to the surface of the glass and, unlike most natural proteins, it is extremely rugged and can withstand extended exposure to acids (e.g., 1 M HNO_3) and repeated exposure to hydrogen peroxide without any noticeable change in its binding characteristics once it is returned to its optimal binding mode of being fully reduced and in a solution whose pH is greater than 6.

12.3.7 Breakthrough Discussion

"Breakthrough curves" and "strips" are commonly used for column characterization and data analysis for "on-line" systems. A complete breakthrough curve represents the change in the amount of metal in the column effluent as the amount introduced to the column is increased until, finally, the column reaches equilibrium capacity for that particular influent concentration.

Figure 12.5a is a typical breakthrough curve shape for Cd introduced into a PLC-CPG microcolumn. A metal solution is passed through the column and the metal in the effluent is recorded as a function of volume passed through the column. The presence of a baseline signal is indicative of quantitative binding (i.e., nearly 100% of the metal introduced to the column is bound) and suggests chelation to strong sites. The saturation of these binding sites results in "breakthrough," indicated by an increase in effluent concentration above the baseline level. The sloped region reveals weaker binding where only a fraction of the metal introduced is bound to the column. When the effluent concentration is equal to the influent concentration, the column has reached equilibrium capacity and the total amount of metal bound can be calculated from the shaded area in Figure 12.5a. The data collected from this type of experiment can be converted to binding isotherms in a relatively straightforward manner:

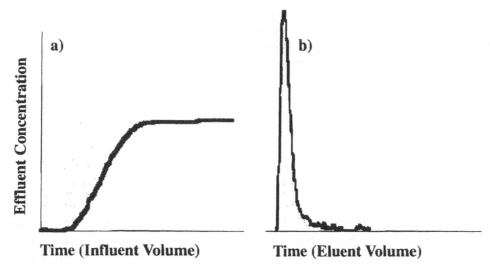

FIGURE 12.5
Typical breakthrough curve (a) and strip peak (b).

$$C_s = C°V - \int_0^V C_m dV \tag{1}$$

where C_s is the metal bound to the stationary phase after a volume V is passed through the column, $C°$ is the influent concentration, and C_m is the mobile phase effluent concentration. A plot of C_s vs. C_m yields the binding isotherm.

The metal can also be stripped from the column with a suitable eluent (e.g., dilute acid works well for PLC), resulting in a stripping peak similar to that indicated in Figure 12.5b. Again, the shaded area in Figure 12.5b represents the amount stripped and, in the case of quantitative recovery, should be the same area as the shaded region in Figure 12.5a.

12.3.8 Batch Studies and K_{eq}

Slow mass transport between the phases in column or slow kinetics can fail to establish equilibrium (Fang, 1993). For accurate determinations of the equilibrium constants for metal-PLC complexes, batch mode studies need to be conducted. To accurately determine the formation constants or, conversely, the dissociation constants of the strong binding sites, competing ligands are used in the bulk solution. For the strongest Cd binding sites on PLC, ethylenediamine tetraacetic acid (EDTA) was employed as the competing ligand. Ethylenediamine (en) was used as a competing ligand for somewhat weaker sites. The equilibrium between Cd and either PLC or the competing ligand, L, is given by:

$$PLC + CdL_n \rightleftharpoons Cd - PLC + nL \qquad K_{eq} = \frac{[Cd - PLC][L]^n}{[Cd_n][PLC]} = \frac{K_f}{K_d} \tag{2}$$

where the following two equilibria must also hold:

$$PLC + Cd^{2+} \rightleftharpoons Cd - PLC \qquad Kf = [Cd - PLC]/[Cd^{2+}][PLC] \tag{3}$$

$$CdL_n \rightleftharpoons Cd^2 + nL \qquad K_d = [Cd - L_n]/[Cd^{2+}][L]^n \tag{4}$$

The formal amounts of metal and ligand in the system are known; the total Cd in the solution C_{sol} can be determined and the Cd bound to the immobilized PLC can be calculated ($[Cd - PLC] = C^0_{Cd} - C_{sol}$) after appropriate correction for dimensions used. Assuming further that $C^0_L \geq n\, C^0_{Cd}$, then the following approximation is reasonable:

$$C_{sol} = [Cd^{2+}] + [CdL_n] \approx [CdL_n] \tag{5}$$

This permits restatement of K_f:

$$K_f = \frac{K(C^0_{Cd} - C_{sol})(C^0_L - C_{sol})^n}{C_{sol}(C^0_{PLC} - C^0_{Cd} + C_{sol})} \tag{6}$$

Since both K_f and C^0_{PLC} are unknown, at least two measurements must be made at different concentrations. However, the use of competing equilibria using a complexing agent with a known K_d circumvents the problem of working with extremely low metal concentrations and attempting to measure extremely small quantities of free metal in the equilibrated solution.

These binding strength studies carried out by Jurbergs and Holcombe (1997) resulted in the determination of four types of binding sites for immobilized PLC. The first site was stronger than Cd-EDTA and its binding constant estimated to be $K_1 > 10^{13}$; and the next was $10^9 < K_2 < 10^{11}$, being stronger than en, but weaker than EDTA. The weak sites were estimated using a non-linear least squares algorithm (Cernick and Borkovec, 1995). The formation constants for the remaining weaker sites ranged from 10^7 to 10^4.

12.3.9   Flow Studies and Establishment of K_{eq}

Cd chelation properties of immobilized PLC were investigated using the flow system described in Figure 12.4 for extraction of Cd from buffered aqueous solutions. The impacts of influent concentration and flow rate on the column were characterized. From the breakthrough curves in Figure 12.6 where different flow rates are studied, it is clear that equilibrium is not established within the column at these flow rates. Effective stability constants were calculated for the moderate and weak sites and a number of strong sites were reported for the flow systems (Jurbergs and Holcombe, 1997). The metal influent concentration has an influence on the measured capacity of the column, as one would expect when weaker sites are present that do not become fully occupied.

As noted previously, the breakthrough curve data can be easily converted to isotherms. These data were analyzed for the flow system using the Levenberg-Marquardt algorithm to obtain binding strengths and capacities. The results were comparable to the true equilibrium values obtained from the batch experiments, especially for the lower flow rate of 0.55 mL/min (Jurbergs and Holcombe, 1997).

Immobilized PLC's metal extraction performance was compared with 8-hydroxyquinoline (8HQ) (Howard et al., 1999), a material well characterized in metal extraction and preconcentration applications (Chow and Cantwell, 1988; Malamas et al., 1984; Marshall and Mottola, 1983). The ability to effectively extract metals from seawater matrices is a property of 8HQ, which has been thoroughly studied (e.g., Willie et al., 1983; Sturgeon, 1981; Seubert, 1995; Fang et al., 1984). The studies showed that both resins (1) perform well for Cd preconcentration and recovery in a seawater matrix, (2) exhibited flow rate dependent extraction efficiency, and (3) exhibited more than one type of binding site. However, the enhanced selectivity of PLC for the soft acid metals permitted distinction from 8HQ for the quantitative extraction of Cd in the presence of a Co and Ni matrix, as shown in Table 12.3.

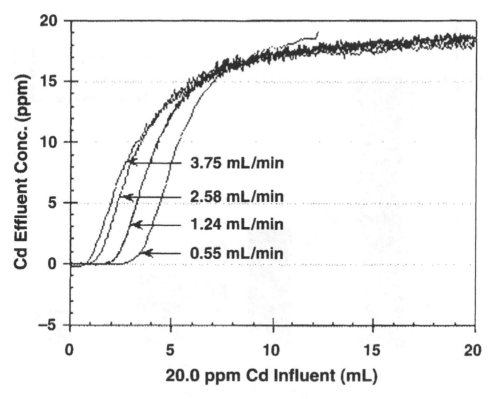

FIGURE 12.6
Breakthrough curves for Cd on PLC at pH 7.

TABLE 12.3

Cd Recovery Efficiencies for PLC and 8HQ (5 min loading at pH 7)

| | 100 ppb Cd in Seawater | | 50 ppb Cd (1.00 mL/min) | |
	1.00 mL/min	5.00 mL/min	500 ppm Co	500 ppm Ni
PLC	108 ± 8%	103 ± 6%	101 ± 3%	102 ± 3%
8HQ	94 ± 8%	101 ± 6%	2 ± 1%	4 ± 1%

In the microcolumn, on-line FIA system, it was suspected that mass transport prohibited the establishment of equilibrium within the column at the flow rates employed (Fang, 1993). This was confirmed by comparing the effective reduction in the strong site capacities for PLC and 8HQ as the flow rate increased, i.e., as the solution contact time with the immobilized chelator decreased. Figure 12.7 shows this comparison. It should be pointed out that the two immobilized chelators are physically different from each other (i.e., PLC is a polymeric chain with a large number of potential binding sites, while 8HQ is a simpler bifunctional ligand) and their binding sites do not resemble each other chemically. Thus, it is unlikely that binding kinetics are sufficiently similar to promote the similarity in flow rate dependence observed (Fang, 1993).

Effective stability constants of Cd-PLC and Cd-8HQ obtained in the FIA system were determined using nonlinear least squares analysis of the Cd binding data and are shown in Table 12.4.

Although 8HQ has a larger number of strong binding sites than PLC, this does not necessarily make it a better choice for remediation, as can be seen in the Co and Ni matrix studies that illustrate the importance of selectivity.

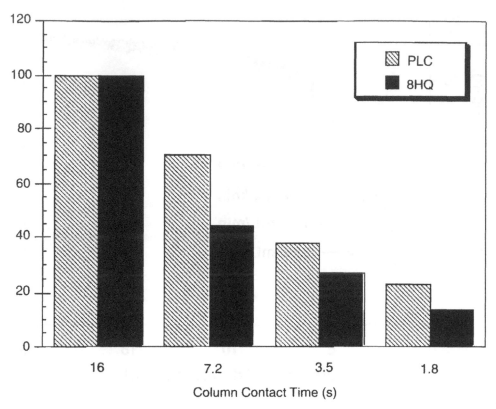

FIGURE 12.7
Flow rate dependence of strong site capacity for PLC and 8HQ.

12.3.9.1 pH Effects

The effect of pH on capacity (Table 12.5) shows that PLC has a greater capacity for Cd in more alkaline solutions. This is due to the greater number of deprotonated SH groups at pHs greater than 6, the pK_a of immobilized PLC, since the more reactive form of the sulfhydryl group is the deprotonated form, the mercaptide ion (S^-) (Torchinskii and Moiseevich, 1994).

12.3.10 Redox Characteristics

Studies of immobilized PLC revealed interesting and potentially useful redox properties. Berger et al. (1956) showed that SH groups of PLC can be oxidized to disulfides or sulfenites by oxygen or hydrogen peroxide as well as oxidizing metals such as Cu(II) (Berger et al., 1956). Although, in homogeneous solution the oxidation of PLC results in a cross-linked, insoluble product (Berger et al., 1956), the oxidation of immobilized PLC has been shown to be fully reversible by reduction of the column with dithiothreitol (DTT) (Howard et al., 1998), and the original binding properties of the column restored. DTT and PLC have a similar pH dependence due to their common reactive thiol groups, and this dependence can be employed to produce intermediate stages of reduction which provide different levels of capacity and binding strength of the PLC (Howard et al., 1998). As shown in Figure 12.8, the capacity of the column can be controlled by oxidizing the column with peroxide and reducing with DTT at pH 4.5 for varied times. Full reduction at this pH takes up to 48 h, but at pH 8.1 the reduction is complete within 10 min.

TABLE 12.4

Conditional Stability Constants and Number of Sites for
Cd-8HQ and Cd-PLC Complexes, pH 7.0

8-Hydroxyquinoline-CPG	Poly-L-Cysteine-CPG
$K_1 = 10^9 - 10^{11}$	$K_1 = 1 \pm 0.5 \times 10^{13}$
$K_2 = 4 \pm 1 \times 10^8$	$K_2 = 10^9 - 10^{11}$
$K_3 = 2 \pm 1 \times 10^6$	$K_3 = 1 \pm 1 \times 10^6$
$K_4 = 2 \pm 0.5 \times 10^4$	$K_4 = 2 \pm 1 \times 10^4$
n_i (mmol/g 8HQ-CPG)	**n_i (mmol/g PLC-CPG)**
$n_1 = 10$	$n_1 = 1 \pm 0.1$
$n_2 = 12 \pm 0.1$	$n_2 = 6$
$n_3 = 3 \pm 1$	$n_3 = 20 \pm 1$
$n_4 = 33 \pm 1$	$n_4 = 10 \pm 3$

TABLE 12.5

Effective Cd Capacities for Reduced PLC-CPG with
Influent Flow Rate of 1.00 mL/min

pH	Cd Bound (mmol/g of CPG)	Cd recovered (%)
4	3 ± 1	111 ± 12
5	15 ± 2	65 ± 2
6	11 ± 1	105 ± 6
7	16 ± 1	110 ± 6
8	22 ± 2	91 ± 6
9	42 ± 3	90 ± 1

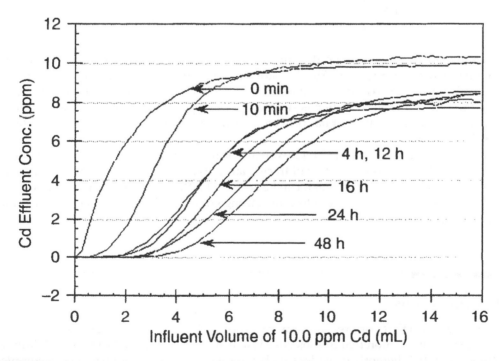

FIGURE 12.8
Cd breakthroughs run at several levels of reduction from the peroxide oxidized state.

TABLE 12.6

Effective Cd Capacities for Oxidized States of PLC-CPG (pH 7.0, 1.0 mL/min)

	Fully Reduced	Partially Reduced	10 min H_2O_2 Oxidation	18 h Oxygen Oxidation[a]	18 h Oxygen Oxidation[b]
Total capacity (μmol/g)	4.8 ± 0.2	4.1 ± 0.2	2.2 ± 0.1	3.1	3.2
Strong site capacity	2.5 ± 0.3	1.2 ± 0.1	< 0.2	0.74	1.3

[a] Oxidation from partially reduced state.
[b] Oxidation from fully reduced state.

TABLE 12.7

Cd, Zn, and Pb Effective Capacities and Recoveries for Peroxide-Oxidized PLC-CPG (pH 7.0, 1.0 mL/min)

Metal	Strong Sites (μmol/g CPG)	Total Sites (μmol/g CPG)	Recovery
Cd	< 0.2	2.2 ± 0.1	100 ± 4
Zn	< 0.1	11	103 ± 2
Pb	2.5	5.0	105 ± 5

TABLE 12.8

Preconcentration of 26 ppb Cd from Na, K, Mg, Ca, and Cu(II) Matrices Using H_2O_2 Oxidized PLC-CPG (pH 7.0, 3 mL samples at a flow of 1.0 mL/min)

Matrix Metal	Cd Recovery (%)
Na (500 ppm)	93 ± 3
K (500 ppm)	92 ± 3
Mg (500 ppm)	93 ± 4
Ca (500 ppm)	91 ± 2
Cu (10 ppm)	100 ± 3

Different oxidizing agents were also studied for their effects on immobilized PLC: hydrogen peroxide, oxygen, and Cu^{2+}. Studies showed the peroxide to be the most efficient oxidant, resulting in full oxidation within minutes, whereas oxygen was much slower to oxidize the PLC. Oxygen oxidation is suspected of being rate limited by the slow kinetics of O_2 bond dissociation (38), but has proven to be much faster in the presence of oxidizing metals like Fe^{3+} or Cu^{2+} (23).

Even in its oxidized form, PLC retains many of its strongest binding sites and can be effectively used for preconcentration of trace quantities as is shown in the following tables (Tables 12.6 and 12.7). In Table 12.8 it can be seen that although capacity is diminished by oxidation, the binding remains selective and strong enough to be very effective for preconcentration applications.

12.4 Conclusions

Any remediation activity or preconcentration is optimally served by a highly selective chelator, since it is uncommon that the objective is to isolate all metals in the sample. Additionally, the selectivity inherently provides the chelator with enhanced performance in complex matrices since competition between the target metal and interfering ions is minimized. Optimal use of chelators for metal extraction further dictates the ability to easily release the target metal for purposes of recovery and ultimate disposal or reuse of the material in a compact, concentrated form.

The potential of immobilized polyamino acids, as illustrated by the initial work with poly-L-cysteine, for such remediation and preconcentration activity has been demonstrated. The chain apparently reaches a free energy minimum for complexation through tertiary structure alteration during binding. This is likely responsible for the strong binding as well as the easy release when this tertiary structure is perturbed, e.g., alteration in pH to convert bonding mode of PLC to a tight random coil with little binding strength or capacity. This provides the stereochemistry afforded by natural proteins and cage-type molecules while affording the opportunity for easy, reversible metal release and ruggedness.

The tertiary structure of PLC responsible for the strong binding is similar to the motif used in nature, although the relative size of the PLC compared to many natural proteins and the absence of a fixed tertiary structure in the absence of the metal does not render the polyamino acids inactive after harsh acid treatment, as is the case for nature's proteins. By immobilizing the polymers, reversible chemistry (e.g., protonation/deprotonation and oxidation/reduction) can be performed without degradation of the activity or loss of binding capacity. Columns subjected to numerous cycles of acid stripping and harsh oxidation over a period of years have retained their binding capacity and strength, thus satisfying the prerequisite for an effective ion exchanger intended for practical remediation, i.e., an extended lifetime.

This preliminary work with PLC illustrates the potential for a new mode of complexation through the use of polyamino acid polymers and copolymers. Since all are polypeptides, the chemistry required for their production is identical, thus eliminating the need for elaborate synthetic projects to construct and test new polymers. The amino acid "building set" also provides an array of cation, as well as anion, ligand functionalities from which to construct the chelator. Strong binding is not "forced" through the geometry built into the system as is the case with fixed cage-type molecules or traditional polydentate ligands. Instead, the long, coiled nature of the polymer and reliance on thermodynamic driving forces to minimize the system's free energy foster coiling around the target metal to achieve optimal stereochemistry for strong complexation.

In particular, PLC has proven to be an efficient chelator for "soft acid" metals such as Cd, Pb, and Zn, and its inherent selectivity enhances its performance in high salt and even some transition metal matrices, which makes PLC a competitive metal extraction tool even when compared to well-characterized and widely used 8HQ. Immobilization on a robust stationary support has enhanced its useful lifetime, although the half-life has not yet been determined since the longest operating column continues to perform as per its initial specifications for 3 years. However, like other chelators it is of limited utility for the rapid extraction of metals from complexes (e.g., metal-EDTA) where dissociation kinetics of the solution complex are rate limiting.

Acknowledgment

This work was supported, in part, by a grant from the Gulf Coast Hazardous Substance Research Center, Beaumont, Texas.

References

Anderson, B., Evaluation of Immobilized Metallothionein for Trace Metal Separation and Preconcentration, Ph.D. thesis, University of Texas, Austin, 1994.

Autry, H.A. and J.A. Holcombe, Cadmium, copper and zinc complexes of poly-L-cysteine, *Analyst*, 120, 2643, 1995.

Bard, A.J., *Chemical Equilibrium*, Harper and Row, New York, 1966.

Berger, A., J. Noguchi, and E. Katchalski, Poly-L-cysteine, *J. Am. Chem. Soc.*, 77, 4483, 1956.

Burguera, J.L., Flow injection atomic spectroscopy, in *Practical Spectroscopy*, Vol. 7, E.G. Brame, Jr., Ed., Marcel Dekker, New York, 1989, 1.

Cernik, M. and M. Borkovec, Regularized least-squares methods for the calculation of discrete and continuous affinity distribution for heterogeneous sorbents, *Environ. Sci. Technol.*, 29, 413, 1995.

Cherifi, K., B.D.-L. Reverend, K. Varnagy, T. Kiss, I. Sovago, C. Loucheux, and H. Kozlowski, Transition metal complexes of L-cysteine containing di- and tripeptides, *J. Inorg. Biochem.*, 38, 69, 1990.

Chow, P.Y.T. and F.F. Cantwell, Calcium sorption by immobilized oxine and its use in determining free calcium ion concentration in aqueous solution, *Anal. Chem.*, 60, 1569, 1988.

Cleland, W.W., Dithiothreitol, a new protective reagent for SH groups, *Biochemistry*, 3, 480, 1964.

Crist, R.H., K. Oberholser, N. Shank, and M. Nguyen, Nature of bonding between metallic ions and algal cell walls, *Environ. Sci. Technol.*, 15, 1212, 1981.

Edsall, J. and J. Wyman, *Biophysical Chemistry*, Vol. 1, Academic Press, New York, 1958.

Elmahadi, H.A.M. and G.M. Greenway, Immobilized cysteine as a reagent for preconcentration of trace metals prior to determination by atomic absorption spectrometry, *J. Anal. At. Spectrom.*, 8, 1011, 1993.

Fang, Z.-L., *Flow Injection Separation and Preconcentration*, VCH Publishers, New York, 1993.

Fang, Z.-L., J. Ruzicka, and E.H. Hansen, An efficient flow-injection system with on-line ion-exchange preconcentration for the determination of trace amounts of heavy metals by atomic absorption spectrometry, *Anal. Chim. Acta*, 164, 23, 1984.

Hamer, D.H., Metallothionein, *Ann. Rev. Biochem.*, 55, 913, 1986.

Harrison, P.M., *Metalloproteins*, Verlag Chemie, Weinheim, 1985.

Howard, M., H. Jurbergs, and J.A. Holcombe, Effects of oxidation of immobilized poly-L-cysteine on trace metal chelation and preconcentration, *Anal. Chem.*, 70, 160, 1998.

Howard, M.E., H.A. Jurbergs, and J.A. Holcombe, Comparison of 8-hydroxy quinoline and poly-L-cysteine for trace metal preconcentration, *J. Anal. At. Spectrom.*, 14, 1209, 1999.

Jurbergs, H.A. and J.A. Holcombe, Characterization of immobilized poly(L-cysteine) for cadmium chelation and preconcentration, *Anal. Chem.*, 69, 1893, 1997.

Kagi, J.H.R. and B.L. Vallee, Metallothionein: a cadmium and zinc-containing protein from equine renal cortex, *J. Biol. Chem.*, 236, 2435, 1961.

Kagi, J.H.R. and Y. Kojima, Chemistry and biochemistry of metallothionein, in *Metallothionein II*, Vol. 52, Kagi, J.H.R. and Kojima, Y., Eds., Birkhauser Verlag, Basel, 1987, 25.

Li, N.C. and R.A. Manning, Some metal complexes of sulfur-containing amino acids, *J. Am. Chem. Soc.*, 77, 5225, 1955.

Li, T.-Y., A.J. Kraker, C.F. Shaw, and D.H. Petering, Ligand substitution reactions of metallothioneins with EDTA and apo-carbonic anhydrase, *Proc. Natl. Acad. Sci.*, 77, 6334, 1980.

Majidi, V., J.D.A. Laude, and J.A. Holcombe, Investigation of metal-algae bonding nature by 113Cd nuclear magnetic resonance, *Environ. Sci. Technol.*, 24, 1309, 1990.

Malamas, J., M. Bengtsson, and G. Johansson, On-line trace metal enrichment and matrix isolation in atomic absorption spetrometry by a column containing immobilized 8-quinolinol in a flow-injection system, *Anal. Chim. Acta*, 160, 1, 1984.

Marshall, M.A. and H.A. Mottola, Synthesis of silica-immobilized 8-quinolinol with (aminophenyl)-trimethoxysilane, *Anal. Chem.*, 55, 2089, 1983.

Masoom, M. and A. Townshend, Determination of glucose in blood by flow injection analysis and an immobilized glucose oxidase column, *Anal. Chim. Acta*, 166, 111, 1984.

Rayson, G. and L. Drake, Plant-derived materials for metal ion-selective binding and preconcentration, *Anal. Chem.*, 68, 22A, 1996.

Sela, M., S. Fuchs, and R. Arnon, Studies on the chemical basis of the antigenicity of proteins, *Biochem. J.*, 85, 223, 1962.

Seubert, A., G. Petzold, and J.W. McLaren, Synthesis and application of an inert type of 8-hydroxy-quinoline-based chelating ion exchanger for seawater analysis using on-line inductively coupled plasma mass spectrometry detection, *J. Anal. At. Spectrom.*, 10, 371, 1995.

Shibata, *Chelates in Analytical Chemistry*, Vol. 4, Marcel Dekker, New York, 1972.

Sigel, H. and A. Sigel, *Metal Ions in Biological Systems*, Marcel Dekker, New York, 1989.

Stillman, M.J., C.F.S. Shaw, and K.T. Suzuki, *Metallothioneins: Synthesis Structure and Properties of Metallothioneins, Phytochelatins and Metal-Thiolate Complexes*, VCH Publishers, New York, 1992, 1.

Sturgeon, R.E., S.S Berman, S.N. Willie, and J.A.H. Desauiniers, Preconcentration of trace elements from seawater with silica-immobilized 8-hydroxyquinoline, *Anal. Chem.*, 53, 2337, 1981.

Torchinskii, Y.M. and I. Moiseevich, *Sulfhydryl and Disulfide Groups of Proteins*, Consultants Bureau, New York, 1974.

Voet, D. and J.G. Voet, *Biochemistry*, Wiley, New York, 1990.

Willie, S.N., R.E. Sturgeon, and S.S. Berman, Comparison of 8-quinolinol-bonded polymer supports for the preconcentration of trace metals from sea water, *Anal. Chim. Acta*, 149, 59, 1983.

Winge, D.R., K.B. Nielson, N.R. Gray, and D.H. Hamer, Yeast metallothionein. Sequence and metal-binding properties, *J. Biol. Chem.*, 260, 14464, 1985.

13

Effects of Natural Zeolite and Bentonite on the Phytoavailability of Heavy Metals in Chicory

László Simon

CONTENTS

13.1 Introduction

Clay minerals have high cation sorption and cation exchange capacity and large surface area (Adriano, 1986; Alloway, 1990; Kabata-Pendias and Pendias, 1992). Natural and synthetic zeolites (sodium aluminum calcium silicates) have high affinity to sorb and to complex trace elements, particularly heavy metals (Mineyev et al., 1990; Baidina, 1991; Gworek, 1992, 1994; Obukhov and Plekhanova, 1995; Chlopecka and Adriano, 1996; Shanableh and Kharabsheh, 1996) and caesium (Campbell and Davies, 1997). Similar properties of bentonite (montmorillonite, sodium aluminum magnesium hydrosilicate) were observed by Krebs and Gupta (1994). Besides other soil additives (e.g., lime, phosphate, apatite, iron oxide, manganese oxide, organic matter) natural zeolites and bentonites have the potentiality of immobilizing heavy metals in contaminated soils and of preventing their accumulation in agricultural plants (Mench et al., 1998).

During recent decades the environmental consequences of industrialization were neglected in Hungary. In many cases the environmental pollution was not revealed, e.g., the contamination of kitchen garden soil with heavy metals in the neighborhood of a former galvanization plant in the city of Nyíregyháza (Northeastern Hungary) was discovered only in 1991. The transition to a market economy is pressing a clean-up of the most critical contaminated sites of the country. The Hungarian government therefore initiated a new environmental remediation program in 1996. Hungary is rich in natural zeolites and bentonites, therefore these relatively cheap clay minerals may play an important role in the soil remediation program.

Considering the above facts the objectives of our study were to evaluate (1) concentration of heavy metals in a soil contaminated with galvanic mud, (2) phytoavailability, transport, and distribution of heavy metals in chicory plant grown in this contaminated soil, (3) heavy metal immobilization capacity of three different natural zeolites and a bentonite with the help of chicory indicator plant.

13.2   Materials and Methods

13.2.1   Soil Sample Collection and Characterization

The uncontaminated (control) soil used in this experiment originated from the demonstration garden of College of Agriculture, Nyíregyháza. The soil samples contaminated with heavy metals were collected in a kitchen garden located near a former galvanization plant (Nyíregyháza, Vasgyár Street). Several basic characteristics of the soils (pH_{KCl}, clay + silt [<0.02 mm particles] content, organic matter %, cation exchange capacity [CEC], macroelement concentrations) were determined according to Hungarian standards. Basic characteristics of the uncontaminated soil used in the experiment were the following: pH_{KCl}: 6.6; clay + silt content: 15.8%; organic matter: 1.3%; CEC: 18.1 meq/100 g, P 0.9 g/kg, K 4.3 g/kg, Ca 34.3 g/kg, and Mg 7.6 g/kg. The soil contaminated with heavy metals had the following basic properties: pH_{KCl}: 6.8; clay + silt content: 16.0%; organic matter: 1.1%; CEC: 8.2 meq/100 g, P 1.4 g/kg, K 2.4 g/kg, Ca 53.3 g/kg, and Mg 4.8 g/kg (all elements were determined in HNO_3/H_2O_2 extracts). Both soils were slightly acidic loamy sands and had brown forest soil character.

The plots were sampled with stainless steel gauge auger. Two parallel soil samples were taken from 0 to 25 cm depth combining 20 subsamples. After air drying the soil samples were screened on a 2-mm sieve. To determine exchangeable or "plant available" fraction of heavy metals 5 g of soil samples were extracted with 50 cm^3 of 0.01 M $CaCl_2$ or Lakanen-Erviö solution (0.02 M H_4EDTA in ammonium acetate buffer, pH 4.65), respectively. To extract "total" amount of heavy metals 2-g samples were digested with cc. HNO_3 and H_2O_2 (3:1 v/v) prior to elemental analysis. Samples were shaken for 2 h.

13.2.2   Characterization of Natural Zeolites and Bentonite Used in the Experiment

Three different types of natural zeolites (RBZ clinoptilolitic rhyolite tuff, MHZ mordenite rhyolite tuff, and MSC clay mixed clinoptilolite [clinoptilolite altogether with H-montmorillonite] and a bentonite (MHB montmorillonite) sample originated from Healing Minerals Geoproduct Ltd. (Mád, Hungary) and were mined in Zemplén Hills, Hungary.

The RBZ type is a medium-hard microporous zeolite. Composition: clinoptilolite ≈50%, mordenite 0-10%, altered volcanic glass 20-30%, montmorillonite 10-15%, quartz 0-5%, feldspar 0-5%, and limonite 1-2%. Characteristics: density ≈2.05 g/cm^3, wet surface pH: 7.8, NH_4^+-ion exchange capacity: 118 meq/100 g, Ag^+-ion binding capacity: 177 meq/100 g. The

MHZ type is a hard porous light zeolite. Composition: mordenite 40-60%, altered volcanic glass 20-30%, montmorillonite 10-15%, quartz 0-5%, feldspar 0-5%, and limonite 1-2%. Characteristics: density ≈ 1.82 g/cm^3, wet surface pH: 7, NH$_4^+$-ion exchange capacity: 70 meq/100 g, Ag$^+$-ion binding capacity: 176 meq/100 g. The MSC type is a rare and worldwide unique soft zeolite formation. The main components are clinoptilolite and H-montmorillonite which are present at the same ratio. Characteristics: density ≈ 1.50 g/cm^3, wet surface pH: 5-6, NH$_4^+$-ion exchange capacity: 65 meq/100 g, Ag$^+$-ion binding capacity: 52 meq/100g. The MHB type bentonite is a gray, wet, soft, and plastic clay. It contains, besides calcium potassium montmorillonite, illite, crystoballite, kaolinite, and amorphous volcanic glass. Physicochemical properties: swelling in water, absorption and ion-exchange of Na$^+$-cations, tixotrophy, formation of stabile suspension-gel (characterization and analytical data are courtesy of Tibor Mátyás from Healing Minerals Geoproduct Ltd., Mád, Hungary).

13.2.3 Growth Chamber Pot Experiments with Chicory

Growth chamber pot experiments were conducted with chicory (*Cichorium intybus* L., var. *foliosum* Hegi, cv. Wild) to study the phytoavailability of heavy metals in the galvanic mud contaminated soil (Experiment 1), and their immobilization with zeolites and a bentonite (Experiment 2). Larger amounts of uncontaminated and contaminated soil were collected during 1995 and 1996 as described above; the soils were air dried, homogenized, and screened on a 2-mm sieve before utilization.

Experiment 1: Chicory was grown in uncontaminated basic soil mixed (or replaced) with 0, 10, 25, 50, 75, or 100% (m/m) of heavy metal contaminated soil (collected in 1995) in a growth chamber experiment. A completely randomized experimental design with six replications was used. In one plastic pot (16 cm diameter) with 1500 g of soil three uniformly sized 4-week-old seedlings were planted. After 8 weeks of growth (temperature: 25 ± 2°C, illumination: 5000 lux for 8 h daily, watering: regularly with distilled water to maintain constant field capacity moisture content) the plants were harvested and separated into underground parts (roots and rhizomes) and shoots. Plant samples were washed in tap water and rinsed in three-times-changed deionized water. After the determination of dry weights (70°C, 14 h) the samples were ground (<1 mm) and digested with cc. HNO$_3$ and H$_2$O$_2$ (3:1 v/v) prior to elemental analysis. The experiment was done between February and April 1995.

Experiment 2: Chicory was grown in uncontaminated soil or in heavy metal-contaminated soil (collected in 1996). The heavy metal-contaminated soil was amended with 5% (m/m) of three different types of zeolite (RBZ clinoptilolite; MHZ mordenite; MSC clinoptilolite altogether with H-montmorillonite) or MHB montmorillonite (bentonite), respectively. The zeolite and bentonite samples were dried (105°C, 3 h), ground, and sieved. The <0.25-mm fraction was thoroughly mixed with heavy metal-contaminated soil and moisturized with distilled water. After 14 days of soil incubation three plants (6-week-old seedlings with three to five true leaves) were planted in one pot containing 1500 g of soil–zeolite/bentonite mixture. All other growth conditions were as described in Experiment 1; the plants were processed after 8 weeks of growth. After harvesting of plants the soil-zeolite/bentonite growth medium was sampled in three replicates from every pot. The samples were sieved (<0.5 mm) to remove root debris, air dried, and extracted with CaCl$_2$, EDTA in ammonium acetate buffer, or cc. HNO$_3$ and H$_2$O$_2$ (3:1 v/v) to determine the exchangeable, "plant available," and "total" concentration of heavy metals in growth medium at the end of the experiment. The experiment was done between February and June 1996.

TABLE 13.1

Heavy Metal Concentrations in Uncontaminated (Control) and Contaminated Soil (determined in cc. HNO_3–H_2O_2 extracts, Nyíregyháza, Hungary, 1995)

Element (µg/g)	Cd	Cr	Cu	Ni	Zn
Uncontaminated soil	0.3	12.2	15.0	14.9	41.0
Contaminated soil	10.1*	228.5*	119.0*	56.2*	107.0*

Note: Student's t-test. Data are means of three replications. Statistically significant at *P <0.001 level.

13.2.4 Elemental Analysis of Soil and Plant Samples

The elemental composition of soil and plant samples was determined by the inductively coupled argon plasma emission spectrometry (ICAP, model Labtam 8440M, Australia) technique in triplicate at the Central Chemical Laboratory, Debrecen University of Agricultural Sciences, Debrecen, Hungary. For validation of plant analysis CRM 281 rye grass (Commission of the European Communities, Community Bureau of Reference, Brussels) certified reference material was used.

13.2.5 Statistics

Statistical analysis of the experimental data was done by Student's t-test using Statistix 4.0 software (Statistix, 1992).

13.3 Results and Discussion

13.3.1 Phytoavailability of Heavy Metals in a Galvanic Mud-Contaminated Soil

Careless handling of galvanic mud resulted in contamination of the kitchen garden soil with cadmium (Cd), chromium (Cr), copper (Cu), nickel (Ni), and zinc (Zn) (Table 13.1). The heavy metal concentrations in the uncontaminated control soil were in common range. The Cd, Cr, and Ni concentrations in the contaminated soil exceeded the Hungarian and international regulatory values (Kádár, 1995; Adriano, 1986; Alloway, 1990; Kabata-Pendias and Pendias, 1992). The regulatory limits for heavy metal concentrations in arable soils are 1 to 3 mg/kg for Cd, 75 to 100 mg/kg for Cr, 75 to 100 mg/kg for Cu, 50 mg/kg for Ni, and 200 to 300 mg/kg for Zn in Hungary, depending on soil properties (Kádár, 1995).

Chicory has been demonstrated as an indicator plant for Cd contamination in soils or nutrient solution (Simon et al., 1996), and chicory indicated the excess of Cd, Mn, and Zn in municipal sewage sludge compost used as a soil amendment (Simon et al., 1997). This sensitive indicator plant was grown in the contaminated soil to study the phytoavailability, accumulation, and distribution of the above heavy metals. With increasing ratio of the contaminated soil vs. uncontaminated soil, linearly increasing amounts of Cd, Cr, Cu, and Zn were detected in chicory roots and rhizomes and shoots (Figures 13.1 and 13.2).

Comparing the cadmium, chromium, copper, and zinc concentrations in roots and rhizomes with shoots a close correlation was found in their accumulation (Figures 13.1 and 13.2). It means that roots and rhizomes of this species do not act as a "barrier" against the accumulation of these trace metals in shoots. This phenomenon could useful when chicory (cultivated or wild form) is used as a phytoindicator of heavy metal contamination in soils.

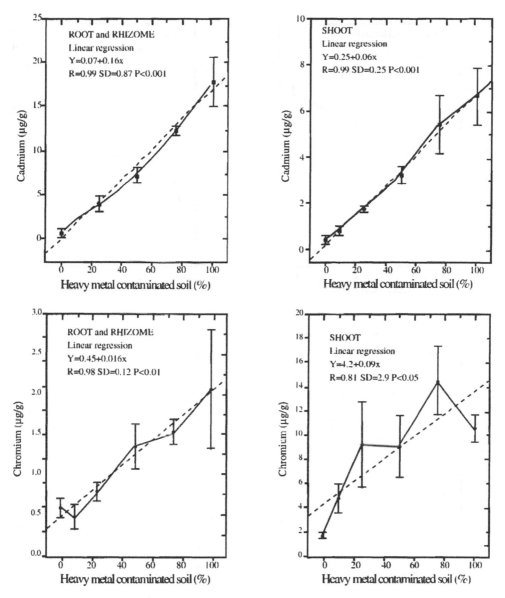

FIGURE 13.1

Cadmium and chromium accumulation in roots and rhizomes and in shoots of chicory grown in a galvanic mud-contaminated soil (pot experiment, Nyíregyháza, Hungary, 1995).

The accumulation of nickel was negligible and its concentrations in chicory roots and rhizomes or shoots did not reflect the level of soil contamination (Figure 13.3). Although 5 to 10 mg/kg Cd, 1 to 2 mg/kg Cr or 15 to 20 mg/kg Cu in shoots may cause phytotoxicity in sensitive plant species (Kabata-Pendias and Pendias, 1992), no phytotoxicity symptoms were observed in chicory and the dry matter accumulation of plants was undisturbed (data not shown).

13.3.2   Immobilization of Heavy Metals with Natural Zeolites and Bentonite

Heavy metal concentrations in different extracts of the uncontaminated and contaminated soils are shown in Table 13.2. Elevated levels of heavy metals were detected in the

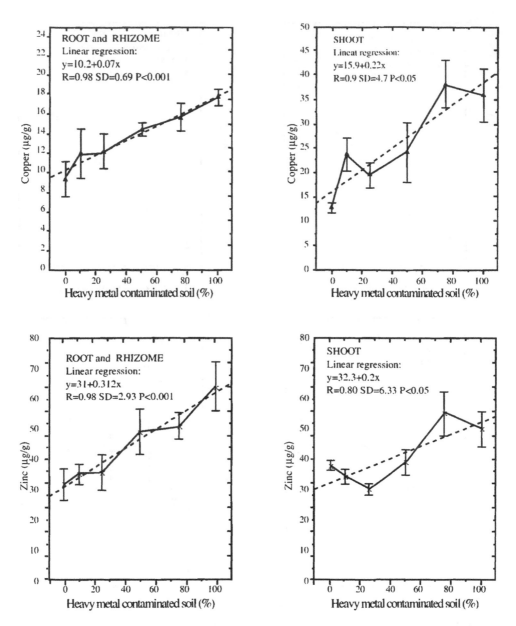

FIGURE 13.2
Copper and zinc accumulation in roots and rhizomes and in shoots of chicory grown in a galvanic mud-contaminated soil (pot experiment, Nyíregyháza, Hungary, 1995).

exchangeable and "plant available" fractions of the contaminated soil; this predicts the danger of entering these heavy metals into the biosphere (Table 13. 2). The heavy metal concentrations were higher in samples collected in 1995 than in 1996 (compare Tables 13.1 and 13.2); this indicates the heterogeneity of soil contamination in the kitchen garden.

Dry matter accumulation of chicory was unaffected by heavy metal contamination of the soil or by zeolite or bentonite treatment of this soil (Table 13.3). No visible phytotoxicity symptoms were observed. The water absorption or macronutrient (P, K, Ca, Mg) uptake of plants was undisturbed by zeolite or bentonite application (data not shown).

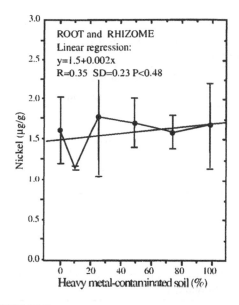
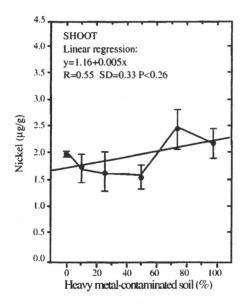

FIGURE 13.3

Nickel accumulation in roots and rhizomes and in shoots of chicory grown in a galvanic mud-contaminated soil (pot experiment, Nyíregyháza, Hungary, 1995).

TABLE 13.2

Heavy Metal Concentrations in Different Extracts of an Uncontaminated (Control) and Contaminated Soil in Nyíregyháza, Hungary, 1996

	Cd	Cr	Cu	Ni	Zn
	µg/g				
Extractant	CaCl₂				
Uncontaminated soil	<0.01	<0.01	0.07	<0.01	<0.01
Contaminated soil	0.10*	<0.01	0.20*	0.14*	<0.01
Extractant	H₄EDTA in ammonium acetate buffer (Lakanen-Erviö)				
Uncontaminated soil	0.6	0.9	8.6	0.9	14.7
Contaminated soil	48.9*	36.7*	93.7*	25.0*	77.7*
Extractant	cc. HNO₃ – H₂O₂				
Uncontaminated soil	0.9	28.3	17.4	19.9	55.7
Contaminated soil	53.5*	327.3*	125.0*	95.3*	139.7*

Note: Student's t-test. Data are means of 3 replications. Statistically significant at * P <0.05 level.

The zeolite and bentonite additives reduced the accumulation of zinc in chicory roots (Table 13.4). A slight reduction in Ni, Cr, and Cu accumulation was also observed in several cases but the rate of this has not proved to be statistically significant (Table 13.4). Considering the heavy metal concentrations in the whole plants (roots and rhizomes + shoots) a similar tendency was observed; in the zeolite- or bentonite-treated cultures 16 to 25% less Cr, Cu, and Zn was measured than in controls. Observations of other authors based on similar pot experiments are controversial, indicating in most cases zinc immobilization by zeolites in contaminated soils. Baidina (1991) found that zeolite application decreased Zn and Pb

TABLE 13.3

Dry Matter Accumulation in Chicory Grown in Heavy-Metal Contaminated Soil Treated with Zeolite and Bentonite (pot experiment, Nyíregyháza, Hungary, 1996)

Treatment	Root and rhizome	Shoot
	(g/plant)	
Uncontaminated soil (1)	0.12	0.42
Contaminated soil (2)	0.13[ns]	0.39[ns]
2 + 5% RBZ clinoptilolite	0.10[ns]	0.39[ns]
2 + 5% MHZ mordenite	0.17[ns]	0.39[ns]
2 + 5% MSC clinoptilolite with H-montmorillonite	0.10[ns]	0.38[ns]
2 + 5% MHB montmorillonite (bentonite)	0.16[ns]	0.38[ns]

Note: Student's t-test. Data are means of six replications.
ns: Statistically not significant as compared to treatment 1.

TABLE 13.4

Heavy Metal Accumulation in Chicory Grown in Galvanic Mud-Contaminated Soil Treated with Zeolites and Bentonite (pot experiment, Nyíregyháza, Hungary, 1996)

Treatment	Cd	Cr	Cu	Ni	Zn
			(μg/g)		
			Root and Rhizome		
1	1.4	6.1	17.3	4.2	43.8
2	27.6	109.7	59.5	11.0	43.9
3	25.3	79.5	48.5	8.6	27.1*
3	23.1	90.9	53.7	9.0	30.2
5	21.1	66.5	38.6	17.4	32.7*
6	22.3	92.3	50.0	8.2	28.9*
			Shoot		
1	2.0	1.8	9.3	2.2	29.9
2	40.8	5.2	13.2	4.0	34.9
3	39.7	4.6	13.4	3.8	36.2
4	43.1	3.0	13.3	3.5	36.5
5	41.0	3.6	14.2	4.3	34.9
6	41.4	2.9	11.7	2.5	34.4

Note: 1: control uncontaminated soil
2: contaminated soil
3: 2 and 5% clinoptilolite
4: 2 and 5% mordenite
5: 2 and 5% clinoptilolite with H-montmorillonite
6: 2 and 5% montmorillonite (bentonite)
Student's t-test. Data are means of three replications. Statistically significant at $\sqrt{P}$<0.1.
*P <0.05 level as compared to treatment 2.

mobility in a zinc-smelter contaminated soil, but heavy metal accumulation of beet was not diminished. Mineyev et al. (1990) found that natural zeolite decreased the concentration of mobile zinc in Zn, Cd, and Pb contaminated soil, but the negative effects of heavy metals were not reduced in barley. In another pot study natural zeolite (clinoptilolite) significantly reduced the Zn concentration in maize and barley grown in soil artificially contaminated with Zn (Chlopecka and Adriano, 1996). Obukhov and Plekhanova (1995) got similar results; application of zeolite reduced the mobility of Pb and Zn in a contaminated soil, and their uptake in maize and barley. Synthetic zeolite introduction to a contaminated soil reduced the Zn content of lettuce leaves by 36 to 65% (Gworek, 1994). In a sewage sludge or heavy metal salts-contaminated soil, however, zeolite application had no effect on heavy

TABLE 13.5

Heavy Metal Concentrations in Different Extracts of Uncontaminated (Control) Soil and of Zeolite/Bentonite Treated Contaminated Soil after 8 Weeks of Chicory Growth (pot experiment, Nyíregyháza, Hungary, 1996)

Treatments	Cd	Cr	Cu	Ni	Zn
			(μg/g)		
Extractant	*CaCl₂*				
1	<0.01	<0.03	0.05	<0.02	<0.01
2	<0.01	<0.03	0.1	<0.02	0.05
3	<0.01	<0.03	0.23*	<0.02	<0.01*
4	<0.01	<0.03	0.24*	<0.02	<0.01*
5	<0.01	<0.03	0.08	<0.02	<0.01*
6	<0.01	<0.03	0.03*	<0.02	<0.01*
Extractant	*H₄EDTA in ammonium acetate buffer (Lakanen-Erviö)*				
1	0.3	0.1	9.7	2.3	9.8
2	35.7	11.0	68	15.2	41.2
3	33.0	11.8*	68	16.6*	41.9
4	32.9	11.8*	68	16.0*	41.9
5	32.6	12.0*	74*	16.2*	43.2
6	32.4	11.5*	68	15.2	40.9
Extractant	*cc. HNO₃ – H₂O₂*				
1	1.1	21.9	17.6	16.9	47.3
2	35.5	218	103	90	112
3	33.4	218	103	91	116
4	36.8	241*	113*	98	119
5	35.0	223	126*	91	116
6	37.0	230	105	87	108

Note: 1: control uncontaminated soil
2: contaminated soil
3: 2 and 5% clinoptilolite
4: 2 and 5% mordenite
5: 2 and 5% clinoptilolite with H-montmorillonite
6: 2 and 5% montmorillonite (bentonite)
Student's t-test. Data are means of three replications.
Statistically significant at *P <0.05 level as compared to treatment 2.

metal concentrations in chicory (Scotti et al., 1995). Sodium montmorillonite mixed with zinc contaminated soil reduced zinc-concentration in plants without causing physiological deficiency of this element (Krebs and Gupta, 1994).

In Table 13.5. heavy metal concentrations are shown in different extracts of zeolite or bentonite-treated contaminated soil after 8 weeks of chicory growth. Zeolite or bentonite amendments significantly reduced the exchangeable (CaCl₂ extractable) Zn content in the contaminated soil. "Plant available" (Lakanen-Erviö solution extractable) or "total" (cc. HNO₃ – H₂O₂ extractable) concentrations of zinc, however, remained unaffected by zeolite or bentonite application. Since the exchangeable form of Zn in soil is considered as the most bioavailable to plants (Kabata-Pendias and Pendias, 1992), the definitively lower exchangeable Zn in zeolite or bentonite-treated soils may explain the lower Zn uptake in chicory roots. Our results confirm the findings of Chlopecka and Adriano (1996); zeolite (clinoptilolite) ameliorant significantly reduced the exchangeable form of Zn in a flue dust-contaminated soil, and Zn uptake in maize or barley test plants.

13.4 Conclusions

The soil of a kitchen garden located near a former galvanization plant was found to be contaminated with Cd, Cr, Ni, Cu, and Zn. Chicory indicator plant grown in this soil accumulated elevated levels of Cd, Zn, Cu, and Cr in its roots and rhizomes and in shoots. Dry matter accumulation of chicory was unaffected by heavy metal contamination of soil or by zeolite or bentonite application. Natural zeolites and bentonite reduced the accumulation of Zn in chicory roots and rhizomes, and a slight decrease was also detected in Ni, Cr, and Cu uptake of the test plants. The decrease in Zn uptake of chicory may be related to decrease in exchangeable (bioavailable) form of zinc in contaminated soil after zeolite or bentonite application. We plan open-field experiments with crop plants to study and verify the long-term heavy metal immobilization effects of Hungarian natural zeolites and bentonites in contaminated soils.

Acknowledgments

This research was supported by the Hungarian Scientific Research Fund (project F016906), the Foundation for the Hungarian Higher Education and Research (project 681/96), and Hungarian Soros Foundation. Valuable help was provided by Prof. Dr. Zoltán Györi (Central Chemical Laboratory, Debrecen University of Agricultural Sciences, Debrecen, Hungary) and his co-workers, Drs. József Prokisch and Béla Kovács, in elemental analysis. The help of Tibor Mátyás (Healing Minerals Geoproduct Ltd., Mád, Hungary) in zeolite and bentonite characterization is gratefully acknowledged.

References

Adriano, D.C., *Trace Elements in the Terrestrial Environment*, Springer-Verlag, New York, 1986, 1–95.
Alloway, B.J., Ed., *Heavy Metals in Soils*, Blackie & Son, Ltd., Glasgow and London, 1990, 29–39.
Baidina, N.L., The use of zeolites as heavy metal absorbents in technogenically contaminated soils (in Russian), *Izvestiya Sibirskogo Otdeleniya Akademii Nauk SSSR., Seriya Biologicheskikh Nauk*, 6, 32, 1991.
Campbell, L.S. and B.E. Davies, Experimental investigation of plant uptake of caesium from soils amended with clinoptilolite and calcium carbonate, *Plant and Soil*, 189(1), 65, 1997.
Chlopecka, A. and D.C. Adriano, Mimicked *in situ* stabilization of metals in a cropped soil: bioavailability and chemical form of zinc, *Environmental Science and Technology*, 30(11), 3294, 1996.
Gworek, B., Lead inactivation by zeolites, *Plant and Soil*, 143, 71, 1992.
Gworek, B., Zeolites of the 3A and 5A type as factors inactivating zinc in soils contaminated with this metal, *Roczniki Gleboznawcze*, 44 (Suppl.), 95, 1994.
Kabata-Pendias, A. and H. Pendias, *Trace Elements in Soils and Plants*, CRC Press, Boca Raton, FL, 1992, 3–66.
Kádár, I., *Contamination of the Soil-Plant-Animal-Man Food Chain with Chemical Elements in Hungary* (in Hungarian), KTM, MTA-TAKI, Budapest, 1995, 86–371.
Krebs, R. and S.K. Gupta, Mild remediation techniques for heavy metal contaminated soils (in German), *Agrarforschung*, 1(8), 349, 1994.

Mench, M., J. Vangronsveld, N.W. Lepp, and R. Edwards, Physico-chemical aspects and efficiency of trace element immobilization by soil amendments, in *Metal-Contaminated Soils: In Situ Inactivation and Phytorestoration*, Vangronsveld, J. and S. D. Cunningham, Eds., Springer-Verlag, Berlin, and R.G. Landes Company, Georgetown, TX, 1998, 151–182.

Mineyev, V.G., A.V. Kochetavkin, and B. Nguyen-Van, Use of natural zeolites to prevent heavy-metal pollution of soils and plants, *Soviet Soil Science*, 22 (2), 72, 1990.

Obukhov, A.I. and I.O. Plekhanova, Detoxication of dernopodzolic soils contaminated by heavy metals: theoretical and practical aspects (in Russian), *Agrokhimiya*, 2, 108, 1995.

Scotti, I. A., E. Lombi, and G.M. Beone, Influence of natural materials on assimilation ability of heavy metals (in Italian), *Annali della Facolta di Agraria, Universita Cattolica del Sacro Cuore Milano*, 35(1-2), 69, 1995.

Simon, L., J. Prokisch, and B. Kovács, Chicory (*Cichorium intybus* L.) as bioindicator of heavy metal contamination, in *Contaminated Soils: 3rd Int. Conf. on the Biogeochemistry of Trace Elements*, Paris, May 15-19 1995, Prost, R., Ed., CD-ROM, D:\data\communic\066.PDF, Colloque n°85, INRA Editions, Paris, France, 1997.

Simon, L., H.W. Martin, and D.C. Adriano, Chicory (*Cichorium intybus* L.) and dandelion (*Taraxacum officinale* Web.) as phytoindicators of cadmium contamination, *Water, Soil and Air Pollution*, 91, 351, 1996.

Shanableh, A. and A. Kharabsheh, Stabilization of Cd, Ni, and Pb in soil using natural zeolite, *Journal of Hazardous Material*, 45, 207, 1996.

Statistix 4.0, *User's Manual*, Analytical Software, St. Paul, MN, 1992.

14

Heavy-Metal Uptake by Agricultural Crops from Sewage-Sludge Treated Soils of the Upper Swiss Rhine Valley and the Effect of Time

Catherine Keller, Achim Kayser, Armin Keller, and Rainer Schulin

CONTENTS

14.1 Introduction

Application on agricultural lands is a popular method for the disposal of sewage sludge, as it represents at the same time a low-cost fertilizer. However, if excessive loads of pollutants are introduced with the application of low-quality sludges, this practice may adversely

TABLE 14.1

Quantities of Heavy Metal Present in Sewage Sludge and Their Transfer to Agriculture in 1989 and 1994 and Average Heavy Metal Concentrations Measured in 1989 in Switzerland

Metal	Quantity in Sewage Sludge			Concentrations in Sewage Sludge		Soils Guide Values (g·t⁻¹ DM)
	t·yr⁻¹	t·yr⁻¹	%used in agriculture	Weighted Mean (g·t⁻¹ DM)	Limit Values (g·t⁻¹ DM)	
	1989	1994	1994	1989	1992	
Mo	1.5	1.2	52	7.0	20	5
Cd	0.9	0.5	42	4.0	5	0.8
Co	2.2	1.7	54	10	60	—
Ni	9.1	8.5	44	43	80	50
Cr	27.4	17.8	49	129	500	50
Cu	82.9	82.0	50	388	600	40
Pb	49.5	28.0	57	232	500	50
Zn	293.6	234.4	56	1378	2000	150
Hg	0.6	0.4	51	2.6	5	0.5

Note: Swiss limit values for sewage sludge and guide values for soils are given for comparison.

From Candinas, T. and A. Siegenthaler, Grundlagen des Düngung: Klärschlamm und Kompost in des Landwirtschaft, _Schriftenreihe der FAC Liebefeld_, 9, Liebefeld-Bern, 1990, SFSO (Swiss Federal Statistical Office) and SAEFL (Swiss Agency for the Environment, Forests and Landscape), The Environment in Switzerland 1997, EDMZ, Bern, Switzerland, 1997, 372; Keller, T. and A. Desaules, Flächenbezogene Boden-belastung mit Schwermetallen durch Klärschlamm, Schriftenreihe des FAL, 23, 1997. With permission.
*OIS, Ordinance Relating to Impacts on the Soil, 1st July 1998, SR 814.12, applicable to mineral soils (<15°° organic matter) extraction 2 M HNO₃.

affect soil fertility, threaten groundwater quality, and lead to food chain poisoning. Consequently, over the past 20 years, governments have imposed limits either for maximum heavy-metal loads in soils or for amounts of sewage sludge and heavy metal concentrations in sewage sludge applied to soils.

In Switzerland, the first regulations concerning the use and the quality of sewage sludge were issued in 1981 (sewage sludge ordinance) and revised in 1992 (Table 14.1). Though the total amounts and heavy metals concentrations of sewage sludges have decreased considerably after these regulations were enforced (SFSO, 1997) (Table 14.1), mass flux analyses show that heavy metals still accumulate in agricultural soils when the tolerance limits for sludge quality and application rates are fully exploited. Moreover, distribution on fields is not uniform and local areas may have received excessive loads. In total, 55% of the sewage sludge produced in 1994 (4 million cubic meters) was used in agriculture, leading to yearly total addition of ca. 200 t of heavy metals (nearly 10% of the total heavy metals added to these soils) (SFSO, 1997). Keller and Desaules (1997) calculated that if the maximum concentrations allowed by the ordinance were applied at the maximum rates tolerated, sludge treated would reach the Swiss guide values for Pb and Cu within 100 years. They estimated that almost 44,000 ha have concentrations above the Swiss guide values for Cu and Zn and almost 65,000 ha for Cd due to application of sludges. Together with the other sources of pollution, contaminated areas could amount to as much as 200,000 ha (Häberli et al., 1991), that is, 15% of the surface used for agriculture and settlements.

Considerable uncertainty exists about the long-term fate of polluting heavy metals. One possibility is that the mobility and bioavailability of soil-polluting heavy metals stabilize or even decrease with time (the so-called "plateau effect") (Dowdy et al., 1994; Smith, 1997; Brown et al., 1998). On the other hand, it is also possible that metals become more mobile, e.g., because of the mineralization of sewage sludge organic matter ("time bomb effect") (Zhao et al., 1997). Field studies covering several decades have produced ambiguous results (Chang et al., 1997; Logan et al., 1997) and led to contradictory conclusions (Chaney and Ryan, 1993; McBride, 1995). The new USEPA (1993) regulations in the United States have induced scientists to reevaluate the results obtained from long-term field experiments and to assess the phytotoxicity and bioavailability of heavy metals added to soils through repeated applications of biosolids (McBride, 1995; Schmidt, 1997). Results of long-term experiments have recently been summarized by Berti and Jacobs (1996), Barbarick et al. (1997), Miner et al. (1997), Sloan et al. (1997), and Zhao et al. (1997). In Switzerland, Krebs et al. (1998) found that after 15 years, heavy metals extracted by 0.1 M NaNO$_3$ (so-called "bioavailable fraction," OIS [1998]) increased with time in soils that had been amended between 1976 and 1984 with sewage sludge. This increase was correlated with a pH decrease and raises the question of stability with time of soil characteristics and sludge residuals including the organic matter content. Indeed, McBride (1995) found that soil characteristics and sludges' inorganic constituents seem to exert an increasing control with time on metal solubility.

The available evidence indicates that the fate of heavy metals in soils and the associated risks may vary considerably, depending on soil properties, cultivation practices, and climatic factors. This means that an extensive data set covering a wide range of conditions is necessary to enable predictions of the metal availability in the long term.

In this chapter we present the results of an experiment which was started in 1969. In the first years, massive doses of sewage sludges from various origins were applied repeatedly on plots of conventionally farmed arable land. We were interested in the effects of these treatments on plant uptake of the polluting metals and the development of phytoavailability over time.

14.2 Materials and Methods

14.2.1 Geographic and Climatic Conditions at the Experimental Site

The experimental site was located at the leveled floor of the Rhine Valley of eastern Switzerland. The valley descends smoothly in a north-northeasterly direction and repeatedly broadens up to 12 km. The climate is relatively mild, permitting productive agricultural activities. Salez is situated at an altitude of 430 m. Mean average temperature is 8.6°C and mean rainfall is 1300 mm with a maximum during summer (stations Vaduz and Saxerriet, respectively [SMA, 1995]). The valley bottom is covered by alluvial deposits, mainly carbonatic clays lying on top of sand or gravel (de Quervain et al., 1963). Soils are generally rich in mineral nutrients. Fluvisols and cambisols are most common and some histosols can be found in former wetlands.

14.2.2 Experimental Setup and Crop Chronology

The experimental plots were first set up in the Rhine Valley in Buchs, northeast of Switzerland, in 1969. Parcels (four treatments, four replicates each) of soils were artificially contaminated with heavy metals from biosolids over a period of 7 years (von Hirschheydt, 1987). Apart from controls with no waste or sludge application, treatments consisted of (a) application of composted municipal waste from a nearby incineration plant; (b) same as (a), but in addition application of various types of highly contaminated sewage sludges; (c) same as (b), but with a double dose of sewage sludges (Table 14.2).

TABLE 14.2

Composted Waste and Sludges Characteristics

a) Amounts of composted waste and sludges applied during contamination period

Treatments	Origin of Soil	Composted Waste	Sludge
Salez	Salez	—	—
Buchs	Buchs	—	—
BW	Buchs	150 m^{3}ha^{-1}a^{-1}	—
BWS1	Buchs	150 m^{3}ha^{-1}a^{-1}	150 m^{3}ha^{-1}a^{-1}
BWS2	Buchs	150 m^{3}ha^{-1}a^{-1}	300 m^{3}ha^{-1}a^{-1}

b) Type and origin of the sludges applied[a]

Year	Sludge Type/Origin
1969	Galvanic industry
1970	Galvanic industry; wood tar
1971	Paint production + neutralization treatment
1972	Acetone production
1973	Galvanic industry
1974	Paint production residues
1975	Galvanic industry

[a] The composted waste was produced by the Buchs waste incineration plant.
From von Hirschheydt, A., Zur Wirksamkeit von Schwermetallen aus Müllkomposten auf Ertrag und Zusammensetzung von Kulturpflanzen. Teil I und II. Studienreihe Abfall-Now. Abfalltechnisches Labor mit Anhang am Institut für Siedlungswasserbau, Wassergüte- und Abfallwirtschaft der Universität Stuttgart, Bandtäle 1, Stuttgart, 1987. With permission.

The original design was a 4 × 4 Latin square with plot sizes of 12.5 m². In 1987, the plots were moved to their present location in Salez, approximately 15 km to the north, because the Buchs site was claimed for construction purposes (Stenz, 1995). The topsoil (25 cm depth) of each plot was translocated separately to Salez, where the experiment was re-established. In addition to the soils originating from Buchs, a set of four replicate plots with local topsoil from Salez was installed. The Salez soil, which has different characteristics with respect to some soil parameters, was included in the experiment, as it was also used as subsoil in the plot setup.

The experimental setup of Salez represented a fully balanced factorial design with four replicates of each of the following five "treatments" of soil and waste/sludge applications:

S   Salez soil with no waste or sludge application

B   Buchs soil with no waste or sludge application

BW   Buchs soil with only composted municipal waste application

BWS1   Buchs soil with composted municipal waste + single dose of sewage sludge application

BWS2   Buchs soil with composted municipal waste + double dose of sewage sludge application

Plot size was 1.8 m², totalling an experimental area of 36 m².

Between 1989 and 1993, the crops listed in Table 14.3 were grown. In 1994 and 1995 the site lay fallow. In 1996 beets were grown once more: this time two cultivars were tested, all plots were divided into two halves, and each half was planted with one cultivar. Plots were treated uniformly with respect to fertilization and application of pesticides, regardless of the crop. Until 1993 they were fertilized with NH_4NO_3 + Mg, Colzador, and Tresan Bor.

TABLE 14.3

Crop Rotation from 1989 to 1996

Year	Crop Type	Strain (Cultivars)
1989	String beans (*Phaseolus vulgaris*)	Felix
1990	Maize (*Zea Mays*)	Blizzard
1991	Sugar beet (*Beta vulgaris*)	Brigadier
1992	Potatoes (*Solanum tuberosum*)	Bintje
1993	Lettuce (*Lactuca sativa*)	Soraya
1993	Spinach (*Sinacia oleracea*)	Polka F1
1994 and 1995	Fallow	
1996	Sugar beet (*Beta vulgaris*)	Brigadier + Monofix

TABLE 14.4

Physical and Chemical Parameters of Soil Samples Collected in July 1990

	pH	C_{org}	$CaCO_3$	Sand	Silt	Clay	CEC_{pot}	Al_{am}	Fe_{am}	Al_{cryst}	Fe_{cryst}
				(%)			(meq·kg⁻¹)		(g·kg⁻¹)		
Salez	7.5	2.6	13.2	22.1	60.4	17.5	177	0.7	4.0	1.9	12.5
Buchs	7.3	2.1	17.5	18.8	41.9	9.3	125	1.2	6.1	2.0	9.3
BW	7.2	2.5	13.1	50.3	39.8	9.9	135	1.8	5.5	2.6	10.2
BWS1	7.2	2.7	16.5	49.8	40.2	10.0	131	1.7	4.4	2.5	10.2
BWS2	7.1	2.8	16.0	49.7	40.6	9.6	130	1.6	5.3	2.6	10.2

Al_{am}: amorphous Al; Fe_{am}: amorphous Fe; Al_{cryst}: crystalline Al; Fe_{cryst}: crystalline Fe

TABLE 14.5

Total Heavy Metal Concentrations (Average of Four Replicates ± Standard Deviations) in Soil Samples (0–20 cm depth) Collected in July 1990 (HNO_3/$HClO_4$/HF-Extracts)

	Cd	Cr	Cu	Ni	Pb	Zn
				(mg·kg⁻¹)		
Salez	0.3 ± 0.1	7 ± 2	750 ± 2	68 ± 6	33 ± 1	112 ± 2
Buchs	4 ± 1.8	83 ± 10	174 ± 30	71 ± 10	353 ±73	691 ± 73
BW	4 ± 0.7	95 ± 5	243 ± 21	81 ± 13	587 ±124	1020 ± 53
BWS1	30 ± 2.9	259 ± 42	250 ± 25	181 ± 48	520 ± 65	1420 ± 98
BWS2	65 ± 4.6	464 ± 90	211 ± 7	347 ± 63	695 ± 89	1968 ± 124

In 1996, NH_4NO_3 + Mg and $(NH_2)_2OC$ were used for N-fertilization. The herbicides used were Gesaprim® and Alipur®. Ridomil-Fortex® was applied to avoid fungal infections.

14.2.3 Soil and Plant Analysis

Topsoils (0 to 20 cm) were sampled in spring 1989, summer 1990, and fall 1990, 1993, and 1996. In 1989 samples from replicate plots of the same treatment were bulked on site. In all other sampling campaigns, composite replicate samples were taken per plot. In 1996 samples were taken from one plot of each treatment every 10 cm along the soil profile. UFAG Laboratories (Sursee, Switzerland) carried out soil analysis for 1989–1993; the samples of 1996 were analyzed in our lab. Selected soil properties and total heavy metal contents of the topsoils are listed in Tables 14.4 and 14.5.

Soil samples were oven dried at 40°C, crushed, and sieved to 2 mm with a nylon sieve. Soil pH was measured in 0.01 M $CaCl_2$ (FAC, 1989). Carbonate content was determined with a Poisson apparatus by measuring the CO_2 volume produced (FAC, 1989). Organic carbon content was determined using a modified version of the $K_2Cr_2O_7$-method (UFAG, internal method). Total N was measured after Kjeldahl digestion following the DIN 19684 procedure, and total P was determined colorimetrically after smelting in $KNO_3/NaNO_3$ and digestion in boiling HNO_3/H_2SO_4 (FAC, 1989). Cation exchange capacity and base saturation were determined by $BaCl_2$-triethanolamine extraction at pH 8.1 (FAC, 1989). Iron- and Al-oxides were determined after extraction with cold (amorphous forms) or boiling (amorphous + crystallized forms) NH_4-oxalate (FAC, 1989). "Total" heavy metal concentrations were determined in duplicate after digestion in $HNO_3/HCLO_4/HF$ (Ruppert, 1987), "pseudo-total" heavy metal concentrations with boiling 2 M HNO_3 (FAC, 1989), and "soluble" heavy metals were extracted with 0.1 M $NaNO_3$ (FAC, 1989). The distinctions between "total," "pseudo-total," and "soluble" were made after Gupta et al. (1996) and according to their biological relevance (Gupta and Aten, 1993).

Plant samples were rinsed thoroughly under tap water, oven dried, preground in an ultra centrifuge, and ground in an agate ball mill. For heavy metal analysis 1-g samples were either oven-digested in a 1:1 mixture of boiling HNO_3 (65%) and H_2O_2 (30%) or 0.5-g samples were microwave-digested in 2 mL HNO_3 (65%), 2 mL HF (48%), and 1 mL H_2O_2 (30%). Flame and graphite furnace atomic absorption spectrometry (AAS) and inductively coupled plasma atomic emission spectrometry (ICP-AES) were used for the chemical analysis of extracts.

14.3 Results

14.3.1 Heavy Metal Distribution and Migration in Soil

14.3.1.1 Effects of Sewage Sludge Treatments on Soil Properties — Aging Effect

The Salez soil differs markedly in most of the investigated soil properties from the Buchs soil (control and treatments). No major differences were found on the soil properties of the Buchs plots, except for a slight increase in organic matter and a decrease in pH with increasing load of biosolids (B<BW<BWS1<BWS2) (Figure 14.1). While the organic carbon and carbonate contents and the texture remained constant, pH increased between 1990 and 1993 in all soils, in particular in the soils treated with compost and sewage sludge. Soil pH was always lower in the soils treated with biosolids than in the controls.

14.3.1.2 Effects of Sewage Sludge Treatments on Heavy Metal Concentrations and Binding — Aging Effect

In topsoils, 2 M HNO_3 concentrations of all heavy metals increased with increasing load of biosolids. Consequently, metal concentrations were highly correlated with each other. For example, a strong correlation was found between total Cd and Zn concentrations ($r = 0.88$). In order to assess the significance of the differences observed between treatments, systematic replications of sampling within plots, soil extractions, and measurements were made in 1996 and compared to the variation between treatments. The coefficients of variation for Cd and Zn are shown in Table 14.5 and Figures 14.2a and 14.2b. It was about 7% for replicate analysis of the soluble zinc concentrations (including replicate extractions). Spatial variability was 25% between single four replicate cores within a plot. Bulked soil samples showed about 28% of variation in average between replicate plots of the same treatment. Although only four replicate plots were available for each treatment, treatment effects were significant in spite of this large background variability due to spatial and analytical effects: the coefficient of variation pooled for all treatments was about 90%.

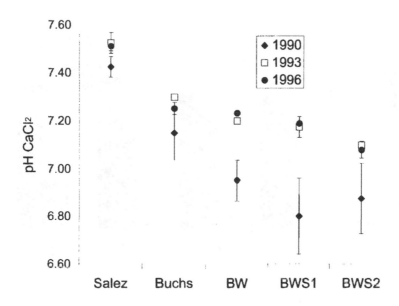

FIGURE 14.1
Soil pH for the two controls and the three treatments in the samplings of 1990, 1993, and 1996.

In comparison to the high total metal concentrations measured in the treated soils, the $NaNO_3$-extractable Cd and Zn concentrations were low, which can be attributed to the high soil pH. Again, the different metals showed high correlation, e.g., the coefficient of correlation between concentrations of $NaNO_3$-extractable Cd and Zn was $r = 0.89$. Moreover, when soluble metal concentrations from 1990 were considered, the log-transformed soluble and total concentrations for cadmium and zinc were correlated with correlation coefficients of 0.78, resp. 0.75. But these differences in $NaNO_3$-extractable Cd and Zn concentrations were solely due to the difference in doses added in form of biosolids because the ratio between the $NaNO_3$- and HNO_3-extractable metal was similar for the three treatments (after subtraction of the respective concentration of the untreated Buchs soil).

Total heavy metal concentrations did not change during the whole period after the end of the biosolids application (data not shown). But the analysis of variance revealed significant time effects on $NaNO_3$-extractable Cd and Zn concentrations: the "soluble" concentrations of both elements pooled over all sewage treatments decreased significantly (P value < 0.001) between 1990 and 93 for Cd and 90 and 96 for Zn (Figures 14.2a and 14.2b). $NaNO_3$-extractable Cd and Zn concentrations decreased in the same proportions in all treatments, but Zn and Cd decreased more rapidly in sewage sludge treated soils than in the waste treatment (BW) and controls (S and B) (Figure 14.3). Thus there was a reduction of the differences between treatments with time.

Figure 14.2 also shows the $NaNO_3$-extractable Cd and Zn concentrations of the samples from 1987: opposite to the trend described above, $NaNO_3$-extractable Cd and Zn concentrations were higher in 1990 than in 1987. In 1987 the soil samples were collected just prior to the translocation on the Salez site.

14.3.1.3   Migration of Heavy Metals through the Soil Profile

Heavy metals profiles were sampled in 1996 to evaluate any possible vertical transfer. As shown by the total content, the contaminated layer was on average restricted to the first 30 cm, which corresponds to the original establishment of the plots. All plots have similar low

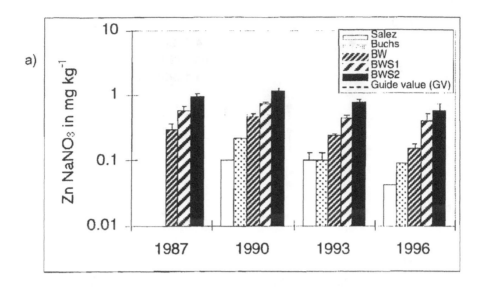

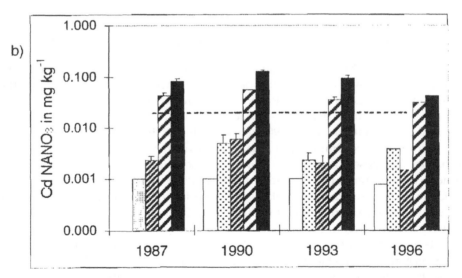

FIGURE 14.2

NaNO₃-extractable Zn and Cd between 1987 and 1996 for the two controls and the three treatments. Data from 1987 are shown for comparison.

concentrations below 40 cm (Table 14.6). All metals follow the same pattern. However, the depth of the contaminated layer which was not always exactly 30 cm, combined with a systematic 10-cm sampling procedure, could explain the abrupt decrease in Cd and Zn concentrations along the profile of treatment BWS1 (pattern different from the other profiles).

Although $NaNO_3$-extractable Zn concentrations decreased with depth, they were still higher in the waste and sludge-treated soils than without these treatments. Also, there was no correlation between the total and the $NaNO_3$-extractable Zn concentrations over depth for the biosolids-treated soils. Whereas the organic carbon content was approximately constant over the soil profiles, the pH showed a tendency to increase with depth in all treatments in positive correlation with decreasing $NaNO_3$-extractable Zn concentrations ($r^2 = 0.62$), indicating that zinc availability was controlled by pH.

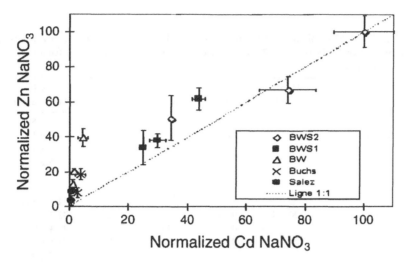

FIGURE 14.3

Relationship between the NaNO$_3$-extractable Zn and Cd. Concentrations have been normalized with Cd and Zn from the BWS2 treatment set to 100. The three points of each treatment correspond to the 3 years 1990, 1993, and 1996 with decreasing concentrations.

14.3.2 Plant Uptake of Heavy Metals

14.3.2.1 Plant Uptake of Heavy Metals and Effects on Crop Production

The heavy metal concentrations found in the crops generally reflected different levels of soil pollution. However, variability between replicates was high in all treatments and heavy metal uptake differed greatly between plant species (Table 14.7). For the Salez soil, Cd and Zn concentrations in plant tissues were always lowest, compared to the nontreated and treated Buchs soils.

Beanstalks contained low to normal concentrations of Cd and Zn when planted on the Buchs and Salez soils, but concentrations in plant tissues increased significantly with higher levels of soil contamination. The most pronounced increase was observed for Cd in the sewage sludge-treated plots BWS1 (10-fold) and BWS2 (40-fold). The concentrations did not differ significantly between plants grown on reference B and treatment BW because the municipal waste (W) did not add significant amounts of Cd to the soil. Zinc concentrations varied less (1.3-fold and 1.8-fold increase, respectively) but the increase was still consistent. In contrast to the stalks, concentrations of both Cd and Zn in bean pods were much lower, especially in the sewage sludge-amended soils. For Zn, no response to the total concentrations in soils was found, whereas for Cd, concentrations in the plant tissues increased more than 14-fold from Buchs soil to treatment BWS2, while still remaining in the range of normal content (Sauerbeck, 1989).

Like in the beans, heavy metal concentrations in maize were different in the different plant tissues studied. Both Cd and Zn concentrations were higher in the leaves. As in beans, an increase was observed with higher soil heavy metal concentrations, but this effect was less pronounced (max. 5-fold for Cd). Nevertheless, concentrations of both metals in all tissues were in a normal range.

In sugar beet, concentrations of Zn and Cd were highest of all plants used in the experiment. In the leaves, Cd content was elevated even in the reference Buchs soil and increased drastically in the BWS1 and BWS2 treatments (9-fold). The concentrations measured were well above critical levels (Sauerbeck, 1989). The same pattern, but to a lesser extent, was

TABLE 14.6

Heavy Metals, pH, and Organic Matter Profiles Measured in October 1996 (after 7 Years of Compost and Sludge Application Followed by 22 years of Conventional Agriculture) for the Two Controls and the Three Treatments

	Depth (cm)	Salez	Buchs	BW	BWS1 (Mean ± sd)	BWS2
Zinc	0.20	42	792	1239	1496 ± 56	2098
(HNO₃-extractable),	20–30	40	237	88	1331 ± 130	296
mg kg⁻¹	30–40	41	293	44	150 ± 15	285
	40–50	42	92	92	65 ± 2	53
	50–75	40	86	52	47 ± 2	61
Cadmium	0–20	0.54	6.73	3.52	24.2 ± 5.00	82.3
(HNO₃-extractable),	20–30	0.25	2.22	0.64	32.5 ± 4.75	12.0
mg kg⁻¹	30–40	0.26	0.36	0.33	2.54 ± 0.88	8.32
	40–50	0.29	0.27	0.66	0.51 ± 0.04	1.18
	50–75	0.26	0.03	0.38	0.37 ± 0.19	0.72
Zinc	0.20	0.04	0.14	0.24	0.33 ± 0.06	0.37
(NaNO₃-extractable),	20–30	0.05	0.05	0.07	0.24 ± 0.10	0.13
mg kg⁻¹	30–40	0.04	0.05	0.07	0.24 ± 0.10	0.13
	40–50	0.04	0.05	0.07	0.09 ± 0.02	0.1
	50–75	0.04	0.05	0.07	0.08 ± 0.02	0.08
pH	0–20	7.6	7.6	7.6	7.6 ± 0.05	7.6
	20–30	7.6	7.6	7.5	7.6 ± 0.09	7.5
	30–40	7.5	7.7	8.2	7.7 ± 0.02	7.5
	40–50	7.6	7.8	7.6	7.7 ± 0.03	7.6
	50–75	7.7	7.7	7.7	7.8 ± 0.04	7.7
OM, %	0–20	3.8	2.8	3.6	3.6	3.7
	20–30	3.5	3.7	3.8	3.6	3.4
	30–40	3.4	4.0	3.7	3.7	3.2
	40–50	n.d.	n.d.	n.d.	4.3	n.d.
	50–75	n.d.	n.d.	n.d.	3.4	n.d.
Clay, %	0.20	29	13	12	6	7
	20–30	17	18	16	10	14
	30–40	20	20	17	15	15
	40–50	n.d.	n.d.	n.d.	15	n.d.
	50–75	n.d.	n.d.	n.d.	16	n.d.

n.d.: not detected

observed for Zn in the leaves. However, in the 1991 planting season, Zn concentrations showed no difference between BWS1 and BWS2, whereas in 1996 concentrations were generally lower and were significantly different. In sugar beet roots, Zn and Cd contents were generally much lower, but revealed the same pattern as the one seen for the leaves. In 1996, no difference in the Zn concentrations was observed, regardless of the levels of soil contamination. This means that the plants had a lower heavy metal transfer efficiency to the leaves with increasing heavy metal in the soil.

Almost the same uptake pattern was observed for potato plants. In the leaves, both Cd and Zn concentrations increased from nontreated Buchs soil to BWS2 treatment, whereas in the tubers, concentrations were both much lower and did not relate as closely to the soil treatment levels. For Zn, an excluder-type uptake pattern was observed, as no significant change in plant concentration was measured from Buchs to BWS2 treatment. Concentrations of both metals were generally in a low to normal range in the tubers, and were elevated in the leaves.

TABLE 14.7

Mean Concentrations of (a) Cd and (b) Zn and Standard Deviations Measured in Plant Tissues (mg·kg⁻¹ dry matter)

		Salez		Buchs		BW		BWS1		BWS2	
						Cd (mg kg⁻¹ dry matter)					
Year	Plant	mean	sd	mean	sd	mean	sd	mean	sd	mean	sd
(a) Cadmium											
1989	Bean stalks	<0.1	<0.1	0.4	<0.1	0.3	<0.1	3.4	1.9	14.3	1.0
	Bean pods	<0.1	<0.1	0.1	<0.1	0.1	<0.1	0.6	<0.1	1.1	0.1
1990	Maize leaves	0.1	<0.1	0.5	0.2	0.2	<0.1	1.7	1.2	2.3	1.0
	Maize cobs	0.1	<0.1	0.1	<0.1	0.1	<0.1	0.2	0.1	0.3	0.1
1991	Sugar beet leaves[a]	0.5	<0.1	3.9	0.6	1.1	<0.1	15.6	9.8	35.8	15.6
	Sugar beet roots[a]	0.3	<0.1	1.1	0.4	0.6	<0.1	3.2	1.6	8.2	3.9
1992	Potato leaves	1.6	0.3	2.9	0.6	3.0	1.8	17.5	1.7	32.0	4.8
	Potato tubers	0.3	<0.1	0.5	0.1	0.3	<0.1	1.2	0.2	1.3	0.8
1993	Lettuce	0.8	<0.1	1.4	0.4	3.9	4.0	7.8	1.1	13.3	0.5
1993	Spinach	1.3	0.7	3.4	1.0	1.8	0.5	13.6	4.5	13.5	9.7
1996	Sugar beet leaves[b]	0.4	0.2	3.2	1.1	2.2	0.4	10.2	2.6	31.1	9.4
	Sugar beet roots[b]	0.3	0.1	2.6	1.1	0.9	0.3	6.6	1.9	6.4	1.5
(b) Zinc											
1989	Bean stalks	55	13	53	10	55	10	68	5	98	5
	Bean pods	33	5	30	m.d.	38	5	40	0	40	<0.1
	Maize leaves	20	4	35	4	41	11	57	4	87	38
	Maize cobs	27	3	40	5	37	5	43	4	47	6
1991	Sugar beet leaves[a]	39	5	250	42	255	47	463	102	468	169
	Sugar beet roots[a]	33	5	198	m.d.	95	13	148	22	155	48
1992	Potato leaves	43	5	128	17	170	26	222	5	298	48
	Potato tubers	20	m.d.	28	5	30	<0.1	35	10	36	5
1993	Lettuce	38	4	67	1	78	3	95	4	104	7
1993	Spinach	85	22	193	29	210	88	270	51	266	40
1996	Sugar beet leaves[b]	33	15	157	63	219	58	200	37	316	56
	Sugar beet roots[b]	42	32	127	51	126	25	130	33	124	26

[a] Sugar beet variety Brigadier 1991.
[b] Sugar beet varieties Brigadier + Monofix 1996.
m.d.: missing data

Cadmium concentrations in lettuce were normal in the Salez and Buchs soils, doubled in BW treatment and showed a sharp increase to critical levels in BWS1 and BWS2. A less pronounced, but still evident increase was observed for Zn. However, concentrations remained within the normal range.

In spinach, a relatively similar uptake pattern was observed, except for the direct comparison of the BWS1 and BWS2 treatments. In these plots, Zn and Cd concentrations were almost identical, indicating plateau-type uptake characteristics. For both metals, concentrations in the plant tissues were normal to critical (BWS1 + BWS2).

Coefficients calculated for heavy metal transfer from soil to the plants were small in the artificially contaminated soil (0.01–0.69 and 0.01–0.28 for Cd and Zn, respectively). This shows that despite the increase in observed Zn and Cd concentrations in the plant tissue, in most cases plant uptake did not reflect the difference in heavy metal loads between the treatments. Transfer coefficients in the Salez and Buchs soils were also <1, while for Cd in the Salez soil, values ranged from 0.13 for bean pods to 4.4 for potato leaves. In all cases, the transfer coefficients were thus within the normal range for Cd and Zn (Kloke et al., 1984; Sauerbeck, 1985). Also, no phytotoxic effects were observed on plants: in particular we didn't measure any difference in biomass between treatments.

TABLE 14.8

Coefficient of Variation (C.V.) in Percentage of Cd and Zn Concentrations in Sugar Beet (Cultivars Brigadier and Monofix 1996) for Various Levels of Replication

Sugar Beet Component of Variation	Leaves Cadmium	Roots Cadmium	Leaves Zinc	Roots Zinc
Laboratory analysis[a]	7.9	n.d.	3.2	n.d.
Variation within single plot[b]	39.4	n.d.	32.6	n.d.
Plot replication[c]	33.2	37.6	33.6	37.0
Total variation[d]	130.0	88.0	57.0	45.0

[a] C.V. of 10 single samples from one mixed bulked sample.
[b] C.V. of four means of $n = 7$ within 4 plots, in total $n = 28$ single plant samples.
[c] C.V. of means for each treatment, 8 bulked samples for each treatment, in total $n = 40$.
[d] C.V. pooled for all data, in total $n = 40$.

n.d.: not detected

14.3.2.2 Spatial Variability of Heavy Metal Contents in Plants

We analyzed the variation of the uptake of Zn and Cd in the leaves and the roots of the sugar beets more in detail to evaluate the significance of the results we obtained for the other crops. The variation of the following factors was analyzed statistically: analytical variability of the leaf samples, metal concentrations of the leaves in one specific plot, plot replications, and treatments. For the roots, plot replications and variation between the treatments were analyzed. As for the soil analysis, differences between the treatments were expected, whereas differences between replicates, within plots and between laboratory analyses, were considered to represent random variations and expected to be low. The results are given in Table 14.8.

The coefficient of variation of Cd and Zn concentrations in the leaves was about 8 and 3%, respectively, for the laboratory analyses and about 40 and 32%, respectively, within one single plot. Within the treatments, bulked plant samples of the four plot replicates showed about 30 to 40% of variation in average for the leaves and roots for both metals. The main variation of the cadmium contents was found between the treatments, with a C.V. (coefficient of variation) of about 130% for the leaves and about 90% for the roots, while the total variation of the zinc was about 60% for the leaves and 45% of the roots for the total variation. Despite this rather large variability (analytical, spatial), treatment effects were dominant, in particular for cadmium and for leaves.

14.3.2.3 Changes Over Time

All plants were always harvested at maturity. However, it is known that heavy metal concentrations in plants usually vary with time (Hein, 1988). Cadmium and Zn concentrations were thus followed in 1996 during the growing season of the sugar beet.

The sugar beets were sampled six times between June and November in order to check for seasonal variations of heavy metal concentrations in leaves and roots. Figure 14.4 shows that Cd and Zn concentrations were highly variable, which can be entirely explained by analytical and spatial variability given that only a small number of plants could be collected each time. The concentrations were consistently lower in the roots than in the shoots. However, no trend was observed over time and thus no evolution related to factors such as sugar beet growth, physiology, or sugar buildup in the plant could be found. These factors are known to vary along the growing season (Cook and Scott, 1993).

Again, the evolution or variability between years could only be assessed for sugar beet (see the plant uptake paragraph) which was grown twice. Concentrations measured in 1996

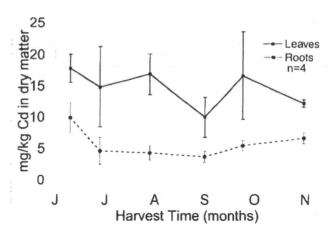

FIGURE 14.4
Cd concentrations in sugar beet leaves and roots (cultivar Brigadier) harvested the June 13, July 1, July 31, September 2, September 25, and October 31, 1996.

were in general lower than in 1990, especially for Zn. This difference was more evident for leaves than for roots. However, at that point we couldn't determine if the difference was due to plant variability, climate, or a difference in metal availability.

14.3.2.4 Plant-Soil Interactions: Influence of Soil Factors on Heavy Metal Uptake by Crops

The general tendency was an increase in Cd and Zn concentrations in plants with increased biosolids application. As presented above, most of the soil parameters did not change either with the treatment or with time, apart from a slight decrease in soil pH with the treatments, and a general increase with time. However, in a multiple regression analysis, pH was never found to have a significant effect on plant uptake. Thus we have chosen to present here only the simple correlation analysis between heavy metal concentrations in plant tissues and in soil.

Pearson correlation coefficients were calculated pairwise for the transformed total and soluble cadmium and zinc concentrations in soil and plant tissues. The results are given in Table 14.9 and Figure 14.5. In general, the metal contents in plant tissues were more closely related to HNO_3- than to $NaNO_3$-extractable metal concentrations. Cadmium and Zn concentrations in plant tissues were correlated with HNO_3-extractable metal concentrations (r between 0.74 and 0.98) and with $NaNO_3$-extractable Cd and Zn concentrations (r between 0.39 and 0.85). In general, these coefficients were larger for Zn than for Cd. The strongest correlation between both $NaNO_3$- and HNO_3-extractable metal concentrations in soil and plant tissues was found for the potato leaves, sugar beet leaves, lettuce, and, to a less extent, maize leaves.

Additionally, it was possible to compare the results obtained for the same crop (sugar beet) grown in 1991 and 1996. As already presented above, Cd and Zn concentrations in leaves were lower in 1996 than in 1991. When correlations were calculated between concentrations in sugar beet leaves and $NaNO_3$-extractable Zn, all sets of data (that is, 1991 and the two varieties grown in 1996) were explained by the same linear regression with the same slope: the decrease in plant concentration seemed thus to be related to the decrease in $NaNO_3$-extractable Zn. $NaNO_3$-extractable Cd concentrations were too low to be taken into account.

We assumed that a better explanation of the soil-plant system (and Cd and Zn distribution in plants) could be obtained when taking into account all the heavy metals present in the soil and added with the treatments. Thus, in addition to the analysis of correlation, we compared the relationships between the heavy metal concentration in soil and plants in a multivariate domain using principal component analysis. The results obtained didn't yield any additional information and are not presented here.

TABLE 14.9

Pearson Correlation Coefficients of the Log-Transformed Concentrations of Cd and Zn Measured in Plant Tissues ($n = 20$) and HNO_3-Extracts («HNO_3») Resp. $NaNO_3$-Extracts («$NaNO_3$») of the Soil

	Cadmium HNO_3		Zinc	
	$NaNO_3$	HNO_3	$NaNO_3$	HNO_3
Maize leaves	0.84	0.77	0.86	0.84
Maize cobs	0.84	0.79	0.82	0.69
Sugar-beet leaves 91	0.89	0.85	0.96	0.71
Sugar-beet roots 91	0.89	0.81	0.81	0.50
Potato leaves	0.97	0.80	0.98	0.89
Potato tubers	0.74	0.64	0.83	0.68
Lettuce	0.90	0.75	0.98	0.78
Spinach	0.76	0.39	0.86	0.65

14.4 Discussion and Conclusion

14.4.1 Impact of the Waste and Sludge Applications on the Soil

As described by von Hirschheydt (1985a), treatments did not affect soil pH immediately. In 1975 soil pH was around 7.2 in all plots at Buchs, as measured in suspensions with distilled water. Apart from metal concentrations, the only immediate effect was found with respect to organic matter content, which had been increased as expected according to the amount of applied biosolids. The evolution of differences in soil pH between treatments, which were found in 1990, thus appears to have been a rather slow process. This process may have been related to the partial mineralization of the introduced organic matter. In 1990 this process must have been completed, as no more changes in total organic matter content were observed between the samplings of 1990 and 1996. The slight increase in soil pH, which was observed in all treatments, must be attributed to other processes.

The level of contamination remained constant for all treatments over the entire study period. Compared to today's legal standards, the pollution introduced with the applied wastes and sludges was severe. But in addition, also the untreated soils were already slightly polluted. According to the Swiss Ordinance Relating to Soil Stresses (OSOL, 1998), the Salez soil exceeded the "guide values" (defining the upper limit of soil considered to be unpolluted) of Cu and Ni, and the Buchs soil those of Zn, Cd, Pb, and Cr. The biosolids applications increased Pb, Cd, and Cu concentrations in all cases above the OSOL "trigger values" (no such trigger value has been defined for Zn), above which health risks have to be assessed by in-depth investigations and land-use may be restricted. The BWS2 treatment even led to Zn and Cd concentrations above the OSOL "remediation values," which means that soil remediation or a ban of land-use for crop production would be required.

In parallel with the total metal loads, also the "soluble" or bio-available concentrations, as determined by extraction with $NaNO_3$, were increased. This finding agrees well with observations of other investigators, e.g., Sloan et al. (1997) and Hamon et al. (1998). Different results, however, have been reported with respect to the temporal evolution of the soluble metal fraction in relation to the total concentration. While Sloan et al. (1997) found a similar decrease of this fraction over time as we did during the 6 years following biosolids application, Krebs et al. (1998) observed the opposite trend, i.e., an increase of $NaNO_3$-extractable Cd and Zn concentrations over a period of 15 years after the last sludge application. In contrast to our results, Krebs et al. (1998) also observed a decrease in soil pH

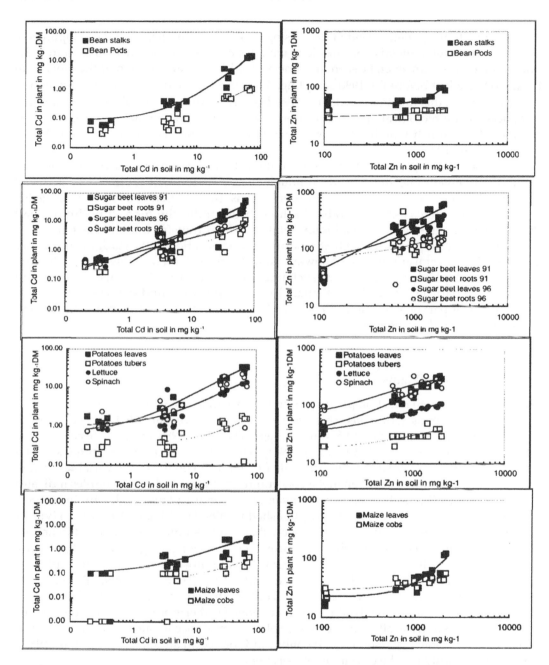

FIGURE 14.5
Relationships between the transformed cadmium and zinc concentrations in soil (HNO₃-extractable) and plants.

between sampling times, whereas in our experiment the decrease in metal availability was accompanied by a slight pH increase. While these differences may be due to different soil and other experimental site conditions, about which we can only speculate here, these results indicate that soil pH probably was the driving variable of the changes in metal solubility. In addition, other factors also may have promoted the "aging effect" of decreasing soluble metal concentrations, e.g., progressive diffusion into microstructures, rearrangement of sorption complexes and co-precipitates, and incorporation and occlusion of metals in insolubles (Alloway and Jackson, 1991; Gupta and Aten, 1993).

In contrast to the general trend of decreasing metal solubility between the samplings of 1990 to 1996, soluble metal concentrations were found to be higher in 1990 than in 1987. This change must, however, be seen in connection with the translocation of the plots from Buchs to Salez just before the 1990 sampling. After the soils had been left uncultivated in the years before, this transfer inevitably caused severe disruption of soil structure and disturbances of soil biological and chemical processes. Unfortunately no measurements are available to establish whether the remobilization of the metals was accompanied by a decrease in soil pH.

Metal mobilization provoked by the translocation of the plots may also have re-accelerated metal displacement into the subsoil. Although such displacement could not be deduced from the depth profiles of total metal concentrations, the fact that $NaNO_3$-extractable Zn concentrations increased measurably in the subsoil under the highest pollution load (BWS2) would agree with such an interpretation. Metal transfer into the subsoil would have been expected primarily the first time after the application of the sludges and wastes as reported by several authors (see Alloway and Jackson, 1991, for a review). Low solubility caused by high pH does not preclude such migration, as transport may also occur bound to suspended particles and colloids. Barbarick et al. (1998) also found an increase in soluble Zn in the lower part of a neutral to alkaline soil after amendment with biosolids with high concentrations of this element.

14.4.2 Impact on Plants

The general treatment effect on plants was an increase in tissue metal concentrations with increasing metal loads in the soils. This effect had been observed already between 1975 and 1983 (von Hirschheydt, 1985b, 1985c) and was again between 1989 and 1996. Because of the close relationship between HNO_3- and $NaNO_3$-extractable concentrations in the soil at a given time, it was not possible to discriminate further between these two variables. In any case, the high correlation between metal concentrations in soil and plant samples indicates that metal availability in the soil was the main limitation for plant uptake. In this respect our findings agree well with previously published results (i.e., Chang et al., 1997; Hyun et al., 1998). Open questions concern the general shape of this nonlinear relationship and its stability over time (Berti and Jacobs, 1996; Brown et al., 1998; Chang et al., 1997). Because of various reasons it is difficult to relate our results to those dealing with the "plateau effect" and "time bomb" (Dowdy et al., 1994; Smith, 1997; Zhao et al., 1997; Brown et al., 1998). The major reason is that we have no soil data with respect to the effects of the sludge applications during and directly after the treatments. Furthermore, except for sugar beets, which were grown twice, all other crops were planted only once in our experiment. In addition, the plateau effect is assumed to be only valid for "clean sludge" (Corey et al., 1987), which was certainly not the case in our experiment.

Zinc and cadmium concentrations exceeded normal values reported for these two elements in the leaves of all crops studied except maize. Cadmium tends to accumulate in leafy vegetables (Alloway et al., 1990) like lettuce and spinach as well as in potatoes leaves. Sugar beet has been reported to accumulate Zn (Davis and Carlton-Smith, 1980 in Alloway and Jackson, 1991). Our results are consistent with these findings, although our transfer coefficients for Cd and Zn were lower than those found by Logan et al. (1997) for corn and lettuce. However, the uptake was not proportional to metal concentrations in soil, because the transfer coefficients decreased with increasing loads of Zn and Cd, as also found by de Villarroel et al. (1993) for Swiss chard.

Because each year a different crop was grown, we could not analyze how the tendency of decreasing $NaNO_3$-extractable metal concentrations in the soil between the samplings from 1990 to 1996 translated into plant uptake, except for a comparison between the 2 years in which sugar beets were grown. The finding that less Zn uptake was observed in 1996 than in 1991 was at least in agreement with our expectation, confirming the results of Hyun et al. (1998). With respect to risks of food chain contamination, this is a fortunate result. How representative it is for other soils and crops remains to be determined.

An ideal extractant to characterize plant availability of metals has not yet been found (Miner et al., 1997). The choice of $NaNO_3$ has been advocated by Gupta and Aten (1993). In the range of near-neutral to alkaline soil conditions, stronger extractants such as DTPA may be preferred (Barbarick et al., 1998; Brown et al., 1998). The fact that we obtained no better relationships between metal uptake by plants with $NaNO_3$- than with HNO_3-extractable concentrations in soil does not speak against using this extractant even under such conditions, as the equal performance of the two extractants in predicting plant uptake can be simply explained as a consequence of the very little variation in soil pH.

Our results show how much metal uptake from the same soil can vary between different crops and within different parts of the same plant. Because of the low availability of the metals in relation to the high total loads, no phytotoxicity was observed, but metal accumulation was still high enough to make most crop products on the highly polluted plots unacceptable for consumption by humans or animals according to current legal standards in Switzerland. Although the trend was going toward a decreasing metal bioavailability in the soil, the process was too slow to expect that this problem would find a "natural solution" by attenuation within the foreseeable future.

Acknowledgment

We would like to thank C. Ludwig and M. Märki for the work done on sugar beets in 1996 and also Werner Attinger and Anna Grünwald for the maintenance of the experiment and the soils and plants analyses.

References

Alloway, B.J. and A.P. Jackson, The behaviour of heavy metals in sewage sludge-amended soils (cf. p. 288), *Sci. Total Environ.*, 100, 151, 1991.

Alloway, B.J., A.P. Jackson, and H. Morgan, The accumulation of cadmium by vegetables grown on soils contaminated from a variety of sources, *Sci. Total Environ.*, 91, 223, 1990.

Barbarick, K.A., J.A. Ippolito, and D.G. Westfall, Sewage biosolids cumulative effects on extractable-soil and grain elemental concentrations, *J. Environ. Quality*, 26, 1696, 1997.

Barbarick, K.A., J.A. Ippolito, and D.G. Westfall, Extractable trace elements in the soil profile after years of biosolids application, *J. Environ. Quality*, 27, 801, 1998.

Berti, W.R. and L.W. Jacobs, Chemistry and phytotoxicity of soil trace elements from repeated sewage sludge applications, *J. Environ. Quality*, 25, 1025, 1996.

Brown, S.L., R.L. Chaney, J.S. Angle, and J.A. Ryan, The phytoavailability of cadmium to lettuce in long-term biosolids-amended soils, *J. Environ. Quality*, 27, 1071, 1998.

Candinas, T. and A. Siegenthaler, Grundlagen des Düngung: Klärschlamm und Kompost in des Landwirtschaft, *Schriftenreihe der FAC Liebefeld*, 9, Liebefeld-Bern, 1990.

Chaney, R.L. and J.A. Ryan, Heavy metals and toxic organic pollutants in MSW-composts: research results on phytoavailability, bioavailability, fate, etc., in *Science and Engineering of Composting: Design, Environmental, Microbiological and Utilization Aspects*, H.A.J. Hoitink and H.M. Keener, Eds., Renaissance, Worthington, OH, 1993, 451.

Chang, A.C., H.-N. Hyun, and A.L. Page, Cadmium uptake for swiss chard grown on composted sewage sludge treated field plots: plateau or time bomb? *J. Environ. Quality*, 26, 11, 1997.

Cook, D.A. and R.K. Scott, Eds., *The Sugar Beet Crop*, Chapman & Hall, London, 1993.

Corey, R.B., L.D. King, C. Lue-Hing, D.S. Fanning, J.J. Street, and J.M. Walker, Effects of sludge properties on accumulation of trace elements by crops, in *Municipal and Industrial Waste 4th Annual Madison Conf. of Applied Res. and Practice on Municipal and Industrial Waste*, Las Vegas, NV, November 13–15, 1985, A.L. Page et al., Eds., Department of Engineering and Applied Science, University of Wisconsin, Madison, 1987, 25.

Dowdy, R.H., C.E. Clapp, D.R. Linden, W.E. Larson, T.R. Halbach, and R.C. Polta, Twenty years of trace metal partitioning on the Rosemount sewage sludge watershed, in *Sewage Sludge: Land Utilization and the Environment*, C.E. Clapp et al., Eds., ASA, CSSA, and SSSA, Madison, WI, 1994, 149.

FAC (Eidgenössische Anstalt für Agrikulturchemie und Umwelthygiene), Methoden für Bodenuntersuchungen, *Schriftenreihe der FAC Liebefeld*, 5. Liebefeld-Bern, 1989.

Gupta, S.K. and C. Aten, Comparisons and evaluation of extraction media and their stability in a simple model to predict the biological relevance of heavy metal concentrations in contaminated soils, *Int. J. Environ. Anal. Chem.*, 51, 25, 1993.

Gupta, S.K., M.K. Vollmer, and R. Krebs, The importance of mobile, mobilisable and pseudo-total heavy metal fractions in soil for three-level risk assessment and risk management (cf. p. 287), *Sci. Total Environ.*, 178, 11, 1996.

Häberli, R., C. Lüscher, B. Praplan Chastonay, and C. Wyss, *L'Affaire Sol: pour une Politique Raisonnée de l'Utilisation du Sol*, Georg Editeur S.A., Geneva, 1991, 192.

Hamon, R.E., M.J. McLaughlin, R. Naidu, and R. Correll, Long-term changes in cadmium bioavailability in soil, *Environ. Sci. Technol.*, 32, 3699, 1998.

Hein, A., Die Ni-Aufnahme von Pflanzen aus verschiedenen Böden und Bindungsformen und ihre Prognose durch chemische Extraktions-verfahren, Ph.D. dissertation, Göttingen, 1988.

von Hirschheydt, A., Field-tests on the influence of heavy metals in refuse compost. I, *Wasser + Boden*, 5, 228, 1985a.

von Hirschheydt, A., Field-tests on the influence of heavy metals in refuse compost. II. Crops and crop yield during the concentration phase, *Wasser + Boden*, 8, 381, 1985b.

von Hirschheydt, A., Field-tests on the influence of heavy metals in refuse compost. III, *Wasser + Boden*, 12, 594, 1985c.

von Hirschheydt, A., Zur Wirksamkeit von Schwermetallen aus Müllkomposten auf Ertrag und Zusam-mensetzung von Kulturpflanzen. Teil I und II. Studienreihe Abfall-Now. Abfalltechnisches Labor mit Anhang am Institut für Siedlungswasserbau, Wassergüte- und Abfallwirtschaft der Universität Stuttgart, Bandtäle 1, Stuttgart, 1987.

Hyun, H., A.C. Chang, D.R. Parker, and A.L. Page, Cadmium solubility and phytoavailability in sludge-treated soil: effects of soil organic carbon, *J. Environ. Quality*, 27, 329, 1998.

Keller, T. and A. Desaules, Flächenbezogene Boden-belastung mit Schwermetallen durch Klärschlamm, *Schriftenreihe des FAL*, 23, 1997.

Kloke, A., D. Sauerbeck, and H. Vetter, The contamination of plants and soils with heavy metals and the transport of metals in terrestrial food chains, in *Changing Metal Cycles and Human Health*, J. O. Nriagu, Ed., Springer-Verlag, Berlin, 1984, 113.

Krebs, R., S.K. Gupta, G. Furrer, and R. Schulin, Solubility and plant uptake of metals with and without liming of sludge-amended soils, *J. Environ. Quality*, 27, 18, 1998.

Logan, T.J., B.J. Harrison, L.E. Goins, and J.A. Ryan, Field assessment of biosolids metal bioavailability to crop: sludge rate response, *J. Environ. Quality*, 26, 534, 1997.

McBride, M.B., Toxic metal accumulation from agricultural use of sludge: are USEPA regulations protective?, *J. Environ. Quality*, 24, 5, 1995.

Miner, G.S., R. Gutierrez, and L.D. King, Soil factors affecting plant concentrations of cadmium, copper, and zinc on sludge-amended soils, *J. Environ. Quality*, 26, 989, 1997.

OSOL, Ordinance Relating to Pollutants in Soil, June 9, 1986, SR 814.12.

OIS, Ordinance Relating to Impacts on the Soil, July 1, 1998, SR 814.12.

de Quervain, F. and D. Frey, Geotechnische Karte der Schweiz im Massstab 1:200 000, mit Erläuterungen. 2. Auflage. Blatt Nr. 2. Hrsg.: Schweizerische Geotechnische Kommission, Organ der Schweizerischen Naturforschenden Gesellschaft. Verlag Kümmerly und Frey, Bern, Switzerland, 1963.

Ruppert, H., Bestimmung von Schwermetallen im Boden sowie die ihr Verhalten beeinflussenden Boden-eigenschaften (Vorschläge). Beilage zum GLA-Fachbericht 2, München 1987 und zum Merkblatt "Bodenkataster Bayern," München 1985. Hrsg. und Verlag: Bayerisches Geologisches Landesamt (GLA), Hessstrasse 128, München, 1987, 3–4.

Sauerbeck, D., *Funktionen, Güte und Belastbarkeit des Bodens aus agrikulturchemischer Sicht*, Kohlhammer Verlag, Stuttgart am Mainz, 1985.

Sauerbeck, D., Der Transfer von Schwermetallen in die Pflanze, in *Beurteilung von Schwermetallkontaminationen im Boden*, DECHEMA Fachgespräche Umweltschutz, Stuttgart am Mainz, 1989, 281.

Schmidt, J.P., Understanding phytotoxicity thresholds for trace elements in land-applied sewage sludge, *J. Environ. Quality*, 26, 4, 1997.

SFSO (Swiss Federal Statistical Office) and SAEFL (Swiss Agency for the Environment, Forests and Landscape), *The Environment in Switzerland 1997*, EDMZ, Bern, Switzerland, 1997, 372.

Sloan, J.J., R.H. Dowdy, M.S. Dolan, and D.R. Linden, Long-term effects of biosolids applications on heavy metal bioavailability in agricultural soils, *J. Environ. Quality*, 26, 966, 1997.

SMA, Annalen der Schweizerischen Meteorologischen Anstalt-Hundertzweiunddreissigster Jahrgang, Schweizerische Meteorologische Anstalt, ISSN 0080-7338, 1995.

Smilde, K.W., P. Konkonlakis, and B. van Luit, Crop response to phosphate and lime on acid sandy soils high in zinc, *Plant and Soil*, 41, 445, 1974.

Smith, S.R., Long-term effects of zinc, copper and nickel in sewage sludge-treated agricultural soil, in *Proc. Extended Abstracts from the 4th Int. Conf. on the Biogeochemistry of Trace Elements*, Berkeley, CA, June 23–26, 1997, 691.

Stenz, B., Schwermetallaufnahme durch Kulturpflanzen auf belasteten Böden, Ein fünfjähriger, praxisnaher Freilandversuch an der Landwirtschaftlichen Schule Rheinhof in Salez. Amt für Umweltschutz, Fachstelle Bodenschutz, Kanton St. Gallen, 1995.

UFAG Internal Method for Organic Carbon Determination, UFAG Laboratories, Kornfeldstrasse 4, CH-6210 Sursee, Switzerland.

U.S. Environmental Protection Agency, 40 CFR Part 257 et al. Standards for the Use or Disposal of Sewage Sludge, Final Rules, *Fed. Regist.*, 58(32), 9248, 1993.

De Villarroel, J.R., A.C. Chang, and C. Amrhein, Cd and Zn phytoavailability of a field-stabilized sludge-treated soil, *Soil Sci.*, 155, 197, 1993.

Zhao, F.J., S.J. Dunham, and S.P. McGrath, Lessons to be learned about soil-plant metal transfers from the 50th-year sewage sludge experiment at Woburn, UK, in *Proc. Extended Abstracts from the 4th Int. Conf. on the Biogeochemistry of Trace Elements*, Berkeley, CA, June 23–26, 1997, 693.

Index